MW00710668

GREEN
ESSENTIALS

GREEN
ESSENTIALS

What You Need to Know

about the

Environment

Geoffrey C. Saign

Mercury House
San Francisco

for Mom and Dad,
who are always there

Copyright © 1994 by Geoffrey C. Saign

Published in the United States by
Mercury House
San Francisco, California

All rights reserved, including, without limitation, the right of the publisher to sell directly to end users of this and other Mercury House
books. No part of this book may be reproduced in any form or by any
electronic or mechanical means, including information storage and
retrieval systems, without permission in writing from the publisher,
except by a reviewer who may quote brief passages in a review.

United States Constitution, First Amendment: Congress shall make no
law respecting an establishment of religion, or prohibiting the free
exercise thereof; or abridging the freedom of speech, or of the press;
or the right of the people peaceably to assemble, and to petition the
Government for a redress of grievances.

Mercury House and colophon are registered trademarks of
Mercury House, Incorporated

Printed on recycled, acid- and
chlorine-free paper, using soy-based inks
Manufactured in the United States of America
Text design and composition by Philip Bronson.

Library of Congress Cataloging-in-Publication Data
Saign, Geoffrey C., 1955–
Green essentials : what you need to know about the environment /
    Geoffrey C. Saign.
        p.   cm.
    ISBN 1-56279-061-7
    1. Environmental sciences—Dictionaries.
GE10.S25   1994
363.7'003—dc20

First Edition
5   4   3   2   1

CONTENTS

PREFACE

This book is designed for anyone who wishes to have easy access to clear information about every major environmental issue. Concerned citizens, environmentalists, students, teachers, and anyone interested in environmental problems can quickly look up specific topics, which are further broken down into subtopics, which are subdivided yet again. This format allows any reader to rapidly answer a particular question or to obtain a quick grasp of a whole topic.

There are three major ways to find information in this book: the table of contents, the index, and the glossary. The table of contents lists the major topics examined in individual chapters. Narrower subjects included within the major topic discussions appear in the index. The glossary defines most of the environmental terms used in the book (and some that are not), in case they are unfamiliar.

Within each chapter on a major topic are subsections outling "Major Sources," "Environmental Impact," and "Human Impact," etc.; within these subsections items are listed roughly in descending order of importance, wherever applicable or possible.

There is extensive legislation on some topics, and the list of environmental laws grows daily at the city, state, national, and international levels. Thus I have been able to mention only a small portion of current legislation, but I have tried to include the most important.

New evidence is arising and new studies are being completed regularly on most topics, and some controversy surrounds a few issues such as global warming, ozone depletion, or the health effects of electromagnetic fields (EMFs). In these cases this book has taken a prudent approach; it would seem wiser to err on the side of safety and health than to commit ourselves to a risky course of action we may later regret.

Acid rain

Definition

Fog, dew, snow, hail, or rain with abnormally high levels of acidity.

This acidity is caused primarily by the release of sulfur dioxide or nitrogen oxides into the air. These gases react with water to form sulfuric acid and nitric acid, which then fall to earth in precipitation. The acids are absorbed into the soil and washed into ponds, lakes, and rivers. Acidic particulates containing sulfur dioxide and nitrogen oxides can also fall to earth as dry soot.

Environmental Impact

Lowered pH: All soils and water have a natural degree of acidity. Acid rain lowers the pH of both soils and water. The stronger the acid, the lower the pH; acid rain has a pH of less than 5.7. Acid rain is harshest on soil, ponds, and lakes whose pH is naturally low, as is the case in the far eastern and western United States; these areas become acidified more easily. The midwestern United States has fairly alkaline soil, which can tolerate acid rain better than soil in most other areas of the world.

Lakes, rivers, and ponds: As small organisms in the water die off, declines in plant and fish life usually follow. When the pH of water is too low, fish may stop reproducing, become weakened, or even die off. Acidity can inhibit the hatching of the eggs of frogs and salamanders that breed in water and can also kill their tadpoles. Sphagnum moss and filamentous algae may form dense mats on lake beds; these plants further increase water acidity by taking materials such as calcium out of the water and releasing hydrogen.

Leaching: Acidic particles combine with nutrients in the soil, causing them to be dissolved and washed out of the soil in a process called leaching. Leaching makes the soil less rich, and plants cannot grow as well in it; acid rain can also leach nutrients out of plant leaves. Acid rain falling on land can cause the leaching of heavy metals such as aluminum, cadmium, zinc, and mercury, which become more active in soils and are also washed into lakes and rivers. Plant and tree growth can be hindered, trees can become poisoned, and fish can die from aluminum that invades their gills. Birds eating insects emerging from acidified lakes may develop high levels of aluminum and may not reproduce as well.

Soils: Acid rain interferes with the decomposition of soil organic matter into nutrients for plants, inhibiting plant growth and ultimately leading to dying forests. Acid rain can also add excess nitrogen to soils, which can have an overall negative impact on tree growth.

Major Sources

Sulfur dioxide (SO$_2$): Coal-burning power plants, which burn coal with sulfur, generate 2/3 of the world's airborne sulfur dioxide. The rest is generated by oil-burning power plants, ore smelters, garbage incinerators, and transportation. One coal-fired power plant can emit as much sulfur dioxide in one year as was emitted in the eruption of Mount Saint Helens: 400,000 tons. Sulfur dioxide is responsible for about 2/3 of northeastern U.S. acid rain.

Nitrogen oxides (NO$_x$): 40%–50% of the world's nitrogen oxides are produced by car emissions; 30% by power plants; and more than 20% from ore smelters, industrial boilers, and biomass burning (such as burning of savanna, trees, agricultural waste, or fuelwood) by humans. The engines and motors of landscaping equipment also contribute to nitrogen oxide emissions. Nitrogen oxides are responsible for about 1/3 of northeastern U.S. acid rain.

Industrial: Smaller amounts of acids are released through industrial chemical use, forming sulfuric or hydrochloric acid. Sulfuric acid is widely used in industry.

Natural sources: Acidic particles (sulfur and nitrogen oxides) are emitted in natural processes, such as volcanic eruptions, lightning, decomposition of organic matter, sea spray, and forest fires. The amount of acidic particles released by human activities is estimated to be up to 4 times the amount released by natural sources.

State of the Earth

Acid rain was first recognized in Sweden in the 1960s. Since then, lifeless lakes and rivers resulting from acid rain have been documented in nearly every country in the world: 14,000 acidified lakes in Canada; 20,000 in Sweden (nearly 1/4 of its lakes); more than 8% of Finland's lakes; and 13,000 square kilometers of Norway's waterways. About 90% of the acid rain in Sweden and Norway comes from other countries. Norwegian Atlantic salmon stocks are virtually extinct owing to acidified river breeding grounds. Liming, which is used extensively in Sweden to restore pH, sometimes helps fish populations survive.

The U.S. Environmental Protection Agency's National Acid Precipitation Project concluded that 10% of Appalachian streams are acidic; out of thousands of studied lakes and streams across the nation, 75% of the lakes and 47% of the streams were found to be acidic. In the Adirondack Mountains, 25% of the lakes, streams, and ponds are already too acidic to support fish life. In more than 300 such lakes all fish have already died. In 1993, Massachusetts and Virginia experienced the highest level of stream and lake acidification since 1983, probably because of heavy, acidic snow melt. Sulfate levels have decreased in some areas in the eastern United States in the past decades, thanks mainly to power plant pollution control; nitrogen levels have only leveled off.

In Ontario, 300 lakes have a pH below 5; in Nova Scotia 9 rivers have a pH below 4.7, a level that cannot support salmon or trout reproduction. The United States is the world leader in both sulfur dioxide and nitrogen oxide emissions. The United States "exported" about 4 million tons of sulfur dioxide, via winds, to Canada in 1990—acid rain knows no boundaries and the particles can travel worldwide.

Air pollutants tend to rise and concentrate at higher elevations. High spruce forests have thus been hardest hit, with thousands of acres already dead in California, North Carolina, Tennessee, and Virginia. Acid rain has damaged trees along nearly the whole length of the Appalachian Mountains; spruce, maple, beech, and oak have all been affected. Cloud acidity there has been measured as low as 2.3—the pH of lemon juice. In Europe, 15% of the forests are damaged; in the United Kingdom, the figure is nearly 65%. Soils in many of Europe's forests are 5 to 10 times more acidic than 50 years ago.

Rain in Brazil has been measured with an average pH of less than 4.5; large tracts of rain forest are threatened by acid rain as a result of biomass burning, which releases nitrogen oxides. Since most tropical insects spend some phase of their life cycle in water (which could become acidified), this acid rain could threaten numerous species. Biomass burning is a serious factor in temperate latitudes as well. Acid rain may be contributing to worldwide declining frog, salamander, and other amphibian populations; 80% of salamander eggs don't hatch when the pH of the water they live in is less than 6.0.

Human Impact

Health: Acidic air particles are believed to have contributed to increases in asthma and bronchitis. The heavy metals that acid rain leaches from soil into water and fish supplies are highly toxic, and if ingested with fish or water, they can cause a variety of serious diseases, including cancer.

Food: Acid rain destroys fish populations and causes billions of dollars in damage to crops in Europe and the United States.

Timber resources: Acid rain kills trees and retards forest growth.

Buildings: Acid rain is dissolving the outer faces of ancient cathedrals and sculptures in Greece, Italy, and nearly every other country in the world. India's Taj Mahal, one of the world's most beautiful structures, is being eroded by acid rain.

Individual Solutions

- Walk or bicycle instead of drive; use mass transit; carpool; combine errands; keep tires inflated and your car tuned.
- Get off the grid (the utility power grid) by becoming energy independent with photovoltaic panels, solar water heaters, or small windmills.
- Practice energy conservation: buy fuel-efficient cars, energy-efficient light bulbs and appliances, and low-flow faucet heads; have an energy audit done on your house.
- Call or write your city, county, state, and federal legislators to demand that utilities practice increased energy conservation and invest in renewables like solar and wind energy; insist on higher miles-per-gallon requirements for cars and scrubbers for utilities.

Industrial/Political Solutions

What's Being Done

- More countries are adding scrubbers to electric utilities.
- Liming (adding calcium carbonate to) lakes and soils to raise their pH levels and improve chemical and biological conditions has had only limited results. Lime can help prevent further destruction and reduce the effects of acid rain. Once a lake is acidified, however, liming can't bring back the whole lake ecology.
- Canada, Europe, Japan, and the United States require all new cars to have catalytic converters. These converters reduce nitrogen oxide emissions by 60% and carbon monoxide and hydrocarbon emissions by 85%. Catalytic converters cannot be used with leaded gas.
- Electric utility plants must purchase electricity developed by small producers (often solar, wind, geothermal, hydroelectric, or other renewable technologies), and must do so at a rate equivalent to avoided cost. This requirement began with passage of the Public Utilities Regulatory Policies Act (PURPA) in 1978.
- The EPA set up a plan in which emission allowances for sulfur dioxide can be bought and sold. The goal is to reduce U.S. sulfur dioxide emissions by 40% by the year 2000. Under the plan, companies can

buy and sell emissions allowances, each of which permits the emission of 1 ton of sulfur dioxide over the course of 1 year. Only a limited number of these allowances will be available each year. By the year 2000 only 8.9 million tons in allowances will be available. About 110 large utilities must have allowances for their emissions by 1995, and about 800 small companies must have them by 2000. This plan will cut about 10 million tons of sulfur dioxide annually from 1980 emission levels. Nitrogen oxide emissions will be reduced by 0.5 to 2 million tons per year starting in 1995. These programs are a result of the 1990 Clean Air Act Amendments.

- The federal government has established tax incentives for solar, wind, and geothermal energy.
- In California, pollution from nonvehicular engines must be cut 46% by 1995, and another 55% by 1999. Federal regulations for August 1, 1996, are aimed at reducing hydrocarbons by 32% and carbon monoxide by 14% by 2003.
- California is focusing on mass transit and phasing in zero-emission (electric) cars ; 200,000 will be in use by 2003. Ten other states have said they will follow California's lead on zero-emission cars.

What Needs to Be Done

- Mandate wet scrubbers on smokestacks at coal- and oil-burning power plants, as well as at ore smelters. Scrubbers remove sulfur dioxide from smokestack emissions, and new research will soon allow them to also remove nitrogen oxides.
- Mandate the use of "clean" coal, which has had the sulfur prewashed out.
- End fossil fuel subsidies.
- Mandate a yearly increasing level of investment for utilities in energy conservation and renewable energy.
- End emissions allowances trading, which allows polluters to buy the right to continue to pollute (which is cheaper than buying scrubbers), and instead focus on pollution prevention.

- Develop international agreements and restrictions on sulfur dioxide and nitrogen oxide emissions.
- Increase investment in energy conservation and efficiency worldwide.
- Encourage worldwide use of automobile catalytic converters; phase in zero-emission cars, like electric cars, or high-miles-per-gallon (150 to 300 mpg) cars such as ultralights with zero-emission converters.

Also See

- Coal
- Nitrogen oxides (NO_x)
- Sulfur dioxide (SO_2)

A

Air pollution

Definition

Gases, liquid droplets, or particles (excluding pure water vapor and droplets) that enter the air from human activities or natural processes.

Air pollution can occur inside or outside of buildings. Indoor air in houses and cars is often more polluted than outdoor air.

Some of the most common air pollutants include carbon dioxide, carbon monoxide, chlorofluorocarbons (CFCs), heavy metals (arsenic, chromium, cadmium, lead, mercury, zinc), hydrocarbons, nitrogen oxides, organic chemicals such as volatile organic compounds (VOCs) and dioxins, sulfur dioxide, and particulates. The most regulated air pollutants are carbon monoxide, sulfur dioxide, nitrogen oxides, lead, particulates, and smog, or ozone.

Environmental Impact

Plants and animals: When particles, liquid droplets, or gases enter the air, some may trap more sunlight, adding to the greenhouse effect; some may accelerate ozone depletion in the stratosphere; some may create smog; and some may create acid rain. Each of the last three problems inhibits plant growth and can thus affect wildlife populations that depend on plant food sources or habitats such as forests. Toxic air emissions eventually precipitate onto soils and water and are consumed by plants, animals, or aquatic life. Toxics can stunt plant growth or make plants vulnerable to disease, pests, or other environmental stresses; in animals or fish, toxics can cause organ, reproductive, and behavior diseases leading to the death of young and adults.

Local buildups: Weather stagnation, temperature inversions, and surrounding mountains can cause pollutants to build to even more dangerous levels in certain areas, especially when there is a lack of wind or the presence of low-lying cloud cover. Also, trees take pollutants out of the air during respiration, so as more trees are burned and cut down, fewer air pollutants are removed from the atmosphere. This process can increase air pollution in local areas.

Transported pollution: Wind currents can transport many air pollutants nearly anywhere in the world.

Major Sources

Automobiles: The burning of oil and gas emits carbon monoxide, VOCs, hydrocarbons, ozone (precursors), nitrogen oxides, peroxyacetyl nitrate (PAN), benzene, toxics, and lead.

Utility power plants: The burning of coal, oil, and gas emits nitrogen oxides, heavy metals, sulfur dioxide, and particulates.

Industry: Manufacturing and processing emit a wide array of toxics, as well as particulates, sulfur dioxide, nitrogen oxides, heavy metals, fluoride, CFCs, and dioxins.

Incineration: Hazardous waste incinerators and municipal solid waste incinerators emit toxics, carbon monoxide, nitrogen oxides, particulates, dioxins, and heavy metals.

Biomass burning: The burning of savanna grasslands and agricultural wastes, fuelwood use, forest fires, and the use of crop residues and other biomass products in power plants emit particulates, carbon monoxide, nitrogen oxides, radon, methane, sulfur, and carbon dioxide.

Small engines: Landscaping mowers, blowers, lawn mowers, hedge trimmers, chainsaws, and other small machines emit nitrogen oxides, hydrocarbons, nitrogen oxides, and toxics.

Catastrophes: Events such as the radiation leak at Chernobyl in the Soviet Union in 1986, the chemical leak at Bhopal, India, in 1984, and the burning of the oil wells in Kuwait in 1991 can lead to emission of toxics, radiation particles, nitrogen oxides, carbon monoxide, sulfur, heavy metals, and particulates.

Mining: Stone-crushing operations, excavation, and processing emit toxics, particulates, nitrogen oxides, radiation particles, and heavy metals.

Erosion: Unpaved roads and farmland emit dust, particulates, dried pesticides, and soil toxics.

Indoor air pollution: Formaldehyde; kerosene heaters; cooking with gas, fuelwood, coal, or dung; leaded paint dust; asbestos dust; radon; and many synthetic indoor products and equipment such as carpet, furniture, drapery, laser printers, and copy machines give off fumes from incorporated and used chemicals.

Natural sources: Volcanic eruptions, blown dust and particles, natural forest fires from lightning, and radiation from background sources emit sulfur dioxide, carbon monoxide, carbon dioxide, chlorine, nitrogen oxides, heavy metals, radon, and particulates.

State of the Earth

Air pollution knows no boundaries, and thus we are all breathing pollutants and air from around the world. In 1991 the U.S. Environmental Protection Agency (EPA) found that nearly 100 major urban areas in the United States exceeded air pollution limits; in 1993 more areas met federal air quality standards, yet 72 major cities and countries still exceeded air pollution limits. The EPA found that indoor air is 5 to 10 times as toxic as outdoor air; indoor air pollution can be exacerbated by tight, well-insulated, newer buildings; in developing countries indoor air pollution from burning fuelwood, dung, and charcoal endangers more than 500 million people.

The United States is the world leader in car manufacturing (and thus exhaust pollution), CFC production, raw material use per capita, and energy and electricity use (and thus related pollutant emissions) per capita. The United States is also first in particulate and hydrocarbon emissions. Cars contribute to more than 50% of air pollution in most cities; motor vehicles are the largest source of air pollution.

The EPA conducted a study that found that using a riding mower for 1 hour creates as much smog pollution as driving a car for 20 miles;

using a gas-powered lawn mower for 1 hour is equivalent to driving 50 miles. The 80 million lawn mowers in the United States produce pollution equivalent to 3.5 million newer cars. Electric mowers are more than 70 times less polluting than gas mowers.

Dangerous air particulates are used in many industries. Silica, used in the silicon chip computer industry, for instance, can be as dangerous as asbestos if breathed in or ingested. In the United States 7 million tons of particulates were released in 1989; taller smokestacks allow pollutants to travel higher, remain in the atmosphere longer, and travel farther.

German scientists reported levels of air pollution across virgin forests in Africa as high as those seen in Europe, probably because of forest fires set by people to clear land. Satellites show that ozone levels in some tropical systems are almost at lethal levels to plants and animals and that carbon monoxide levels are greater than in highly industrialized areas. Biomass burning in tropical and temperate forests worldwide is believed to be responsible for nearly 40% of world tropospheric ozone levels, 30% of carbon monoxide, 40% of carbon particulates, and more than 20% of the world's hydrogen, hydrocarbons, nitrogen oxides, and methyl chloride. Global biomass burning includes mainly the burning of savannas (43% of global biomass burning), agricultural waste after harvesting (23%), forests (18%), and fuelwood (16%); the majority of all fires are started by humans.

For decades trees have been observed dying in central Europe and the United States as a result of air pollution; almost all tree species show some declines at all altitudes. Sulfur dioxide and fluoride emissions at industrial plants in the past have killed and damaged significant stands of trees up to 30 miles away. Heavy metals, acid rain, nitrogen, and other pollutants can act together, synergistically, and place much higher stress on vegetation and trees.

Air pollution reduces crop yields in the United States by as much as 5% to 30% depending on the crop and is estimated to cause losses of $5 billion annually; Sweden estimates harvest losses of 6%; Czech scientists estimate annual harvest losses of $192 million; Europe as a whole suffers billions in crop losses yearly. Some of the worst phytotoxic chemicals are ground-level ozone, sulfur dioxide, nitrogen oxides, flouride, ethylene, and peroxyacetyl nitrate (PAN); ozone is responsible for the majority of crop losses.

In Magnitogorsk, Russia, the Lenin Steel Works put out a million tons of air pollution every year in the early 1990s, more than 2 tons for each of the city's 430,000 inhabitants. Satellite photos showed a stretch of black air and soil 120 miles long and 40 miles wide. The city and central government had no money to revamp the vastly outdated Stalinist factory. In the early 1990s, 90% of the children there had respiratory problems, and people in their 30s died from lung disease. Every day children and older people went to clinics for a few breaths of pure oxygen. In Moscow less than 30% of all factories have air pollution emission controls; in Hungary in the early 1990s it was estimated that 1 out of every 17 deaths was caused by air pollution.

Anywhere from 70,000 to 80,000 chemicals are commonly used in industry worldwide; 80% have not been studied for health or environmental effects. A number of scientists, doctors, and organizations believe that up to 80% of all cancer cases may be related to environmental toxics; in 1991 U.S. industry emitted about 2 billion pounds of air toxics.

In 1984, a Union Carbide plant in Bhopal, India, allowed 30 tons of methyl isocyanate to escape into the air of a slum area and failed to inform the people living nearby. Two hundred thousand were exposed. More than 2,500 died, almost 20,000 were disabled, and thousands are expected to die from related diseases. Thousands of animals died as well. The gas escaped because of poor safeguards, poor maintenance, and poor management. In the preceding 4 years, several accidents had occurred in the same plant, and yet the number of monitoring and safety staff had been reduced. This type of negligent record is common in developing countries that are desperate for industry to provide jobs.

Health: Air pollutants can cause immediate damage to lungs and respiratory systems and can increase asthma, allergies, lung cancer, and early deaths. Eyes, skin, and nasal passages are threatened by exposure to toxic, acidic air. Other illnesses from air pollution are stomach poisoning, heart disease, and headaches, including migraines. Particulates in dust from construction, forest fires, and burning fuels cause the premature deaths of some 50,000 Americans yearly. Small particulates (10 microns, or 1/1000 of a millimeter) can penetrate deep into the lungs and bloodstream and cause respiratory illness and blood vessel diseases. The 135 million Americans under age 17 and over 65 are most susceptible to air pollution because they have weaker immune systems; asthma has been increasing steadily among children (reaching 5% of all children) in some large cities and is less common in developing countries.

Sick-building syndrome: People with health problems related to indoor air pollution are said to have sick-building syndrome. People spend 90% of their time indoors, where air is more concentrated and circulates less than it does outside. Often office or household products, equipment, or materials contain and give off toxic fumes. Symptoms include lung disorders, nasal congestion, skin rashes, inability to concentrate, headaches, fatigue, and irritability. Industrial indoor air can be even more toxic and lead to more severe illnesses, such as chronic lung diseases or cancer.

Indirect exposure: People are indirectly exposed to air pollution when toxics are precipitated out of the air by snow or rain onto soil and crops.

Cigarettes: Cigarettes are the greatest cause of premature deaths in the United States. Tobacco-related diseases and illnesses caused by smoking kill more than 400,000 Americans yearly, and second-hand smoke inhalation kills 53,000 nonsmokers yearly. Smoking has been found to increase the risk of contracting colon cancer, even if a smoker quits smoking. In developing countries smoking rates are higher, and restrictions on smokers are nearly nonexistent. More than 4,000 chemical compounds and more than 40 carcinogens, including carbon monoxide and benzene, have been found in cigarette smoke. The effects of second-hand cigarette smoke on children is many times worse than on adults; in custody cases, more courts are removing children from the home of an indoor smoking parent; the trend is to consider it child abuse to smoke around a minor in the home.

Food and crops: Air pollution reduces food and timber crop yields and threatens fish stocks.

Individual Solutions

- Practice energy conservation: get a house energy audit; drive fuel-efficient cars; keep your car tuned; use mass transit or a bicycle.
- Buy an electric or reel (push) lawn mower; purchase tools with electric motors or hand tools.
- Buy recyclable products; recycle; buy food and other products in bulk to eliminate packaging; buy durable goods.
- Remove indoor toxics; contact the EPA or a local environmental group for a list of household toxics and safe alternatives; purchase nontoxics such as natural carpet with no coatings or harmful backing; avoid petrochemical products. Keep toxics out of the trash; if incinerated they add to toxic air pollution. Check your house for radon.
- Maintain good circulation of indoor air; avoid use of gas stoves and ovens, kerosene heaters, wood stoves, and fireplaces, which are all sources of pollutants; install a home air-cleaning system; be aware that chlorinated swimming pools release chloroform into the air; indoors it is trapped inside and thus builds up.
- Oppose incineration.
- Write or call local, state, and national officials to support zero-emission (electric) vehicles, stronger utility emission controls, and an immediate ban of all ozone-depleting chemicals, including hydrochlorofluorocarbons (HCFCs) and hydrofluorocarbons (HFCs).

Industrial/Political Solutions

What's Being Done

- Air pollution is recognized as a global problem by most countries.
- Catalytic converters are required in Canada, Japan, the United States, and much of Europe.
- Some developing countries are becoming more aggressive in demanding corporate responsibility for pollution and worker safety.
- CFCs are being phased out worldwide; proposed alternatives are HFCs and HCFCs.
- Recycling and source reduction are replacing incineration as approaches to solid waste disposal.
- The 1990 Clean Air Act Amendments address the following: stricter emissions standards on automobile tailpipe pollutants; VOC reductions by 1996; onboard automobile vapor recovery devices for gasoline by 1996; requirements for cleaner fuels and cleaner-running vehicles in the most polluted cities; more controls and limits on pollution emissions at other sources, such as coke ovens; regulations and limits on emissions of almost 200 air toxics; reduction in acid rain through limits on sulfur dioxide and nitrous oxide emissions; phasing out of most ozone-depleting chemicals; and strict enforcement penalties.
- U.S. industries must monitor and report potential health risks caused by routine emissions.
- U.S. utilities must buy emission allowances to emit air pollution.
- By October 1993, the EPA required highway buses and trucks to burn low-sulfur diesel fuel (with 80% less sulfur), to reduce toxic particulate emissions by 90%.
- By November 15, 1993, all states were required to develop emission permit programs to comply with the minimum standards of the 1990 Clean Air Act.
- In California, pollution from nonvehicular engines must be cut by 46% by 1995 and by another 55% by 1999; federal regulations for August 1, 1996, are aimed at reducing hydrocarbons by 32% and carbon monoxide by 14% by 2003.

- Smokers are being restricted, and lawsuits are being filed against tobacco companies for distributing a toxic product. Many fast food chains and other restaurants have banned all smoking.
- Some states are mandating prison sentences for executives of companies that intentionally release serious toxic pollutants.
- Some states are setting their own emissions criteria; California, for instance, is requiring that by 2003, 10% of the cars sold there must be zero-emission cars—that is, electric. Ten other states say they will follow California's lead.

What Needs to Be Done

- Strengthen the Clean Air Act by ending emissions allowances, focusing on pollution prevention, and aiming for zero industrial discharges.
- Require new buildings to be free of petrochemical products.
- Require industries to have plans to limit pollution before they begin a manufacturing process.
- Phase out incineration.
- Make U.S. companies subject to prosecution for pollution negligence in other countries (this step is especially important under the North American Free Trade Agreement, or NAFTA); delete provisions that prevent environmental regulations from being applied in all three NAFTA countries (Canada, Mexico, and the United States).
- Mandate that U.S. cigarette companies include health warnings on all packages sold anywhere; end all government subsidies for the tobacco industry.
- End fossil fuel subsidies.
- End the use of all ozone depleters immediately, including the proposed alternatives: HCFCs, which are also ozone depleters, and HFCs (both are greenhouse gases).
- Maximize technology that already exists, such as cars that can travel more than 100 miles per gallon of fuel and burn much cleaner; phase in zero-emission vehicles.
- Strengthen world efforts to substantially reduce global biomass burning, such as the burning of savannas, forests, agricultural wastes, and fuelwood by humans.

- Help developing countries gain access to family planning education and services. Increases in population raise the amount of energy consumed, electricity used, and raw materials mined and multiply the number of cars used and goods manufactured. Thus population growth adds to the emission of pollutants.

Also See

- Acid rain
- Greenhouse effect
- Ozone (O_3) depletion, stratospheric
- Smog

A

Asbestos

Definition

A group of naturally occurring mineral fibers made of silicates.

Chrysotiles (white asbestos) are soft, long, hollow fibers used in cement, fireproof insulation, textiles, flooring, electronic components, and car brake linings. They are responsible for 90% to 95% of past total asbestos usage.

Amphiboles are small, solid, hard fibers used in gaskets and underground cement pipes. They account for 5% to 10% of past usage, primarily in military and industrial applications.

Asbestos fibers are flexible, poor conductors of electricity, resistant to chemicals, fireproof, soundproof, and nonbiodegradable. They have often been mixed with cement to make it stronger. As any asbestos ages, it begins to crumble into tiny fibers that can remain suspended in the air for hours, days, or almost indefinitely in indoor areas.

Environmental Impact

Dispersal: If pipes burst or if construction is damaged, large amounts of asbestos fibers can be released into the air or water. When discarded, these fibers will be either incinerated or thrown into a landfill. From landfills, mining operations, cleanup sites, and manufacturing sites, asbestos fibers can enter the air and water supplies. It is now believed that removal of newer, nondisintegrating asbestos may release more asbestos into the air than was present before removal.

Major Sources

Environment: Asbestos is found in certain rock formations. Chrysotile is present in Arizona, California, Canada, the Italian Alps, and the Ural Mountains of Russia; amphibole is found in South Africa. The mineral is separated into fine fibers invisible to the eye, which are bound into bundles to be used in products. Asbestos is released into the environment from landfills and during mining, product manufacture, and removal. Asbestos fibers can travel hundreds of miles through water and air before settling.

Construction: Currently the major sources of asbestos exposure are in older buildings that used asbestos in piping insulation, heating insulation, vinyl floor tiles, and roofing shingles. Pipes and ducts are still insulated with asbestos in many public buildings, including schools and hospitals. When asbestos crumbles, the fibers float into the air and can remain suspended for long periods; removal of asbestos fibers can also have the effect of dispersing it.

State of the Earth

Asbestos has been used throughout history. Clay pots in Finland (2500 B.C.) contained asbestos for added strength. Dead pharoahs of Egypt were embalmed in asbestos cloth. The Greeks used asbestos for candlewicks in temples. Charlemagne had an asbestos tablecloth that he cleaned by throwing it into a fire. The Romans noticed that slaves who wove asbestos frequently died of breathing difficulty, most likely as a result of asbestosis (see "Human Impact"). Firemen in Paris used asbestos garments as early as 1853. The widespread use of asbestos for piping and insulation in homes, schools, and office buildings began in the 1930s.

It is estimated that at least 700,000 commercial and public buildings and more than 30,000 schools contain asbestos that has broken down or will eventually break down into its hazardous state. Studies of several hundred school maintenance workers in Los Angeles and New Jersey showed 30% with signs of asbestosis. In New York City it is estimated that as many as 2/3 of the buildings contain asbestos; at 90% of the sites the asbestos is deteriorating. In 1991 there were more than 90,000 court cases related to asbestos disease on file in federal and state courts, which could cost up to $50 billion in restitution from the asbestos companies.

In the United States 65,000 people are estimated to be suffering from asbestosis, with 5,000 to 10,000 asbestos-related deaths annually. Autopsies show that large percentages of urban dwellers may have asbestos fibers in their lungs; in urban areas asbestos is common in the air. Asbestos fibers are more buoyant than dust and may float in the air almost indefinitely. The U.S. Environmental Protection Agency (EPA) has stated that there is no safe level of exposure to asbestos. Many manufacturers of asbestos have moved their companies to developing countries that do not have asbestos regulations, to make and sell their products there. Asbestos has been used in sugar mills and oil refineries in Brazil, China, and Mexico.

Human Impact

Exposure: Asbestos can enter the body through the skin (asbestos miners, asbestos workers, and bystanders to those activities are most susceptible), through the lungs (the most common and most dangerous type of exposure), or through ingestion of asbestos in water, which in the past occurred mainly in municipal water supplies not closely monitored and purified. Ingestion through food is rare. Asbestos in the skin can cause corns and, if exposure is large enough, an arthritic response and clubbing of the fingers (where exposure most often occurs).

Asbestosis: As asbestos fibers age they break down, giving off minute particles that can be inhaled. When inhaled, the fibers lodge in the lungs and bronchial tubes, causing a crippling lung disease known as asbestosis. Eventually the victim cannot breathe. This disease is thought to be mostly due to an amphibole fiber called crocidolite (blue asbestos), but many experts believe all asbestos presents a danger of cancer.

Cancer: Asbestos can cause cancer in the lungs, gastrointestinal tract, stomach, intestines, and rectum, as well as cancer of the inner lining of the chest (mesothelioma—always fatal, often due to crocidolite). Once in the body, asbestos fibers can remain there indefinitely and can move from the lungs to the brain and almost any other organ in the body. One very small exposure to asbestos can lead to cancer even 40 years later. Cigarette smokers exposed to asbestos may increase their risk of lung cancer 50 to 100 times over that of unexposed nonsmokers.

Worker safety: In the past most victims have been workers in asbestos factories, industry, and other areas where asbestos is used daily. Old buildings that are torn down can be hazardous to construction workers if the buildings contain asbestos in their materials.

Individual Solutions

- Before doing any construction or modifications in older buildings, or on even relatively new ones, be sure to find out if asbestos was used anywhere in the construction. If uncertain, have the facility (pipes, ceilings, cement) inspected before you begin, so you can take precautions.
- Make sure anyone who performs asbestos removal for you is a certified professional.
- If considering buying an older building, find out if asbestos was used in any part of its construction; have any crumbling asbestos removed before the purchase, or include the cost of removal in the purchase price. Have an asbestos expert assess any health risks for removal or retaining asbestos materials, and use that information to guide your decision to purchase.

Industrial/Political Solutions

What's Being Done

- Europe has partial bans on asbestos use; blue asbestos is banned in many countries.
- The EPA will phase out 95% of all asbestos uses by the late 1990s.
- The EPA has set standards for asbestos removal in schools.

A

What Needs to Be Done

- Ban selling of asbestos to developing countries.
- Phase out all asbestos uses.
- Document where existing asbestos is located in older construction, and fund its gradual removal, beginning with the areas that are most hazardous to human health.

Also See

- Hazardous waste
- Toxic chemicals

Biomass energy

Renewable energy generated from plant material or manure by burning it to produce heat or electricity or by changing it into a gas or liquid fuel with heat, fermentation, or anaerobic digestion with bacteria. In general, biomass energy is derived from solar energy captured by plants during photosynthesis.

Biomass material is often called biofuel either before or after processing. Fossil fuels are a special form of biomass, since they were formed from decomposed and compressed plant and animal material over millions of years; these sources are finite and not renewable. The rest of this discussion is related to biomass energy that is or can be renewable.

Biomass is the most diversified form of energy among the renewable energies, mainly because different forms of biomass material grow or are available in different locations and climates. No one type of biomass is cheaper or easier to grow or obtain than all others at all places on the planet.

Depending on which biomass energies are used and how they are used, biomass energy sources can range from being nonpolluting and very cheap to polluting and costly. Plant biomass is considered renewable only if it is regrown; it is sustainable only if it is grown without artificial fertilizers, pesticides, or other chemicals.

Municipal solid waste and methane from landfills are also considered sources of biomass energy; however, they are not considered renewable sources of energy, since the creation of solid waste results in the use of more raw materials, which are not replenished (see sections on Incineration, Methane (CH_4), and Solid waste).

Environmental Impact

Air pollution: All air emissions from burning biomass are far less than those produced by burning fossil fuels such as coal and oil; toxics, and sulfur emissions that lead to acid rain, are often negligible; nitrogen oxide emissions are significant and can add to acid rain and smog problems. Burning biomass such as wood or plant crops releases carbon dioxide, carbon monoxide, nitrogen oxides, methane, hydrocarbons, radon, sulfur, and particulates such as soot and ash. But if burned biomass crops are replaced with new crops, the net emissions of carbon are negligible, and thus there is little addition to the greenhouse effect and global warming. Burning trees for fuel without replanting, however, adds significantly to greenhouse gases. Also, particulates in the atmosphere can worsen stratospheric ozone depletion and add to local pollution; emissions such as particulates can be controlled, but worldwide there has been little effort in this area as yet. Biomass produces far less ash than coal, and the biomass ash is much cleaner and can be used as a fertilizer.

Deforestation: Burning trees for fuel is a major cause of deforestation. This practice is causing habitat loss for plants and animals in many areas of the world, especially in developing countries.

Soil degradation: Removing logging or crop residues for biomass energy can deprive soils of necessary nutrients, which are created when such residues decay. Pesticides used to grow biomass crops can pollute soils.

Sewage cleanup: Some biomass plant crops can help convert nontoxic sewage and sludge into plant material and fertilizer. This approach is being taken in some human-created wetlands, which recycle sewage nutrients, keeping them out of rivers and streams.

Cleaner fuels: The use of biomass-derived ethanol and methanol in automobiles reduces pollutants such as nitrogen oxides, benzene, carbon monoxide, other toxics, and unburned hydrocarbons; on the other hand, biomass fuels increase aldehyde emissions, such as formaldehyde. In general, biomass energy sources are much cleaner than fossil fuels since they release no sulfur and much lower levels of other emissions.

Toxics: Burned biomass crops can release heavy metals, toxics, and even radioactive substances that they might have taken up in the soils in which they were grown. Heavy metals can enter the food chain and affect the reproductive abilities of predators, such as hawks.

Major Sources

Fuelwood: Fuelwood is used mainly in developing countries; it is often distinguished from a tree crop, which is planted, cut down, and regrown.

Biogas: Biogas is a mixture of gases (a roughly 60:40 mix of methane and carbon dioxide) emitted from organic matter, manure, or sewage decomposing anaerobically in a biogas digester. The residue is used as fertilizer (superior to raw manure) or sometimes as fish food.

Biomass fuels: Ethanol, methane, and methanol can be made from decomposed or fermented plant material or manure, sometimes with the assistance of biotechnology. Methanol made from fossil fuels is not considered renewable.

Biomass crops: Biomass crops include trees, corn, and perennial plants such as algae, ocean seaweed, and switchgrass. Some can be dried and burned; others are converted into fuel.

Crop waste: Nut shells, corncobs, cornstalks, and sugarcane residue (bagasse) can be burned or converted into fuel.

Forest industry waste products: Wood, paper, pulp scraps and leftovers are usually burned as fuel.

Peat: Peat is partially decomposed vegetation, which can be burned.

Paper: Pelletized or shredded paper with inorganic residue removed can be burned as fuel.

B

State of the Earth

Currently 1/2 the people on earth rely on some type of biomass for their energy needs; roughly 15% of the world's energy needs are met with biomass (and 35% to 40% of the energy needs of developing countries, mostly with fuelwood). Biomass use supplies 4% to 5% of the U.S. energy needs; 15% to 30% is possible. It is estimated that by 2015 15% of U.S. electricity could be supplied from biomass grown on 50 million of 84 million acres of farmland currently kept out of production.

Gasohol is a blend of 10% ethanol and 90% gasoline and is used in nearly 10% of U.S. gasoline; there are 50 fuel-ethanol facilities in the United States, producing 3 billion liters of ethanol a year. Biomass fuels will be price competitive with fossil fuels in the United States by 2000. Ethanol is currently made from corn, sugar beets, and sugarcane, and perhaps in the future it will be made from even more plants. In the United States, corn protein is converted to feed, while corn starch is converted to ethanol. Ethanol would require 40% of the U.S. corn crop to supply U.S. fuel needs. In Albuquerque, New Mexico, 80% of all gasoline contains ethanol; in Des Moines, Iowa, buses have replaced 25% of their diesel fuels with ethanol. Biodiesel (vegetable oil or blends of oil and vegetable oil) is nearly sulfur free, reduces carbon monoxide and hydrocarbons, and yields 2/3 less particulates; Austria is mandating its use in Alpine areas, and it is being tested in France, Germany, Italy, and Switzerland. Currently 1% of transportation fuels come from plant matter from more than 60 biorefineries.

Bagasse supplies 60% of electricity on the island of Kauai, Hawaii, and 40% on the island of Hawaii. In Brazil more than 50% of fuel has come from sugarcane-produced ethanol since 1986; 9 out of 10 cars in Brazil have run on ethanol since 1983. Brazil's ethanol industry employs more than 100,000 workers; it is the largest biomass program in the world. More than a dozen countries currently produce ethanol, and others are considering it.

Dung, wood, crop waste, and charcoal serve as the main cooking fuel for half the world's population; dung and crop waste supply more than

800 million people. Millions of small biogas digesters are in use in India and China (over 6 million there). India's cows generate nearly 600 million tons of manure each year that can be used in biogas digesters. Healthy animals produce 4 to 5 times their body weight in dry dung each year. China is now developing larger systems: In one city, Nanyang, 20,000 households will replace their cooking coal with methane generated from 2 large biogas digesters. The University of Illinois Swine Research Center is studying the use of farm manure in biogas digesters to supply biogas as a supplementary fuel for Illinois farms.

Overall, biomass sources produce energy less efficiently than fossil fuels. Since more biomass is needed to supply equivalent energy needs, biomass is bulkier to transport. Thus local biomass use is best. Biomass used in open fires is very inefficient; villagers in Kenya were given 180,000 wood stoves by one international organization to limit fuelwood consumption. Ten percent of U.S. homes still use fuelwood for some of their heating.

The U.S. forest industry gets more than 50% of its power from wood wastes, and other wood wastes are sold to utilities that use cofiring (burning wood with coal). A number of other companies have used everything from rice hulls and peanut shells to cherry pits to satisfy their fuel needs over the past several decades.

Elephant grass, native to China and Japan, is being tested as an energy source in Germany; it can grow 3 meters in one season and yield 30 tons per hectare in one year. No pesticide is necessary, and it uses fertilizers efficiently. Finland and the United States have been conducting studies of efficient biomass plants and trees. Peat is used for fuel in Ireland, the Netherlands, Scandinavia, Scotland, and the former Soviet Union; peat land is disappearing in Europe but is currently protected in the United States.

Human Impact

Particulates: Fuelwood burned in open air releases particulates that cause and aggravate severe respiratory problems; burning crop biomass in power or heating plants with particulate containment equipment may minimize these effects.

Cleaner fuels: Biogas, from biogas digesters, burns cleanly and causes little pollution. Using crop plants for biomass energy creates less acid rain and air pollution than fossil fuels do and thus causes less crop damage. Ethanol and methanol used in place of gasoline reduce smog, which is responsible for many diseases, including respiratory and skin problems.

Secondary effects: If biomass crops are grown with pesticides or fertilizers, they increase the possibilities of polluting water with nitrates, nutrients, phosphates, and pesticides, which can ruin drinking water, kill fish, and cause cancer. Land use for biomass material may further limit agricultural land use in developing countries or compete with natural habitat.

Economic benefits: Biomass energy can supply more jobs, more incentives for wise land use (for biomass crops), and less dependence on other countries for fuel sources.

Individual Solutions

- Use ethanol or gasohol in your vehicle, if possible.
- Recycle and avoid disposable products, especially paper and tissue products.
- Support funding for renewable and sustainable biomass research.
- Support reforestation programs and sustainable management of forests.

Industrial/Political Solutions

What's Being Done
- Worldwide research is being conducted in most areas of biomass energy.
- U.N. and nongovernmental organization programs are working to help developing countries reforest and use biomass in a sustainable

manner. One example is agroforestry, in which trees are grown with food crops; the trees provide biomass fuel and enhance the crop yield.

- Researchers in Europe and the United States are studying fast-growing hybrid biomass plants.
- The Public Utilities Regulatory Policies Act (PURPA) of 1978 requires public utilities to purchase electricity from small producers and must do so at rates equivalent to avoided cost of replacement or new generation.
- Solar ovens are replacing some fuelwood needs in developing countries.
- The 1992 National Energy Policy Act gives production tax credits for closed-loop biomass (biomass grown only for fuel).
- Use of ethanol and gasohol blends has been mandated in severely polluted areas to lessen carbon monoxide pollution from auto and truck emissions.

What Needs to Be Done

- Reduce particulates from biomass burning.
- Increase research on the releases of heavy metals and toxics in biomass burning.
- Fund biogas converter research for farms.
- Use more ethanol and gasohol to minimize air pollution.
- Replace peat burning with sustainable energies such as solar and wind.
- Minimize fuelwood use worldwide, and create sustainable harvesting and use of biomass sources.

Also See

- Deforestation
- Renewable energy

B

Biotechnology

Definition

Altering living organisms by using cloning, genetic engineering, biochemistry, or other organisms.

Cloning (that is, tissue culture) involves taking a small piece of tissue or cells from one organism and growing other genetically identical organisms from them. Genetic engineering involves using enzymes to "cut" a piece of DNA from one organism and "paste" it to the DNA of another organism. DNA is found in the cells of all living creatures and holds the genetic information that determines growth, cell function, and many other characteristics of an individual. Performing this type of DNA cut-and-paste procedure is also known as transgenic manipulation. Sometimes the terms biotechnology, genetic engineering, recombinant DNA (rDNA) technology, and transgenic manipulation are used interchangeably.

In genetic engineering, the piece of DNA selected to put into another organism usually controls one specific trait. For instance, to make tomatoes grow bigger and faster, the gene controlling large fruit size in one plant might be put into a plant that grows fast, with the hope of ending up with a plant that grows fast with bigger fruit. If the tomato doesn't taste good, researchers might also look for a "taste" gene from another plant. In many animals only a small percentage of genes have been identified. And some traits, like milk production in cows, are governed by many different genes.

Biotechnology, often called biotech, can have different impacts on the environment and humans depending on how it is used; many of the long-term health and environmental risks have not yet been adequately assessed. Animal rights have sometimes been sidelined by biotech operations. Also at issue is the ownership of the genetic material that is used in biotech products.

Environmental Impact

New species: With genetic engineering, new species of animals and plants are being created that could never occur in nature. Biotech plants that are grown with new genetic traits, such as pesticide resistance, better nitrogen fixation to grow faster, or drought and acid resistance, could transfer these traits through pollen to weeds, making them more difficult to control. Biotech plants may also alter natural ecosystems. Bacteria and viruses that are genetically engineered might escape into the environment and self-replicate, causing harm that cannot be foreseen; bacteria used to clean up an oil or toxic spill might also kill valuable organisms in the soil or water or cause eutrophication in the water, which could kill native plants and fish. Bacteria designed by biotechnology to destroy crop pests might also destroy nontarget species, just as pesticides do. Genetically engineered animals such as farm animals or fish might act like exotic species and compete with native species, potentially upsetting the balance of an entire ecosystem. For example, if cows are engineered to be resistant to tsetse flies (which carry sleeping sickness in Africa), these cows might then take over the grazing land of many wild animals, which would lose their habitat.

Pesticide use: Using biotechnology to create herbicide-resistant plants will lead to continued use of herbicides (a form of pesticide). Many herbicides are toxic to wildlife and can enter the food chain. They are washed off the soil into water supplies and are taken up by fish, predators, and smaller organisms. Herbicides can cause cancer, reproductive problems, behavior problems, and the death of young and adult animals; even biodegradable herbicides can break down into more toxic components.

Animal welfare: Growth hormones such as bovine growth hormone (BGH) (also called bovine somatotrophin or BST) have had adverse effects (such as infections) on the cattle they have been used on to increase milk production. In initial and current biotech research other kinds of animals have been genetically manipulated and have suffered deformities, diseases, and crippling effects. There are also animal welfare concerns for the animals being used in medical research. Mice have been given human genes that cause cancer in them, so that researchers can study the cancer.

Extinction: If the focus of biotechnology is to create "perfect" animals and plants to be mass-produced for human consumption, natural native varieties may be lost through lack of agricultural use.

Major Sources

Major biotech products, or areas of research, are:

- Animals (such as transgenic mice) that have cancer and other diseases for medical research.
- Growth hormones for animals (cows, pigs, fish) that improve digestion; increase muscle mass or decrease fat; and increase size, rate of growth, or rate of milk production. Also being researched are farm and aquaculture animals that are resistant to viruses or other contagious diseases.
- Pharmaceutically useful human protein production in animals, such as mice, fish, and rabbits; in cow and goat milk; or even in algae, which can then be harvested.
- Different food crop plants, trees, and algae so they can fix their own nitrogen; grow bigger faster; or are resistant to drought, heavy metals, frost, salt-laden soils, acid soil, herbicides, or insects that might damage them.
- Animal vaccines, by modifying viruses.
- Human medicines and diagnostics, by creating biotech variations of natural plant chemicals; other examples are the growth hormone for dwarfism, insulin for diabetics, erythropoietin for renal failure, vaccines, and antibodies.
- Biological controls such as insects, bacteria (bacterial pesticides), viruses, or plants that will prevent crop losses to pests or weeds.

- Neutriceuticals, which are genetically engineered food products used for a specific health effect, such as psyllium fiber taken to reduce cancer risks and garlic extract to reduce blood cholesterol.
- Bioremediation, which uses microorganisms like bacteria, fungus, or others to break down specific toxic materials such as polychlorinated biphenyls (PCBs), hydrocarbons, and oil in air, water, or soil. Natural organisms can be encouraged to grow with aeration or nutrients, or genetically altered organisms can be introduced.
- Food-processing enzymes and microbes for cheese, fruit juice, food flavors and enhancers, yeast hybrids, animal feed, and conversion of starch.
- Industrial enzyme catalysts, which speed up chemical reactions.
- Biotech bacteria that help the fermentation or anaerobic digestion of a biomass energy resource to convert it to a usable fuel like ethanol or methane.
- Microbes that recover metals; this process is called biohydrometal-lurgy.
- Disease organisms for biological warfare.

State of the Earth

Since the first gene was cloned in 1973, human genes have been implanted in animals, animal genes into plants, animal genes into other animals, and plant genes into other plants. Biotech products can be patented, if artificially produced and not a product of nature. The U.S. patent office has received many patent applications; many of these are pending and are for genetically engineered mice and farm animals.

Mice have been implanted with human genes so they produce human enzymes that can be harvested. Fish, such as carp and salmon, are being engineered to grow twice as large as normal. A goat-sheep hybrid has been created by combining the embryos of a goat and a sheep; chicken genes have been put into potatoes; cow genes into chickens; and fish genes into fruits and vegetables.

The first U.S. field trials for recombinant (biotech) plants were held in 1986 and involved tobacco; 94% of all trials since have been held in the

United States (37%), Canada (36%), France (9%), Belgium (7%), and the United Kingdom (5%). Most experiments have involved rapeseed, potatoes, tomatoes, tobacco, corn, and flax; herbicide tolerance is the predominant trait field tested up to 1992. The U.S. Environmental Protection Agency (EPA) approved the first biopesticide for sale in the United States in 1991.

Researchers at Montana State University are developing herbicide-resistant safflower plants; herbicide-resistant cotton seeds have already been developed. A hairy potato has been tested successfully on 5 continents and will be introduced in 30 to 40 countries in 1994—its sticky hairs trap insects. Hybrid poplar trees that grow 4 to 7 feet a year are being studied as a biomass source for ethanol or paper. Hybrid elms are being bred to resist Dutch elm fungus, elm leaf beetles, and elm yellow diseases.

About 1,200 biotech companies are now in business; by the year 2000 Japanese biotech sales should reach about $39 billion. Neutraceuticals are big business in Europe, Japan, and the United States. Researchers are close to creating livestock drug factories, where farm animals will produce medicines, proteins, and other useful products in their milk, which will then be harvested. This idea is projected to reach commercialization by 2001. Protein from oysters is being grown for use in detergents; it is biodegradable.

The Green Revolution was labeled thus because it increased agricultural food production by using hybrid plants that grow fast and increase crop yields. Hybrids, however, are more vulnerable to pests than natural plants and need more irrigation. These plants also need more fertilizer and pesticides, which increase costs to farmers. Pesticides reduce soil fertility as well. Thus increased production from the Green Revolution has peaked, and pesticide use is recognized as detrimental to human health and soil quality. Biotechnology seeks to correct these problems with more expensive technology, which could lead to even greater environmental problems; sustainable agriculture can give sustained crop yields as great as or greater than use of either of the above technologies, without any of the associated risks or detriments, and it enriches the soil for future generations.

Plants are being given bacterial genes of *Bacillus thuringiensis (B.t.)* so they can create proteins that are toxic to moths, fly larvae, and other pests; B.t. is already used as a bacterial pesticide. Limited use of B.t. has been effective for 25 years, but if B.t. genes are used in plants, insects will develop immunity, just as they do with current pesticides. Already the Indian meal moth has developed resistance to B.t.

March 1994 research showed that pieces of viruses injected into a crop plant's genetic code to protect it from viral disease can combine with newly introduced viruses to form a new virus, and thus a new plant disease, which may be harder to control than the original virus.

BGH/BST is created from a natural cow pituitary hormone using biotech techniques; it mimics a natural process in the cow and increases milk production by 10% to 20%. These cows require more feed; BGH can cause mastitis in cows, which may then need more antibiotics. Residues of antibiotics can end up in the milk and can render people susceptible to allergies and resistant to antibodies. Trypanosomiasis— sleeping sickness spread by tsetse flies—is present in 38 African countries and puts up to 60 million cattle at risk. Creating cattle that are immune to the disease is expected to help increase cattle production, which will also reduce wild animal ranges. A genetically engineered virus for cattle immunity to rinderpest is also being developed.

Bioremediation has been used around the world to clean up petroleum spills, stains on concrete and road surfaces, and drilling waste in soils. Researchers are also studying the use of bioremediation to consume nuclear wastes such as uranium, toxics such as PCBs, organic chemicals, and heavy metals. Most uses involve natural organisms; experts believe it will not be until the year 2000 that genetically altered organisms will be widely used in bioremediation. Microbe-enhanced fertilizers were used to fight the 1989 *Exxon Valdez* oil spill off the coast of Alaska.

A dozen large chemical and petrochemical corporations control the global seed market and biotech markets. Some of them are offering African farmers the use of their genetically altered crops in exchange for royalties on crop yields.

Biologists are seeking genetically engineered plants that can increase crop yields in areas with high soil acidity, low moisture, or high heat. They believe that continued research such as this will allow more marginal land to be cultivated to meet the continuing increases in population, especially in developing countries.

The Human Genome Project involves taking samples of blood and hair from 25 people in each of the 600 recognized population groups in the world. Researchers will analyze the DNA in an attempt to see relations between the origins and past migrations of peoples and to correlate diseases and susceptibility to diseases with certain genetic makeups. The human genome is made up of about 100,000 genes and 3 billion base pairs; thus far about 0.5% of the human genome is known. In the United States the National Institutes of Health and the Department of Energy are hoping to finish mapping the U.S. population by 2005. Animal embryos have been cloned for some time; recently human embryos have also been cloned.

Human Impact

Health: Biotechnology is already used in some food processing, but no one knows the long-term health effects of consuming other proposed plant and animal products that have been genetically altered, especially if protein, carbohydrate, and oil percentages and types are changed or if genes from animals end up in plants. Herbicide-resistant plants would lead to continued use of herbicides, which can cause cancer, skin problems, and liver and kidney failure. Most herbicides have not been tested for long-term health effects. Genetically engineered vaccines, injected into animals, can be contagious to humans; contagion has already occurred in a field trial in Argentina. Use of more antibiotics for farm animals due to BGH or other biotechnology applications could lead to other health problems for consumers.

Food production: Production of food is expected to increase, especially in developing countries with marginal land, where genetically engineered crop plants and livestock can tolerate poor soil or other adverse regional conditions. Poor soil conditions, however, are often human-created. As a first step it would be better to correct the source of these problems instead of relying on an expensive new technology as a "fix." Giving hybrids greater genetic strength by splicing in genes from wild varieties would help ensure better crop yields.

Unknowns: It is unknown whether biotech animals and plants, if released into nature, would alter the environment or food chains in a way that is detrimental to humans. Mass use of hybrid plants and animals may accelerate extinction of natural varieties of plants and animals.

Medicine: Medicines from rain forest plants, altered and mass-produced using biotech procedures, will make large quantities of chemicals available to many people who otherwise might not have access to them. It is believed there may be thousands of cancer-healing drugs in the rain forests. Already biotech is used to create numerous proteins, enzymes, vaccines, diagnostics (such as antibodies and recombinant DNA (rDNA), for treating infectious disease, marking tumors, drug monitoring, and blood screening), and other products that are used in medical technology. In the future, biotechnology may be able to isolate the specific genes that cause specific diseases and alter them through gene therapy to prevent the disease.

Fuels: Production of some biomass fuels and other industrial processes rely on biotechnology.

Farmers: Small and mid-size farm losses will accelerate owing to increased dependence on expensive biotech applications. Multinational corporations will have more control over what foods are produced. Farmers in developing countries may become dependent on herbicide-resistant crops and may face continuing high costs, soil degradation, and environmental pollution.

Bioremediation: Bioremediation may be of major benefit in reducing serious hazardous pollutants from industrial wastes, oil spills, hazardous dumps, Superfund sites, or other areas.

Costs: Biotechnology is expected to further raise the cost of high-input farming already characterized by the use of pesticides, artificial fertilizer, and hybrid crop seed. It will also extend the cost of high-input chemical farming to future generations, in terms of soil degradation. Biotech could help farmers in developing countries avoid pesticide and fertilizer expenses if genetically altered crop seed that can be grown without additives on marginal land is in itself not expensive.

Discrimination: As human gene mapping progresses, people with certain genotypes may face discrimination by employers and especially by insurance companies that believe them to be high medical risks.

Genetic control: There is concern that large corporations, through patenting, will gain control of the genetic stock of the planet. Plants or animals with altered genes might contain traits that already exist in other plants, but patents could prevent public access and use of such traits for research. Farmers in India protested in 1993 because, as a result of rules set by the General Agreement on Tariffs and Trade (GATT), they could not keep seed from their crops to grow future crops but instead had to purchase new seed from suppliers every year. Biotech plant products sold on the market may also outcompete and replace the original plant; such a development would economically devastate regions that are dependent on a few major crops. Examples of vulnerable crops are sugar, cocoa, and palm oil.

Individual Solutions

- Buy organic foods from food cooperatives or farmers practicing sustainable agriculture; they do not use pesticides and often use natural varieties of plants.
- Eat less red meat. Less meat consumption could increase food production worldwide more than any new super-plant, since cattle consume huge amounts of grain and require large acreage.
- Support laws limiting the release of genetically engineered plants, animals, or microorganisms into the environment until each has been thoroughly studied and its possible environmental impacts are well understood and found to be benign (see *Superpigs and Wondercorn,* by Michael W. Fox).
- If you are concerned about animal welfare and biotechnology, contact

People for the Ethical Treatment of Animals (PETA) or the Humane Society.

- Write your legislators and demand long-term health testing on biotech products before they are sold commercially; also demand labeling for all biotech foods and BGH/BST-produced milk so that consumers have a choice.

Industrial/Political Solutions

What's Being Done

- The Convention on Biological Diversity states that nations have rights to genetic resources in their territories and can regulate access to them or require payment for use of them; 156 nations have already signed it, including the United States in 1993.
- The international Biological Weapons Convention, signed by the United States in 1972, prohibits research on and use of biological weapons.
- Some pharmaceutical companies are paying rain forest countries service fees and royalties to research their genetic plant material for possible medicines; interest in biotech products may help stave off some rain forest destruction.
- Major corporations are spending billions seeking patents and doing research to create "perfect" and more productive produce and farm animals.
- The three major U.S. federal agencies regulating biotechnology products are the Food and Drug Administration (FDA), the Environmental Protection Agency (EPA), and the Department of Agriculture (USDA) acting under the Federal Policy on Biotechnology, established in 1984. The U.S. General Accounting Office has stated that the USDA has no clear policy and has done little to educate the public about the risks and benefits of biotechnology. All three agencies have poor records on regulation.
- The U.S. FDA is not requiring long-term health testing for genetically engineered foods, has approved BGH/BST, and has decided that BGH/BST-produced milk does not require labeling. The FDA does

not require biotech plants to be labeled but does require that biotech foods containing genes from allergic foods, such as fish and nuts, be labeled.

- The EPA is spending $10 million on innovative cleanup technologies, the majority of which involve bioremediation; the EPA, the Department of Defense, and the Department of Energy together spent $83 million in 1993 for research on environmental biotechnology; almost 400 companies make bioremediation products. The EPA has stated it supports integrated pest management instead of heavy use of chemical pesticides.

- Corporations are funding biotech research at universities, which are also working with the U.S. Department of Agriculture and the U.S. Forest Service to develop herbicide-resistant crops and trees.

- Biotechnology has already been used in many applications: in creating new strains of flowers and crops, in food processing and additives, in microbes and mice for research and industry, in biomass energy production, and in research for "pharming" of genetically engineered drugs in goats, cows, and sheep for human use.

- The U.S. Department of Defense is doing research involving genetic engineering to develop highly sensitive tests to detect biological weapons.

- The National Institutes of Health have applied for hundreds of patents on human genes.

- The first genetically engineered food (slow-rot tomatoes) were introduced to the marketplace in spring 1994.

- A number of major grocery chains have said they will not stock milk produced from BGH-injected cows; seventy dairies and food companies, including Ben & Jerry's Ice Cream, Borden's, and several major infant formula manufacturers, have pledged to boycott BGH-produced milk.

- Allow public debate and involvement in policy making regarding biotechnology applications that pose unknown risks to health and the environment.
- Create an independent interdisciplinary ethics advisory board for the biotech industry, made up of scientists, industry representatives, farm representatives, wildlife biologists, and public representatives.
- Label all biotech foods, such as BGH-produced milk, so consumers have a choice.
- Increase support for small and mid-size farmers; expand subsidies and support for organic and sustainable farmers.
- Mandate that biotech food products receive long-term health testing before they can be sold to consumers.
- End invertebrate and vertebrate animal patenting to prevent monopolies on genetic stock.
- Encourage all countries to declare sovereignty over their genetic resources and to regulate the collection of genetic material. Developing countries should control industrial nations' access to their genetic diversity by restricting transnational corporations, international agencies, and scientists working for either from using genetic material unless royalties, fees, or other arrangements are agreed on beforehand.
- Conduct thorough research to establish a zero-risk assessment and engage in public debate before releasing any genetically engineered plants, animals, or microorganisms into the environment. Genetically engineered microorganisms should not be released into the environment unless they are bred to remain within the host plant or animal and bred to die within a given period of time.
- Ban all pesticides and herbicides until they are tested for long-term health effects.
- Decrease meat production worldwide to make more grain available and to free more land for agriculture instead of grazing.
- Support sustainable agriculture; educate farmers on how to farm sustainably in all countries.

B

- Push the EPA to end support for the use of herbicide-resistant crop plants.
- Ban any agreements or laws that prohibit farmers from using crop seed to grow new crops. Such a step will prevent a few major international seed companies, such as Pioneer, Sandoz, Dow, Imperial Chemical Industries, Monsanto, Shell, Ciba-Geigy, and Atlantic Richfield, from gaining complete control over global crop seed genetic stock.

Also See

- Exotic species
- Pesticides

Chlorine (Cl)

A yellow-green gas, liquid, or crystal.

Chlorine is highly toxic and combines with water vapor to form hydrochloric acid and with organic material to form chloroform. Chlorine is used in numerous products, such as plastics, and in chemicals such as chlorofluorocarbons (CFCs) and polychlorinated biphenyls (PCBs).

Ozone depletion: When CFCs reach the upper atmosphere, chlorine (Cl) reacts with ozone (O_3) in the following way: $Cl + O_3 = O_2$ (oxygen) $+ ClO$ (chlorine monoxide), then $ClO + O = O_2 + Cl$. The chlorine molecule continues to break apart ozone molecules, up to 100,000, before it finally breaks down itself. This process can be accelerated when particulates or ice crystals are present in the air. Ozone destruction increases the ultraviolet radiation that reaches the earth; ultraviolet inhibits plant growth and can destroy phytoplankton in the ocean, disrupting the food chain.

Greenhouse effect: CFCs act as powerful greenhouse gases.

Toxic chemicals: PCBs, dioxins, other organochlorines, and pesticides with chlorine are found worldwide in animal populations, water, and soils. They are persistent in the environment and bioaccumulate in the food chain. They are very toxic to animals and can cause reproductive failure, neurological damage, birth defects, cancer, and death in both young and adult animals.

Biocide: Chlorine in water kills fish, aquatic plants, and small organisms like bacteria, fungi, and algae.

C

Major Sources

Natural sources: Chlorine is found in seawater and on land in deposits of sodium or potassium chloride; electrolysis of seawater and salt yields chlorine. About 1,500 organochlorines occur naturally, produced in very tiny quantities by bacteria, fungi, and sea creatures. Chlorine can be released by volcanic eruptions, by evaporation of seawater, and from biomass burning (which accounts for nearly 1/4 of global emissions).

Industry: Chlorine is a common chemical in industrial wastewater. It is also used to create a number of well-known chemicals, such as CFCs, chlorinated hydrocarbons (also called organochlorines) such as PCBs and DDT, industrial solvents, and plastics (polyvinyl chloride, or PVC). Chlorine is common in a number of household products, such as table salt (sodium chloride), bleach, and cleansers. Chlorine is used by the paper industry to bleach paper.

By-products: Some chlorinated chemicals (those that have chlorine attached to them), like dioxin, are by-products of processes such as incineration of plastic and paper in municipal solid waste.

Disinfectants: Chlorine is added to drinking water, swimming pools, and wastewater as a disinfectant.

State of the Earth

The U.S. Environmental Protection Agency (EPA) has estimated it is likely that 100% of the U.S. population already has some chlorinated hydrocarbons in their bodies; it also estimates that 40% of the hazardous wastes incinerated in the United States are chlorinated and thus release dioxins. Environmental toxics such as organochlorines are suspected of causing up to 80% of current increases in breast cancer; some experts believe that 80% of all cancer cases may be due to environmental toxics like DDT, PCBs, dioxins, furans, and other synthetic organic chemicals. When Israel cut organochlorine pesticide use, its abnormally high breast cancer rates dropped.

The Seventh Biennial Report on Great Lakes Water Quality was issued in February 1994 by the International Joint Commission (IJC a treaty organization that monitors Great Lakes water quality issues for the United States and Canada). The report states that mounting evidence in the United States and elsewhere shows that exposure to chlorine-based chemicals has resulted in lower sperm counts in men, dwarfed reproductive organs in males, epidemic rates of breast cancer, and declining learning performance and increasing behavior problems in schoolchildren. Some specific examples of evidence are: women with breast cancer in New York were found to have significantly higher levels of dichlorodiphenyldi-chloroethylene (DDE (see glossary)), a pesticide breakdown product; male children of mothers who were exposed to PCB-contaminated cooking oil in Taiwan in the early 1980s have abnormally small penises; in the past 30 to 50 years sperm counts in men have declined by half in the United Kingdom, while male reproductive tract problems have doubled; and leukemia and bladder cancer rates have increased in Cape Cod communities that were exposed to solvents leaching from the vinyl liners of drinking water pipes. In reference to wildlife, the report states: Great Lakes bald eagles are failing to reproduce, and their eggs and young contain PCBs and DDE; Great Lakes gulls exposed to DDT show abnormal development of sexual organs in both sexes, with significant feminized development in males; alligators in Florida exposed to pesticides had penises 1/2 to 1/3 normal length, abnormal testes and ovaries, and elevated estrogen levels; turtles in Florida with the same exposure produced hatchlings with malformed reproductive organs and abnormal hormone levels. In both wildlife and human examples, the study believes chlorine-based toxics interfere with fetal development and alter sexual characteristics in part because the toxics mimic the female hormone estrogen.

Organochlorines have been found in wildlife populations around the world and in soils and ocean sediments. Some predators and other animals with high levels of these chemicals, studied in different areas of the world over the past decades, have plummeted in number because of thinned eggs, neurological and reproductive problems, and weakened immune systems that have led to diseases such as cancer and to increased deaths.

Of the thousands of pesticides in use, most have not been tested for long-term health or environmental effects. The EPA suspects a number of pesticides of being carcinogens but has not removed them from the market. Chlorinated hydrocarbon pesticides, which are banned in the United States, were still being sold to developing countries by U.S. companies in 1993.

Vinyl chloride monomer (VCM), polyvinyl chloride (PVC), and ethylene dichloride (EDC) are the three largest users of chlorine. Some 11,000 organochlorine compounds are produced or used in roughly 1/2 of all processes in petrochemical production. Twelve million tons of chlorine are produced in the United States every year: 5% goes to water treatment, 12% to inorganic chemicals such as hydrochloric acid, 15% to bleaching wood pulp, 30% to chlorinated organic chemicals, and 30% to PVC (the fastest growing use of chlorine).

The ozone layer has thinned far faster than early estimates predicted. For every 1% decrease in ozone, there is a 2% increase in ultraviolet radiation. A 12% to 15% thinning of the ozone layer has been detected over North America; 20% over Alaska, northern Canada, Greenland, Norway, and Siberia. The EPA estimates this depletion could mean 200,000 deaths and 12 million cases of skin cancer over the next 50 years. Chlorine levels from CFCs are leveling off, thanks to CFC phaseouts, and it is hoped that large-scale depletions may be averted by discontinuing use of CFCs.

The greenhouse effect is responsible for keeping the planet warm; greenhouse gases, including human-created chemicals like CFCs, hold heat around the earth. It is believed that global warming is caused by an increase in greenhouse gases, such as CFCs, which are raising the temperature of the planet; over the past 100 years the temperature has risen 1°F.

Human Impact

Direct contact: Chlorine is toxic and is a strong irritant to the eyes, mucous membranes, and respiratory passages. Chlorine gas (Cl_2) is very toxic, and storage leaks can be dangerous.

In water: When chlorine is combined with water as hydrochloric acid (HCl), it is a strong external irritant and a powerful poison if taken internally. Chlorine can also combine with organic chemicals, such as decaying vegetation in water supplies or industrial organic chemicals, to form trihalomethanes (THMs), which cause cancer; chloroform is an example of a THM. Chlorinated drinking water has been shown to increase the risk of rectal and bladder cancer in communities where it has been studied.

Other toxic chemicals: Bleaching paper and burning plastics produce dioxin, which is a toxic chemical. Dioxin interferes with the immune and reproductive systems and causes cancer. Pesticides, DDT, PCBs, and other chlorinated chemicals have been shown to be powerful carcinogens; some mimic hormones and enzymes and thus disrupt cell chemistry, causing kidney failure, birth defects, liver damage, immune problems, neurological problems, breast cancer, and death. Specifically, some chlorine-based chemicals can mimic the activity of the hormone estrogen, increasing the risk of cancer.

Ozone depletion: Chlorine depletes the earth's ozone layer, increasing skin cancer and cataract rates and inhibiting the growth of food crops; it may affect ocean food chains by depleting Antarctic phytoplankton.

Individual Solutions

- Minimize use of plastic; don't buy PVC (used in vinyl siding, gutters, window frames, pipes, and floor tiles) or vinyl furniture, clothing, or toys.
- Oppose incineration.
- Recycle Freon in your air conditioners.
- Find out if your water system uses chlorine; protect yourself with a water filter. Large amounts of chlorine can be absorbed through the skin in chlorinated water showers, baths, and pools. Use a solar swimming pool purifier, or purifiers other than chlorine.
- Garden organically, and avoid pesticide use.
- Avoid using bleach or other chlorine products; call a local environmental group or the EPA for a list of safe household alternatives.

- Write or call your national representatives to support immediate bans on all ozone-depleting chemicals, including hydrochlorofluorocarbons (HCFCs) and hydrofluorocarbons (HFCs).

Industrial/Political Solutions

What's Being Done

- There are alternatives for 99% of all chlorine products.
- CFCs will be phased out worldwide by the mid-1990s; HCFCs (another ozone depleter) and HFCs will be used until early next century.
- In 1992 a 13-nation Paris Convention for Prevention of Marine Pollution on the North Sea called for a chlorine ban; Mediterranean countries have agreed to phase out discharges of toxic organochlorine compounds into the Mediterranean Sea by 2005.
- The majority of uses for PCBs are banned in Europe and the United States.
- Paper manufacturers in Europe are using oxygen-based instead of chlorine-based chemicals.
- A major European PVC flooring manufacturer is phasing out PVC use; PVC packaging is also being reduced in Europe.
- Austria, Germany, and Sweden are moving toward bans on chlorine use. An Austrian pulp mill phased out chlorine and uses ozone for bleaching. A Swedish water purification method using ultraviolet light and reverse osmosis removes 95% of all chemicals and compounds.
- The International Joint Commission on the Great Lakes (IJC) of Canada and the United States has proposed banning all industrial use of chlorine; it concluded that organochlorines in the Great Lakes are responsible for a broad array of human and wildlife diseases that are especially threatening to fetuses and infants. British Columbia, Canada, is complying with the IJC proposal; one of the largest pulp-producing areas in the world, it is phasing out chlorine discharges by pulp and paper mill industries by 2002, with organochlorine reductions starting in 1995.
- The EPA will institute stricter emission standards for dioxins in 1994.

- The American Public Health Association (APHA), an organization of 30,000 doctors and public health officials, has called for the elimination of many industrial uses of chlorine.
- A number of communities are taking chlorine out of their drinking water and using ozone, ultraviolet radiation, or other safer alternatives to disinfect the water; California is conducting a 2-year test using ozone and hydrogen peroxide as a drinking water disinfectant.

What Needs to Be Done

- Substitute alternatives for chlorine use in drinking water.
- Support zero chlorine, waste, and toxic discharge by industry into sewage systems and waterways.
- Separate sewage from industrial wastewater.
- Accelerate the ban on CFCs and the alternatives, HCFCs and HFCs (HCFCs threaten the ozone layer, and both are destructive greenhouse gases).
- Phase out incineration; ban the burning of plastic in incinerators.
- Prevent power plants from flushing water systems with chlorine.
- Limit the use of plastic in packaging.
- Mandate the testing of all pesticides for long-term health and environmental effects; ban use of a pesticide until it has been shown to be safe. All pesticide testing should be done by neutral parties.
- End the sale of banned pesticides by U.S. companies to developing countries.
- Ban the use of chlorine in paper manufacturing; phase out all industrial use of chlorine as a feedstock.
- Require all government agencies to purchase only chlorine-free paper.

Also See

- Dioxins
- Ozone (O_3) depletion, stratospheric

C

Coal

A fossil fuel created through the compression of organic material (largely plant remains) in the layers of the earth's crust over millions of years. It is made up primarily of carbon and hydrogen, as well as varying amounts of mineral matter.

Coal ranges from soft (bituminous—more than 80% carbon) to hard (anthracite—more than 90% carbon). Soft coal is the least valuable and has the highest moisture content, most impurities, most metals, and most volatile gases. Low-grade, soft coal also has the largest amounts of sulfur compounds. Soft coal yields about 10% smoke and dust pollutants when burned. One ton of hard coal yields the same energy as 3 tons of soft coal. Sulfur can be prewashed out of coal, which is then called clean coal.

Peat is the first stage in the formation of coal. Lignite (60% to 70% carbon) is a grade of decomposed material halfway between peat and coal. Natural gas, methanol, and a variety of synthetic fuels (synfuels) can be made from coal. A small percentage of coal is used by the petrochemical industry to make chemicals.

Environmental Impact

Burned coal: Burned coal gives off heavy metals such as mercury, arsenic, lead, and cadmium, as well as creosote, carbon dioxide, radon, particulates, sulfur dioxide, and nitrogen oxides. Thus burned coal is a major contributor to acid rain, air pollution, the greenhouse effect, smog, and possible global warming. Acid rain, air pollution, and smog harm plants and animals. Acid rain destroys lakes, forests, and fish and animal populations. Synthetic fuel and methane produced from coal have many of the same effects on the environment. Heavy metals, toxic to plants and animals, bioaccumulate in the food chain. Sulfur dioxide can damage plants by causing leaf injury and discoloration. Coal fly ash can be toxic with heavy metals and radioactive particles, and when placed in landfills contaminates soils and threatens nearby water sources.

Mining: Coal mines (strip mines and underground tunnels) destroy natural habitats and cause erosion and leaching of acidic particles into nearby soils and water. Coal-mining wastes can lead to air and water pollution. Even though mining companies are required to return land to its original state through reclamation, their efforts are rarely completely successful, especially in areas with low rainfall.

Ozone depletion: Burning coal produces sulfur dioxide, which changes to sulfates such as sulfuric acid in the atmosphere. Sulfates are particulates that can increase ozone destruction by increasing chemical reactions between chlorine and ozone. Ozone blocks ultraviolet rays from the sun; thus thinning ozone can lead to damaged plants. Ultraviolet also damages plankton in Antarctic waters, which may affect the ocean food chains. Sulfates may also block incoming ultraviolet, and the net effect is unclear for localized areas.

Petrochemicals: Chemicals made from coal are often toxic and persist in the environment. They can enter the food chain and harm the health and reproduction of birds and other animals. These chemicals bioaccumulate in predators, which are especially susceptible to their harmful effects.

Major Sources

Mining: Coal can be found deep under ground or very close to the surface, in which case strip-mining is used. Coal beds can be a thin film only inches thick, several feet thick, or up to 50 feet thick. Often coal beds are covered by tens to hundreds of feet of sandstone, shale, or other soil and rock material. Strip-mining is the most effective way to mine coal and recovers 85% to 90% of the coal in a site; it is used for 60% of all coal recovery. About 10% of all mined coal is burned at the mine mouth to provide electric power.

Transport: Coal can be transported by barge or rail or as a slurry—in which it is pulverized, mixed with water, and sent through a pipe. The Black Mesa coal slurry pipeline is 273 miles long and transports 5 million tons of coal annually from Arizona to Nevada, using 2,700 gallons of water per minute. Environmental and water concerns have ended further construction of slurry pipelines.

Use: Coal is the primary source of electric power and is used mainly in power plants. It is also used in developing countries for home heating and cooking.

State of the Earth

World coal reserves are expected to last at least 130 to 200 years. Nearly 3/4 of recoverable reserves are soft coals, which are high in sulfur; this is the type of coal used in most electric utility plants. In the United States only 1% of coal reserves are hard low-sulfur coal, and most of these reserves are located in the Rocky Mountain region, where environmental concerns have prevented mining. Other large reserves are in Appalachia and the central United States.

The 1992 World Energy Conference estimated recoverable bituminous coal reserves: the United States has 29.3% of world reserves; the former Soviet Union, 19.8%; China, 13.5%; India, 8.5%; South Africa, 7.8%; Australia, 6.9%; Poland, 4.2%; Germany, 3.4%; Indonesia, 1.1%; and Canada 0.8%. For lignite, the former Soviet Union has 30.5% of world reserves; Germany, 17.1%; Australia, 12.8%; the United States, 9.7%; Indonesia, 7.3%; China, 5.7%; Yugoslavia, 4.6%; Poland, 3.5%; Turkey, 2.1%; and Bulgaria, 1.1%. China is now the largest producer of coal worldwide and relies on coal for 3/4 of its energy needs; China also experiences extensive crop damage. Many homes in China and other developing countries still burn coal indoors for cooking and heat, contributing to high rates of lung disease.

Centralized coal power plants lose 2/3 of the available coal energy as waste heat; cogeneration can increase efficiency to 60% to 90%. Coal provides about 1/3 of U.S. energy needs and 55% of its electrical needs, and roughly the same share of world energy and electrical needs. In the United States 87% of coal is used in electric utilities. Coal plants use millions of gallons of water for cooling; dry-cooled coal plants are being developed in response to water scarcity pressures. U.S. utilities are planning to open about 30 new coal power plants between 2001 and 2011; the number could triple.

A typical utility power plant may burn 10,000 tons of coal each day; with even 5% waste (a low estimate), that leaves 500 tons of solid waste to dispose of each day. The coal industry generates 90 million tons of waste annually, most of which goes into landfills. Coal ash can contain toxic metals, trace contaminants, and high radon levels. Researchers are studying the feasibility of using coal ash in construction or as a soil enhancer for crops like apple orchards.

Lifeless lakes and rivers and dying forests (caused by acid rain, largely the result of burned coal) have been documented since the 1960s in nearly every country in the world. Abandoned Appalachian underground coal mines have leached acids into thousands of miles of streams, rivers, and aquifers; many of the streams can no longer support aquatic life. Carbon dioxide represents up to 40% to 50% of human-created greenhouse gases, and much of it is produced from burned coal.

Human Impact

Health: Burned coal results in smog, released sulfur dioxide, heavy metals, and particulates, which cause lung, respiratory, and other health problems. Heavy metals can cause cancer, heart disease, kidney problems, nausea, and headaches. Such problems are especially severe in countries with few pollution controls. Some coal is used to manufacture petrochemicals. Petrochemicals are some of the most toxic humanmade chemicals and can cause a number of health problems, including neurological problems and cancer. Ozone depletion, enhanced by burned coal particulates, increases the incidence of skin cancer and cataracts.

Food: Burning coal results in acid rain, air pollution, and smog, all of which damage crops. Acid rain can also reduce fish resources. Acid rain leaches heavy metals into water; metals from acid rain and air pollution are taken up by fish; ingesting the heavy metals in fish can cause cancer and other health problems.

Worker safety: Underground coal miners develop black lung disease (coal worker's pneumoconiosis) after breathing coal dust for many years.

The dust builds up in their lungs and causes coughing, wheezing, lung inflammation, and eventually lung damage leading to death. Several hundred mine workers are killed annually in South Africa's coal mines, which are the world's deepest; miners there often die as a result of methane explosions. More than 9,600 coal miners died in 1992 in Chinese mines.

Individual Solutions

- Consider solar and wind energy for home water heating and independent electrical needs.
- Practice energy conservation to minimize use of fossil fuels.
- Support strict air emission standards for all fossil fuel use.
- Write your congressional representatives to mandate increased use of energy conservation and renewable energy by power utilities.

Industrial/Political Solutions

What's Being Done
- Some countries, including Canada, Japan, European countries, and the United States have pledged to reduce carbon dioxide emissions by 2000.
- The Antarctic Treaty of 1961 (amended in 1991) prevents any mineral extraction in Antarctica for at least 50 years, until 2041.
- In Germany, topsoil and subsoil are saved in strip-mining, so they can be replaced after mining to make reclamation easier and more complete.
- Electric utility plants must purchase energy developed by small renewable producers (such as wind and solar), and must do so at a rate equivalent with avoided costs of new or replaced generation. This policy began with passage of the Public Utilities Regulatory Policies Act (PURPA).
- The U.S. Department of Energy has a $4.6 billion Clean Coal Technology Program to reduce sulfur dioxide, nitrogen oxides, and carbon dioxide from coal.

- Technology is being developed to use acoustic agglomeration (sticking together) of fine particles in coal stacks.
- The 1990 U.S. Clean Air Act mandates that sulfur emissions be cut by 10 million tons (from 1980 levels) by 2000, 1/2 of that amount by 1995; nitrogen oxide emissions are to be cut by 2 million tons by 1996.
- The U.S. Coal Mine Health and Safety Act of 1969 reduced risks to miners by requiring better ventilation and safer tunnels; it was amended in 1977 by the Mine Safety and Health Act.
- The U.S. Surface Mining Control and Reclamation Act of 1977 requires coal-mining companies to restore strip-mined areas to their initial conditions as nearly as possible.
- To end the dumping of coal ash in landfills, utilities are looking at construction uses for the ash.
- In Minnesota and a number of other states, utility regulators must include environmental costs when deciding whether to approve new power plants, and they can only approve the least expensive option presented.

What Needs to Be Done

- Install scrubbers on all coal plant air stacks.
- Mandate increasing utility use of energy conservation and renewable energies.
- Increase investments in solar, wind, and other renewable energies, and energy conservation.
- Obtain world agreement on reducing emissions of sulfur dioxide and nitrogen oxides.
- Help developing countries reduce air emissions from coal burning; the United States has a pilot program with Poland to do this.

Also See

- Fossil fuels
- Greenhouse effect

D

Deforestation

Destruction of forests by disease, burning, cutting, flooding, erosion, or pollution.

Forests are classified mainly as either temperate (having a moderate climate) or tropical. There are a number of different types of temperate forests worldwide. Coniferous forests are mainly needle-leaved trees and include evergreens such as spruce, pine, redwoods, cedar, hemlock, sequoia, and fir. These northern forests often make up what is known as old-growth forest.

Old-growth forests contain trees that can be several thousand years old and when cut down would take at least 100 years to regrow to maturity; the complete ecosystem of the forest might never revive. Such a forest has never been harvested, is extremely productive, and is diverse in trees over 200 years old. It has fallen logs and is home to a wide number of species of animals, birds, amphibians, and other life, most of which are interdependent. Old-growth forests are too complex to "regrow." Second- and third-growth forests often have trees about the same age and are more dense and dark; old-growth forests are more open, have trees of varying heights, and support many more species of plants and animals.

Deciduous forests, also temperate, have broad-leaved trees that shed their leaves seasonally and include oak, elm, maple, birch, aspen, and chestnut. Closed temperate forests are dense, with thick undergrowth; open forests have more space between the trees and continuous grass cover. Temperate coastal rain forests exist in areas such as the Pacific

Northwest of the United States and Canada; they are always on a coast near mountains and have high rainfall.

Tropical forests include arid forests, such as savanna and open forests, and moist forests, such as seasonal or monsoon forests, mangroves, and rain forests (50% of all tropical forests).

Tropical rain forests occur near the equator. They have rainfall of 80 to 400 inches each year, remain green all year (like temperate evergreens), and include ebony, mahogany, and teak. Most of the nutrients in a tropical rain forest are held by the living vegetation, not the soils. Tropical rain forests have a nearly continuous upper canopy of leaves that perpetually block most sunlight from reaching the ground.

Environmental Impact

Deforestation results in:

Extinction: The majority of the world's species live in tropical rain forests.

Soil erosion: Tree roots prevent soil from washing away.

Water pollution: Eroded soils enter rivers and lakes and inhibit plant growth, harm fish populations, and destroy coral reefs. Nutrients and heavy metals in the eroded soils also enter the water; nutrients may cause eutrophication, which kills fish and plants; heavy metals bioaccumulate in plants and animals.

Greenhouse effect: Destroyed trees release the carbon dioxide stored in their tissues; carbon dioxide is a major greenhouse gas. Living trees also remove carbon dioxide from the air for photosynthesis. Thus fewer trees result in more carbon dioxide remaining in the atmosphere. Forest soils absorb methane; when a forest is burned, the methane is released with the carbon and adds to the greenhouse effect. Termites also thrive in destroyed forests and produce large amounts of methane in their digestive tracts.

Air pollution: Burning trees adds particulates to the air, which can speed up ozone depletion, create clouds, and affect local weather patterns. Forests also help to filter pollution out of the air, a capability that is lost when they are destroyed.

D

Water shortages: Trees and vegetation help soil to hold water, which percolates down to groundwater supplies. Without the vegetation some soils can become hard, and the water simply runs off. Trees provide shade, and when they are gone more water is lost to evaporation.

Flooding: When trees and vegetation are cut down, flooding increases. Again, without trees, rain washes silt, nutrients, and pollution that are harmful to fish into streams and rivers; erosion is also accelerated.

Major Sources

Agriculture: Farmers clear forested land by cutting or burning to grow food crops or for plantations of rubber and palm.

Livestock production: Ranchers clear forested land to raise cattle or other livestock.

Timber industries: Timber companies cut forests to obtain raw and exotic woods or to establish tree farms and plantations.

Fuelwood: Local residents cut trees to produce firewood for cooking or heating.

Air pollution: Toxics, particulates, acid rain, smog, and other pollutants inhibit tree and plant growth. All can be transported on the wind and reduce tree resistance to diseases and pests.

Aquaculture: Mangrove swamps, especially in Africa, Asia, and Central and South America, have been cut or are being cut along coastlines for the purpose of growing commercial finfish, shrimp, and shellfish.

Development: Forests are cut down to make room for urban development, roadways, and airports.

Mining pollution: Forests are often destroyed by mining for fossil fuels and minerals; pan mining for gold is also adding toxics to rain forests worldwide.

Dams: Reservoirs of large dams have flooded existing forests.

Diseases: Diseases often kill forests. Some are exotic strains introduced from elsewhere, like Dutch elm disease and chestnut blight. Tree resistance to native diseases is also lowered by pollution.

Population increases: Many of the problems already mentioned increase as the world population increases. Demands for more wood products, more fuelwood, more agricultural land, and more meat intensifies pressure to cut down more forests. As more products are created, more pollution is generated, which adds to the stress of forests everywhere.

Industrial countries: Forests are being destroyed in every country in the world. Even though the Amazon in Brazil and other tropical rain forests have gained most of the publicity for deforestation, some industrial countries, like Canada and the United States, are still going through rapid deforestation. Also industrialized countries are creating much of the demand for beef, exotic rain forest wood, and wood products, which is a major cause of deforestation.

State of the Earth

As of 1990, experts estimated that 8 to 10 billion acres of the earth were still forested, that is, 1/4 to 1/3 of the earth's land surface. Roughly 50% lies in developing countries, and 50% lies in industrial countries. Worldwide, more than 60% of all forested lands have been cleared, and 4 billion trees are cut yearly to produce paper.

Nearly 1/3 of the United States, 750 million acres, is still forested, most of it second growth. This area is about 70% of what was forested in 1600. More than 90% of old-growth forests have been cut down in the United States; many species, including the spotted owl, can survive only in old-growth forests—the spotted owl needs up to 5,000 acres per pair of adult birds. Tree theft by logging outfits is a major problem in America's old-growth forests; the Forest Service has not aggressively pursued thieves, and in the past Forest Service rangers investigating theft have often been told to keep quiet or be transferred. About 214 West Coast salmon runs are near extinction; 90% of the extinctions are due to logging, which has buried spawning beds under silt, stripped shading vegetation, caused floods, and pulled out stream logs that formed sheltering pools.

Siberia, which holds more than 1/2 of the world's evergreen forests and 1/5 of all forested land, is being logged at the rate of 5 million acres a year. Dams are flooding the forests, and logging operations are poaching wildlife and clear-cutting trees. Once logged, the semifrigid land turns to swamp and cannot be replanted. Thirty million people live in the forests, including 24 distinct indigenous groups who are pressing for ecotourism, extractive reserves, and land rights.

Coastal temperate rain forests are rare, covering less than 0.2% of the world's land. They exist in Australia, Europe, Japan, New Zealand, North and South America, Tasmania, and Turkey. More than half of all coastal temperate rain forests have already been cut down. The only large stands left are in British Columbia.

British Columbia is logging the largest and most productive temperate rain forest in North America at a rate of 1/2 million acres yearly. Ninety percent of the logging involves clear-cutting, and less than 40% of the original forest remains. The United States bought more than half of all British Columbia wood products in 1992.

About 1/2 of all tropical rain forests have already been cut down. The remaining rain forests, 4.38 billion acres, cover only 5% to 7% of the world's land. Forty million acres of tropical forests are destroyed annually. In the 1980s, Haiti lost 40% of its forests; Paraguay, 39%; and El Salvador, 36%. In total 385 million acres of rain forest were lost. Thirty percent to 40% of the world's rain forests lie in Brazil, which has lost 12% of its rain forests. About 50 million acres have been burned, and about 100 million acres total have been destroyed. Every year 3.7 million acres are cleared. Still more forest is degraded by the effects of roads, projects, and fragmentation. The Amazon, which covers 60% of Brazil, occupies 1.3 billion acres and extends to 9 countries; 5% to 8% of the total Amazon has been lost.

Rain forests hold 50% to 90% of the world's species. With rain forest destruction, 5,000 to 50,000 species are going extinct each year. Millions of northern birds winter in the rain forests.

Tropical countries supply 15% of total global timber production. About 70% of tropical timber comes from Malaysia and Indonesia and 5% from Brazil; 80% of it goes to Europe, Japan, and the United States. Japan made 24 billion disposable chopsticks in 1991 from Asian hardwood.

Fragmentation of rain forests, where large tracts of land are broken up by roads, development, mining operations, timber cutting, and other intrusions, is worst in Southeast Asia and Oceania (88% of rain forests fragmented), Africa (80%), Central America (66%), and the Amazon (60%); this fragmentation accelerates extinctions. The biggest road in the Amazon, BR-364 Highway, aims to connect the east coast of Brazil to the west coast of South America. Pan mining for gold, which releases large amounts of mercury into forests, is occurring in Brazil, Ecuador, Ghana, Papua New Guinea, the Philippines, and Zimbabwe. Ecuador has lost 7 million acres of forest to large-scale mining, oil spills, and related toxic organic chemicals.

Major landslides and erosion due to deforestation have occurred in Borneo, Brazil, Indonesia, Malaysia, Nepal, and Thailand. In 1988 and 1993, tree loss in Bangladesh, India, and Nepal resulted in flooding in much of Bangladesh. Often rain forest cleared can be used for agriculture for only 2 to 5 years before nutrients are depleted and for cattle for 3 to 10 years before it is overgrazed or eroded. Recent studies have shown some cleared land is still healthy after several generations of use and that soil type and land use are key factors.

Most medicines originate in nature, and 1/4 of those prescribed in the United States originate in the rain forest. The rosy periwinkle, for example, is used in treatments for lymphocytic leukemia and Hodgkin's disease. Rain forest products are also used in drugs for malaria, heart disease, birth control, and surgery. At least 2,000 rain forest plants have been identified by the National Cancer Institute as having anticancer properties; more than 1/2 of the world's people still use natural plant remedies. Rain forests also provide latex, essential oils, coffee, bananas, cinnamon, chewing gum, nuts, and many other products.

Statistically, nearly all of the next 5 billion people born on the planet will be born in developing countries, putting more stress on rain forests and other forests because of fuelwood and farming needs. The search for fuelwood is one of the largest reasons for deforestation in many tropical areas. In some developing countries nearly 1/2 of the people still use wood for fuel, including rural citizens of Brazil, India, Indonesia, and

Madagascar. Fuelwood users total 1 to 2 billion people. Africa has the highest rate of fuelwood use and the highest rate of deforestation. In sub-Saharan Africa 90% of households use fuelwood for cooking; 3 billion use wood or charcoal daily. In the mid-1800s the United States derived 90% of its energy from fuelwood; 10% of U.S. homes still use fuelwood for some of their heating needs. By 2000, 1/2 of the people in developing countries will have insufficient fuelwood.

Land tenure—that is, the right to land ownership—is the main reason so much of Papau New Guinea's land is still forested. Constitutionally, more than 95% of all land is owned by people through the tribes they belong to. Lack of broadly held land tenure had led to deforestation in many countries—people have no stake in the long-term health of the forest. In most developing countries, governments and wealthy people own most of the land; in Brazil, for example, 4.5% of landowners control 80% of the farmland.

Rain forests hold about 100 billion tons of carbon in their tissues; in the past decade Brazil was the fourth largest carbon emitter in the world, releasing several hundred million tons of carbon dioxide each year from deforestation. Deforestation contributes to 25% of global carbon dioxide emissions; world soils hold 1.5 trillion metric tons of CO_2, which can be lost through deforestation. Rain forests are also the largest terrestrial producers of the world's oxygen.

Damage from smoke, sulfur dioxide, or other pollution makes trees more susceptible to infestations from pests such as bark beetles, weevils, and fir lice. Smoke also reduces resistance to frost. Cement dust reduces photosynthesis in lime and elm trees.

Current deforestation is occurring at 10 times the rate of reforestation. In any case, replanting trees cannot bring back the lost animals, plants, and nutrients that a true, diversified forest contains.

Human Impact

Genetic material: Literally millions of possible biotechnology products, such as new medicines, could disappear with rain forest losses. Most potential ingredients are unknown except to indigenous peoples.

Ecosystem knowledge: Scientists understand very little about the complex ecosystems of rain forests or even what other types of knowledge might be available in a rain forest. This knowledge could be useful in ways we do not yet even understand.

Indigenous peoples: As forests are destroyed, indigenous peoples are forced from their forest homelands.

Food: Some tropical forest soils erode quickly, resulting in lower crop yields. Cutting mangrove forests reduces breeding areas for marine fish, which are already under pressure from ocean overfishing.

Fuelwood: Fuelwood is a major use of trees in developing countries. Loss of trees makes it difficult to meet cooking and heating needs and creates an incentive to use dung and other biomass fuels that are otherwise used for fertilizer. The result can be impoverished soils.

Water shortages: With deforestation groundwater is not recharged as quickly; lack of tree cover contributes to increased evaporation and thus can exacerbate local drought conditions.

Energy needs: Trees provide shade. Cutting them increases electrical needs (for air conditioning), leading to increased coal burning and thus air pollution such as acid rain and smog.

Respiratory problems: Burning trees releases particulates that can cause respiratory problems.

Disease: Unidentified viruses trapped in rainforest soils may be transmitted to humans more frequently as deforestation accelerates.

Aesthetic concerns: The extinction of species that accompanies deforestation raises important aesthetic issues.

Global impact: Continued deforestation presents other unknowns, such as the impact on local or global weather patterns, the threat of global warming, and possible ecosystem failure as thousands of species continue to go extinct, especially in the tropics.

Individual Solutions

- Plant trees, especially around your home, to provide shade and lessen the need for air conditioning.
- Use less paper; buy products that you can reuse, such as cloth towels

instead of paper towels; recycle paper; don't use disposable wood products, like chopsticks; build with brick or other materials instead of wood.

- Oppose incineration, which destroys paper.
- Don't buy tropical hardwoods such as tropical teak, rosewood, mahogany, ramin, lauan, or maranti; the United States is one of the largest importers of Brazilian timber. Instead use oak, cherry, birch, pine, or maple; be aware that these types may also be unsustainably harvested.
- Purchase rain forest sustainable products, like Brazil nuts, or products that incorporate rain forest plant products if the sellers are indigenous peoples, not commercial interests, and if other political, social, and economic reforms accompany the projects.
- Eat less red meat.
- Contribute to organizations like the Nature Conservancy, which allows people to "adopt an acre" with donations, causing more rain forest to be set aside.
- Call or write your legislators, the U.S. Forest Service, and the Bureau of Land Management to demand an end to subsidies to timber companies and funding of reforestation. Demand that remaining old-growth forests be protected and logging in them ended. Demand that imports of tropical timber be restricted.

Industrial/Political Solutions

What's Being Done

- Worldwide, indigenous people are organizing to protect their remaining forests. For years the Penan have created roadblocks in Sarawak, Malaysia, to block logging.
- Some farmers in developing countries are practicing agroforestry, which is the practice of planting trees and shrubs among crops to prevent soil erosion and provide fuelwood, nuts, and fruits.
- Paper products are being recycled worldwide in most industrial countries.

- Extractive and biosphere reserves in South and Central America are growing rubber, nuts, and other forest products that are harvested from rain forest land where other development is prohibited.
- Tropical countries have created national parks and reserves to protect forests.
- Solar cook boxes are being used in some developing countries to replace fuelwood for cooking.
- Shaman Pharmaceuticals and other organizations fund individuals to study with local natives to learn medicinal and other uses of tropical rain forest plants and trees.
- Merck and Company, Ltd., is paying Costa Rica's National Biodiversity Institute (INBio) $1.1 million to research species for products. For any products that are developed, INBio will get royalties, 1/2 of which will go to the government for conservation. Indonesia, Kenya, and Mexico are developing similar programs.
- Tree planting programs exist in a number of countries; Global Releaf planted its millionth tree in mid-1993.
- Debt-for-nature swap programs trade money for land conservation with countries that wish to lessen their debts by agreeing to put aside a piece of land for conservation purposes.
- The U.S. Forest Service is helping Venezuela reclaim tropical rain forest in the Guayana region that was damaged by mining; Honduras has stopped logging on the Mosquito coast; Brazil has extended more protection to the Amazon.
- Boycotts have pressed for a halt to the use of rain forest beef in hamburgers and rain forest timber that has been produced from nonsustainable methods. By 1995 B&Q plc, the United Kingdom's largest home improvement chain, will stock only wood products produced sustainably—none that result from clear-cutting of first- or second-growth forests.
- Indigenous people in Ecuador are suing Texaco for dumping oil into hundreds of human-made lagoons, which have seeped carcinogenic toxics into wetlands, rivers, and soils.
- The National Forest Management Act, the Endangered Species Act,

and the National Environmental Policy Act all work to protect forests in the United States.

What Needs to Be Done

- Phase out incineration.
- End timber subsidies and funding of forest road construction. Tax dollars have subsidized timber companies through the U.S. Forest Service for decades; timber has been sold at a loss to timber companies. The U.S. Forest Service is the largest road builder and maintainer in the world, overseeing 340,000 miles of roads in wilderness areas. They plan to upgrade all of it and add 262,000 more miles by the year 2040.
- Fund the U.S. Forest Service independently of logging profits thereby eliminating budgetary rewards to the Forest Service for timber sales.
- Demand that the U.S. Forest Service prosecute timber theft; in the past it has not allowed its personnel to pursue theft aggressively.
- Protect old-growth forests from more development by banning all logging there; support only sustainable logging and end clear-cutting practices.
- Ending log exports will increase jobs.
- Conduct plant research to replace trees as a paper source. For example, kenaf grows fast in 15-foot bamboo-like stalks, requires few chemicals, and needs no chlorine for bleaching; 4,000 acres were grown in the United States in 1993, 5,000 in Europe.
- Strengthen reforestation and tree-planting programs.
- Train foresters worldwide to understand the social and economic factors in forest use.
- Decrease meat consumption.
- Make family planning available to all families in all countries.
- Encourage developing countries to claim rights to their genetic resources.
- Tax virgin wood use for paper to encourage recycling.
- Help developing countries meet their energy needs with energy conservation and renewable energy for cooking, heating, and electricity.

- Mandate labeling of rain forest beef in all foods, pet foods, and other products; end imports of rain forest hardwoods.
- Help developing countries with debt relief and poverty; debt and poverty are driving forces in deforestation. The United States relieved substantial debt to Bolivia, Chile, and Jamaica in 1990; 1/2 of developing countries' debts are owed by 27 countries with 97% of the world's tropical rain forests.
- Instead of purchasing just primary goods, lower industrial country trade barriers to manufactured goods from developing countries, helping the economies of developing countries and lowering incentives to exploit native forests.
- Establish government debt-for-nature swaps; industrial countries can save enormous tropical forest acreage in developing countries by swapping the owed national debts. These swaps should include all debts, so the remaining debt load is not actually increased, as it has been with some swaps. Attention must be paid to economic and social justice and to the land rights of indigenous peoples. Countries need to grant land rights to local communities and indigenous tribes so they can manage the forest resources for their needs; land reform needs to redistribute land equitably from the few to the majority.
- End all international funding by the World Bank and other lending institutions for the building of dams, roads, mines, and other damaging projects in rain forests; focus on educating, funding, and assisting women in developing countries. In many developing countries women are responsible for collecting the fuelwood and thus play a key role in the preservation of their forests.

Also See

- Extinction
- Livestock problems

Dioxins

Definition

A large family of more than 75 closely related organic compounds, the most well known and toxic being 2,3,7,8-tetrachlorodibenzo-paradioxin (TCDD). Furans, also sometimes lumped under the heading of dioxins, are related compounds.

Dioxin is a human-created chemical by-product formed during the manufacturing of other chemicals and during incineration. It was a major component of Agent Orange, the defoliant used during the Vietnam War.

Environmental Impact

Soil and water: Dioxins enter the air from a number of processes and then can precipitate into water and soil. The chemicals can last for years in soils.

Animals: Animals can take in dioxins from air or water or by eating other animals. Dioxins are fat soluble and bioaccumulate in the fatty tissues of animals, especially fish. Studies show that dioxin is the most potent animal carcinogen ever tested, as well as the cause of severe weight loss, liver problems, kidney problems, birth defects, and death.

Major Sources

Dioxins are the by-products of dozens of chlorine-based industrial chemical reactions and processes:

Incineration: Burning of municipal solid waste (paper and plastic), sewage sludge, hazardous waste, and medical waste releases dioxins.

Pulp mills: Dioxins are created during pulp and paper bleaching in paper mills.

Manufacturing: Production of industrial chemicals or products such as polyvinylchloride (PVC), certain types of wood preserving, oil refining, and metal smelting can lead to dioxin emissions.

Herbicide production: Production of herbicides such as 2,4,5-T creates dioxins.

Old chemical dumps: Chemicals can combine to form dioxins.

Paper cartons: Cartons can leach dioxins into milk in trace amounts.

State of the Earth

Dioxins were an ingredient in the defoliant Agent Orange, which was used in Vietnam (20 million gallons on 3 1/2 million acres). Agent Orange ruined thousands of acres of land in Vietnam and caused thousands of birth defects. Thousands of U.S. veterans still claim they suffer health symptoms—including liver damage, endocrine system degeneration, and cancer—related to Agent Orange, which has been linked to a number of different types of cancer and skin disorders.

In 1976 a Hoffman–La Roche chemical plant in Seveso, Italy, exploded. A cloud of dioxin spread over the surrounding countryside, killing animals in the streets. Studies tracking 37,000 people showed those exposed in Seveso have leukemia, lymphoma, and liver cancer rates far above what is considered average.

In 1983, the town of Times Beach, Missouri, was evacuated owing to high levels of dioxin in the soil. TCDD had been mixed with waste oil and sprayed for dust control. There is no clear indication yet of the possible health problems the population of Times Beach may experience due to their exposure. More than 100 hazardous waste sites in the United States have dioxin contamination.

In 1984 the U.S. Environmental Protection Agency (EPA) claimed dioxin was the "most potent animal carcinogen" it had ever evaluated. But later studies claimed otherwise and caused a controversy on the issue of dioxin's toxicity; a 1993 report on the Seveso victims supported the earlier findings of dioxin's toxicity. It is suspected that most Americans have low levels of dioxins in their bodies.

D

Human Impact

Health: Humans can breathe in dioxin, absorb it through the skin, or eat dioxin-contaminated food; 90% of exposure is from such foods as meats, dairy products, and fish. Exposure to dioxin can cause a severe skin problem called chloracne. High doses over prolonged periods cause cancer, and there is growing evidence that low-level dioxin exposure may damage the immune and reproductive systems.

Breast milk: Dioxin accumulates over time in the fatty tissues and thus can be found in breast milk.

Animal testing: Dioxin produces cancer and other serious diseases in animals at lower doses than any other human-created chemical, although this result varies from one species to another. Animal testing is how the effects of carcinogens and dangerous chemicals on humans are evaluated.

Individual Solutions

- Use nonbleached, nonchlorine paper products.
- Oppose incineration.
- Minimize your use of plastic products, especially PVC, used in vinyl siding, gutters, window frames, blister packs, cooking oil bottles, and liquid detergent containers.
- Demand from legislators and the EPA laws that will put an end to dioxin emissions and industrial use of chlorine.

Industrial/Political Solutions

What's Being Done

- Current technology can reduce up to 99% of dioxins in incinerator emissions.
- Further research on the toxicity of dioxins is being conducted.
- Chlorine is being phased out of pulp mills in Canada and several countries in Europe.
- An international Canadian-U.S. organization studying the Great Lakes has proposed banning all industrial uses of chlorine.

- The EPA will be instituting stricter emission standards for dioxins in 1994.
- The U.S. Food and Drug Administration ordered milk carton manufacturers to reduce dioxin levels in cardboard containers by 1992.
- The EPA has issued stricter dioxin emission permits for hazardous waste incinerators.
- The herbicide 2,4,5-T has been banned for most uses except rangeland and rice fields.
- The American Public Health Association (APHA—an organization of 30,000 doctors and public health officials) has called for eliminating many industrial uses of chlorine.
- Researchers are working on effective, safe methods to destroy dioxins.

What Needs to Be Done

- Phase out incineration; require incinerators to use state-of-the-art technology to burn dioxin emissions.
- End chlorine bleaching of paper.
- Mandate strict controls over indirect dioxin production; phase out or change processes that create dioxins.
- Ban 2,4,5-T completely.
- Phase out all industrial feedstock uses of chlorine.
- Investigate dioxin emissions from PVC plants.

Also See

- Chlorine
- Incineration

Ecotourism

Definition

Travel that has a positive effect on the place visited, which is often a natural scenic area, wildlife park, or wilderness area.

Since travel has become easier and cheaper and is done by more and more people yearly than in the past, it has become a major factor in environmental degradation and pollution worldwide; ecotourism seeks to prevent these problems.

Environmental Impact

Habitat: Ecotourism preserves and supports pristine, natural areas in their original state. For developing countries it provides a revenue-producing alternative to logging or destroying natural habitats for other reasons.

Pollution: Ecotourism minimizes the impact of visitors with regard to garbage and other pollution.

Wildlife: With ecotourism species are not threatened and instead are protected and supported.

Major Sources

Organizations: Organizations worldwide are offering eco-tourist programs. Many places that are not easily accessible to individuals, because of location or local government restrictions, are open to tourists through organizations that do research and provide funding to the host country. Programs such as these allow inexperienced and experienced tourists alike to safely venture into foreign terrains, such as rain forests or mountains. Most organizations provide complete preparation, know the local dangers, and can take visitors to the best areas to see and explore. Some programs, such as Earthwatch, encourage teachers to go on their outings, so that they can bring their experiences back to the classroom.

Individual travel: Even without organizations, people can use the principles of ecotourism wherever they travel.

State of the Earth

More than 1/2 billion people now travel; 1 billion will be traveling by 2000. Travel is the largest and fastest-growing industry in the world. Ecotourism makes up 15% to 25% of all travel; 85% of travelers wish to benefit the area they travel to. U.S. ecotourists spend $14 billion a year—7% of total world travel expenses. Wildlife watching is the most popular type of travel: Kenya obtains 1/3 of its yearly foreign exchange from nearly 1/4 million tourists; Costa Rica earns more than $400 million a year; Ecuador's Galápagos Islands earn nearly $250 million. Estimates show, however, that less than 50% of tourism dollars stay in some host countries, and in some countries less than 10%.

Ecotourist dollars in Rwanda helped the government to fund antipoaching patrols and hire farmers as guides to save the mountain gorillas; sporadic local warfare is disrupting these efforts. Excessive numbers of tourists in some wildlife parks have driven animals into hiding.

More than 1/4 million Himalayan trekkers have contributed to the destruction of forests for firewood and lodges, trampled low-lying vegetation, hunted exotic animals, and discarded trash. As a result Nepal has replaced firewood burning with kerosene heaters and efficient solar water heaters, and hands out tree seedlings to trekkers to plant.

Fiji is focusing on ecotourism instead of large tourism developments. It plans to build isolated resorts near pristine areas and offer nature hikes and river rides. Some resorts will allow travelers to visit local villages once a week. The villages will supply fish and other items for the resorts.

Many countries, in the rush to get tourism dollars, are polluting and overcrowding their beaches, mountains, forests, and lakes. Raw sewage is flushed into waters or left in areas where it does not decay; solid waste is thrown into open pits; native plant and animal species are sold as artifacts and curios. Rain forests in Brazil, Costa Rica, and Ecuador are heavily visited; China, Mexico, Nepal, Puerto Rico, and Tanzania are also heavily traveled.

Even places that are very well managed and protected, such as the Galápagos Islands, are under stress as more than 50,000 visitors yearly disturb animals and introduce exotic species of plants. In a number of ecologically sensitive places, resorts have sprung up to cater to wealthy tourists; these resorts inevitably harm the environment that people have come to see. In Kakadu National Park in northern Australia, tourists have vandalized sacred aboriginal rock art, and trampling of soils has led to erosion problems.

Human Impact

- Ecotourism encourages appreciation of and interaction with nature and preserves it for future generations.
- It encourages responsible action in a foreign area.
- It respects the rights of locals and native peoples; it can replace more abusive and nonsustainable tourism.
- It helps build friendly relationships between nations.
- It has the aesthetic value of helping to preserve endangered wildlife and their habitats and leaving a place as it was found.
- Ecotourism preserves ecosystems, which filter air and water pollutants, are vital for groundwater regeneration and soil preservation, and affect the local climate. It helps to educate the public about the importance of natural ecosystems as well as the importance of how we interact with them.
- Ecotourism provides income for the host country and local popula-

tions so both can maintain habitats in their natural states and avoid pressures to destroy them for logging or other purposes.

Individual Solutions

- Never litter; take out any garbage you bring into a natural area.
- Stay on designated trails; don't disturb animals.
- Familiarize yourself ahead of time with the place you are going to visit.
- Always observe local customs and regulations; remember, you are the guest.
- Don't buy coral or other endangered or exotic curios even though they may be legal to purchase. Many plant, animal, and insect species worldwide are threatened by travel and tourist purchases. Check with the Humane Society or the U.S. Fish and Wildlife Service if you have questions.
- Be aware that a developing country might not have sufficient safeguards to protect wildlife sanctuaries; use your own judgment to decide if an activity is harmful to native species or not; consult professionals before you leave. Remember that your actions may be duplicated by thousands, or even millions, of other visitors yearly; it is often the cumulative effect over time that is damaging.
- Avoid disposable products of any type when you travel; be aware that developing countries may not recycle or even dispose of trash in an environmentally safe method. Even in industrial countries, disposables eat up resources, cause pollution, and often are disposed of in a way that harms the environment.
- When choosing a tour operator, find one that observes local customs; financially supports the visited area; seeks to enrich your appreciation of nature, conservation, and the environment; involves local participation; and operates in an environmentally sound and sustainable manner. Be sure you understand the risks for disease, injury, and any other dangers and the limits of liability of your tour operator before you take any trip. Other characteristics of ecotours that might be

desirable are: a small tour, a high ratio of guides to tourists, established operator with good references, and use of local hotels and facilities to support the host country. Contact the Center for Responsible Tourism or the Ecotourism Society for references; they have studied the quality of a number of different ecotourist programs.

Industrial/Political Solutions

What's Being Done

- Governments, private operators, and local populations are restricting tourist travel to safeguard natural areas.
- Some countries are limiting numbers of visitors to sensitive areas, such as coral reefs.
- Some governments, such as Kenya, have involved their own people in most aspects, including ownership, of the tourist business.
- Groups like the Ecotourism Society are helping host countries redesign tourism programs so they are sustainable and environmentally sound.
- Organizations such as Earthwatch, the Nature Conservancy, Conservation International, the Audubon Society, and others offer people opportunities to accompany scientific research expeditions or to visit biopreserves on environmentally sound visits; visitor funds help support the projects and areas visited, as well as local populations.
- Tourist businesses like hotels and restaurants in some areas are realizing that if they pollute their local attractions—that is, the environment —they will eventually lose their business. As a result some businesses are using environmentally sound products, energy conservation, and better waste disposal methods. Businesses are also practicing environmentally sound management to please and attract a growing percentage of environmentally aware patrons.

What Needs to Be Done:

- Regulate travel in pristine or sensitive areas and monitor travel groups that claim to offer environmentally sound travel.

- Include local populations in ecotourism program decisions, operations, and planning, and do not displace them in favor of tourism.
- Ensure that a portion of the tourist dollar goes to the host country, especially the local populations; park fees should support park services.
- Push for all tour operators to develop responsible rules of conduct, observe local regulations, and give financial support to the areas visited.
- Encourage local businesses and governments to include in the tourist expense the costs of pollution control, solid waste disposal, and environmentally sound management. Regulate tourist numbers according to the stress they pose on an ecosystem, so that areas can be sustainably managed and developed and so that high usage does not eventually result in reduced tourism.

Also See

- Indigenous peoples displacement

E

Electromagnetic fields (EMFs)

Nonionizing radiation from electric and magnetic sources; basically an invisible force, with electric and magnetic components, surrounding electric currents and appliances. EFs (electric fields) occur whenever an electric charge is present; MFs (magnetic fields) occur whenever charges, such as an electric current, are in motion.

Nonionizing radiation is caused by long, low-frequency waves of radiation that do not have the energy to ionize an atom (knock an electron out of an atom's orbit and thereby change the atom's charge and possibly break chemical bonds) but can affect living beings in different ways. There are natural EMFs, but most exposure is due to human-created sources.

Other examples of nonionizing radiation are ultraviolet radiation (UV), visible light, infrared radiation, and microwave radiation.

Environmental Impact

Plants: It is unknown how increasing EMFs in the environment, · through power lines, substations, and other equipment, may affect plants.

Animals: Studies of the effects of EMFs on animals show changes in biological rhythms and behavior, impaired early central nervous system development, and avoidance behavior toward EMFs. EMFs also affect the amount of estrogen in animals, increasing susceptibility to breast cancer in females, and affecting the male reproductive system, too. Even weak EMFs have been shown to damage chick embryos.

Human-created sources: Microwaves, clock radios, computer terminals, water bed heaters, electric blankets, home power inlets, power lines, electric power substations, vacuum cleaners, blenders, food processors, electric shavers, cordless telephones, coffee makers, vending machines, fluorescent lights, and hair dryers produce EMFs; all appliances using electricity have EMFs.

Naturally occurring EMFs: EMFs are generated by the earth, lightning, and thunderstorms.

Magnetic fields are measured in a unit called a gauss (G). One gauss equals 1,000 milligauss (mG). A typical home without any appliances may have a natural background magnetic field of 0.5 to 4 mG. Cancer threshholds may begin at 2.3 to 3 mG. Walls, trees, shrubs, and other obstructions can block electric fields but not magnetic—an appliance will send its magnetic field through a wall into an adjacent room.

Power lines are one of the largest sources of human EMF exposure. Studies since the late 1970s have shown links between leukemia deaths in children and power line exposure—including neighborhood power lines. In 1990 a study by the U.S. Environmental Protection Agency (EPA) found a link between EMFs and leukemia, lymphoma, and brain cancer; other studies have found increases in cancer in electrical plant workers. The EPA has said that 60 Hz magnetic fields from power lines and home sources may be a cause of cancer.

A recent Swedish epidemiological study looked at power lines, utility usage, and cancer in 436,000 people for the past several decades. It found that children near power lines have a 25-fold increased risk of getting leukemia. Another Swedish study found that cancer risks increase 3-fold with EMF exposure of 3 mG and higher; power lines were found to increase risks of cancer 4 times over those not exposed. In Stockholm power lines must be removed if they are close to areas that children frequent; Swedish authorities are considering requiring

200-meter rights-of-way on either side of power lines. Another Swedish study linked occupational exposure to EMFs and leukemia in adult men. Studies in the United Kingdom showed that EMF radiation was linked to depression and suicide.

EMFs from appliances can be very strong, but exposure is usually fairly short and their fields rapidly decrease with distance (most appliances emit small EMFs at a distance of 3 feet and thus threaten the user more than they do other people in the household). Power line radiation fields do not diminish as sharply over distance as other sources and may be strong even 100 or more feet away. Magnetic fields are more difficult to shield than electric fields.

Can openers, power saws, electric shavers, hair dryers, vacuum cleaners, and mixers often emit the highest magnetic fields (250 to 20,000 mG); electric blankets and water bed heaters emit weaker fields, but since the person is closer to the field over a long period of time, exposure presents more risk. Shavers and hair dryers pose more risk than other appliances, since they are often used daily and held near the head. In the office, copying machines and laser printers emit high EMFs.

Human Impact

Health: Electric fields usually do not penetrate the body; instead they create charges on the surface of the body. Magnetic fields pass through the body and create electric currents in it. EMFs may interfere with the electrical charge of cell membranes or affect outside signals to and from cells. Overall there is uncertainty about how EMFs affect the body. Many scientists and doctors, however, believe that EMFs are linked to leukemia, Hodgkin's disease, lymphoma, brain cancer, reduced immunity, birth defects, miscarriages, and male and female breast cancer. Spiking or rapid fluctuations of magnetic fields, such as those produced by video display terminals (VDTs), may be the worst danger, which could explain the abnormally high incidence of cataracts in VDT operators and birth defects in their children.

Children: Children seem to be more vulnerable to EMFs than adults, and there is growing evidence that early trimester miscarriages increase for women who use VDTs more than 20 hours a week; brain tumors, developmental delays, and leukemia in infants of exposed mothers also increase. Studies have consistently shown the strongest correlations between EMFs and childhood leukemia.

Broken bones: Exposing broken bones that are not healing to magnetic fields can encourage the repair process.

Individual Solutions

- Test your home for EMFs generated by electrical equipment with a portable hand-held tester (check with local hardware stores or the EPA), or hire a company to do a complete inspection of background charges in your house. Often inexpensive solutions can reduce exposure to home EMFs.
- Keep your distance from all appliances, and limit exposure; keep clock radios, electric clocks, and telephone answering machines 6 feet away from the head of your bed.
- Remain an arm's length away from VDTs; remain 3 to 4 feet away from a neighbor's terminal, especially the side or back, where the EMFs are strongest. Computer shields, external and internal, can block EMFs.
- Avoid water beds (and thus their heaters) and electric blankets, or unplug them before you go to bed; stay at least 6 to 10 feet away from television sets; avoid or minimize use of hand-held electric shavers, hair dryers, and kitchen appliances; use hand-powered tools or appliances when possible.
- If you have children, keep them away from power lines.
- Fluorescent bulbs generate stronger fields than incandescent bulbs (but are much more efficient); don't sit next to them.
- Don't put cribs or children's beds against a wall adjacent to an appliance in another room.
- Buy or rent a home in an area free of major power substations or power lines.
- Use an oven or stove-top to heat food instead of a microwave.

E

Industrial/Political Solutions

What's Being Done

- Some countries are limiting placement of electric power lines, especially near schools and other areas where children are often present.
- Researchers worldwide are studying the relationship of EMFs to cancer and other diseases.
- The National Academy of Sciences is conducting a study on EMFs, which should be ready at the end of 1994; an EPA report on EMFs should be out in 1995.
- A number of companies are marketing EMF-blocking accessories or building them right into the equipment that generates EMFs.
- Florida has adopted magnetic field limits of 150 to 250 mG at the edge of new transmission line rights-of-way; New York has an interim limit of 200 mG for new high-voltage lines.

What Needs to Be Done

- Set limits for EMF generation on all appliances, and build in EMF-blocking devices.
- Encourage research by independent organizations, and not electric power companies, on health problems due to EMFs
- Remove power lines in areas where children are often present.
- Conduct research on ways to limit EMFs in power sources of all types.

Energy conservation

Reducing energy use in all its forms by changing methods of use, patterns of living, and equipment or by increasing energy efficiency.

Most energy is used to create heat or combustion energy for transportation, cooking, heating, manufacturing, and electricity. Fossil fuels and biomass energy are the main sources of heat energy and electricity. Electricity can also be derived from nuclear energy and from renewable souces such as hydroelectric, wind, solar, and geothermal energy.

Currently "new" energy is being created through energy efficiency; saving electricity through efficiency is many times cheaper than building new power plants.

Environmental Impact

Energy conservation would result in:

Reductions in fossil fuel use: A decline in use of fossil fuels would reduce acid rain, air pollution, mining pollution, greenhouse gases and global warming concerns, deforestation, soil degradation, freshwater degradation, ocean degradation, extinctions, wetlands degradation, wildlife habitat losses, fish and wildlife losses, and toxics in the food chains.

Reductions in nuclear energy use: Lowered use of nuclear energy would reduce nuclear waste, radiation accidents, radon increases due to uranium mining, and local increases in backround radiation.

Reductions in hydroelectric energy use: Decreased use of hydroelectric energy would reduce wildlife habitat losses, erosion, extinctions, deforestation, and river fish losses.

Reductions in biomass energy use: Decreased use of biomass energy would lessen particulate and greenhouse gas emissions.

Reduced energy needs: Decreased overall energy needs would allow renewable energies such as solar and wind to fill more of our energy needs; these forms of energy are virtually nonpolluting.

Major Sources

Energy conservation can occur in the following areas:

Transportation: Conservation steps include increasing fuel efficiency in automobiles; switching from internal combustion to electric vehicles or ultralights; using bicycles, light rail, metro rail, and freight rail; and abandoning freight trucks wherever possible.

Power plants and industry: Use of cogeneration, efficient motors and lighting, and renewable energy can save energy. Many outdated manufacturing methods can be streamlined and made more energy efficient.

Lighting and appliances: Energy-efficient appliances are on the market in most areas, especially lighting, household appliances, and motors.

Off-the-grid power generation (independent of a major power utility): Home, commercial, and industrial electricity and energy can be supplied with rooftop photovoltaic cells, solar water heaters, solar heating, backup batteries, small wind turbines, and mini hydroelectric systems. Use of off-the-grid sources is expanding throughout Europe and the United States and is already widespread throughout rural areas in developing countries.

Heating: Solar energy is available for heating water and buildings. Better construction and insulation can be used. More efficient faucet heads and washing machines, which conserve hot water heating energy, are also available.

Cooking: Fuelwood (trees, bushes, and shrubs) used in developing countries for cooking can be replaced with solar cook boxes. Fuelwood sources can be made renewable by planting trees amid other crops (agroforestry).

The United States accounts for 1/4 of world energy consumption; nearly 2/3 of the U.S. expenditure on energy is wasted on unnecessary use of appliances, inefficient lighting and motors, and poor conservation practices. Inefficient lighting wastes 75% to 90% of the used energy; efficient bulbs would reduce electricity needs by 10%. One-fourth of all electrical usage in the United States goes to lighting, and 90% of lighting use is for business and commercial needs. Worldwide, lighting accounts for 15% of total electrical usage. Motors consume 1/2 of all electricity, 3/5 of all industrial electricity, and 2/3 of all industrial energy. Household appliances use 1/4 of U.S. electricity; efficient appliances could save 10% to 20% of the electricity used. Efficient electric motors could save 50%. One-fourth of the world's population does not have electricity.

Home energy use accounts for about 15% of all energy use in industrialized countries; industry and commercial uses, 45%; and transportation, 20% in Europe and 40% in the United States. More than 100,000 people (20,000 homes) in the United States are energy independent— off-the-grid; the number is increasing by 20% a year.

Currently nearly 25% of California's electricity comes from renewable sources: nearly 20% is hydroelectric; 6.5%, geothermal; 3%, biomass; 1%, wind; and 0.03%, solar. Renewables now supply about 8% to 10% of U.S. energy needs and are rapidly increasing output.

The technology already exists to save 75% of the electricity and 80% of the oil used in the United States yearly through efficiency, without lowering our standard of living; greenhouse gases could be reduced by 2/3 through such savings. It is much cheaper to save electricity than to make it. The Rocky Mountain Institute (RMI) found that giving 1 energy-efficient fluorescent light bulb to each citizen in India would be 7 times cheaper than building a new power plant and would provide the same amount of energy to reach the same number of people; incandescent bulbs waste 90% of used electrical energy to keep the tungsten filament hot enough to glow. Improved utility efficiency and customer energy conservation can save as much energy as a new power plant can produce and costs many times less; 134 million compact fluorescent bulbs sold in 1992 worldwide saved the production energy of 10 large coal plants.

Between 1981 and 1993, the United States received 4 1/2 times more energy from conservation than from new supply, and 1/3 of new supply came from renewables. Electricity is the most expensive form of energy to make; nuclear power costs 10 to 20 times more than energy efficiency programs. The energy saved from using recycled instead of virgin material is 95% for aluminum, 85% for copper, 80% for plastic, 74% for iron and steel, 65% for lead, 64% for paper, and 60% for zinc. Centralized coal power plants lose 2/3 of the available coal energy as waste heat—cogeneration can raise efficiency to 65% to 90%.

Electricity is not available in many areas in developing countries, and parts of Asia, Latin America, and other regions often experience shortages. Energy efficiency is poor and energy losses are huge. A typical factory in China uses 7% to 75% more energy than one in an industrial country. China spent $6 billion on industrial energy efficiency in the 1980s; Brazil spent $20 million by 1991 on education and motor and light replacements; Mexico is spending $24 million on a residential compact fluorescent light program; Thailand will spend $190 million between 1994 and 1998 on an efficient appliance incentive program; Western Europe had 50 utility-sponsored light programs from 1987 to 1992 in 11 nations; utilities in Germany are spending $150 million on appliance rebates; other countries are following their lead. In developing countries, solar cook boxes, solar refrigerators, and solar lighting is being used; backup batteries can be powered by small windmills, hydropower, or generators.

Electric cars use 1/10 the energy of gasoline engines and consume no engine oil. In the United States cars are responsible for 2/3 of all oil used; worldwide, transportation accounts for 1/3 all oil usage. Volvo has a prototype car that gets 80 miles per gallon on the highway, seats four, and is safe in a 35-mile-per-hour head-on collision—the U.S. standard for head-on collisions is set at 30 mph.

Energy conservation would result in:

Fossil fuel reductions: Decreased use of fossil fuels would reduce pollutants responsible for respiratory diseases such as asthma and emphysema and toxic-related conditions such as cancer, kidney and liver failure, neurological problems, skin problems, and death. Harm to food crops, fish stocks, forestry, and potable freshwater would be reduced. Less rain forest destruction from mining of fossil fuels would mean more plants are preserved for possible medicinal use. Less fossil fuel use would decrease risk of global warming; greenhouse gases would be stabilized. Aesthetic values would increase, as smog and unsightly pollution like ocean oil spills are decreased.

Nuclear energy reductions: A fall in the use of nuclear energy would reduce radiation accident potential, released radon and other radiation wastes, and background radiation; the result would be fewer cases of cancer, reproductive problems, neurological problems, immune system breakdowns, and death.

Hydroelectric energy reductions: A decline in the use of hydroelectric energy would result in less soil degradation from irrigated and thus salinized soils, less displacement of indigenous peoples, and less loss of biodiversity. Downstream river fish stocks and crops would not be depleted.

Reduced costs: Reducing fuel needs and health problems related to high energy usage, especially of fossil and nuclear fuels, would result in substantial savings. The money saved can be spent on education and other pressing social needs, especially in developing countries.

Individual Solutions

- Minimize home energy usage by getting an energy audit; contact your local power company for suggestions—audits are often free.
- Drive less, walk, bicycle, use mass transit, and carpool. Keep your car tuned and tires inflated. Combine errands. Buy cars with high-mileage performance.

E

- When purchasing appliances, check efficiency ratings.
- Replace light bulbs with high-efficiency bulbs; compact fluorescent and electronic lamps last 20 times as long as incandescent.
- Conserve hot water by using front-loading washing machines and low-flow shower and faucet heads. Set the hot water heater to 120°; insulate the tank; use solar water heaters. Set your thermostat to 65° during daytime and 60° at night in winter, and 78° in summer.
- Plant trees for shade, which can reduce air conditioner cooling needs by 50%.
- Keep all switches off when not in use. Use hand-powered tools and rechargeable batteries with a solar recharger. Use a dishwasher only when full. Service air conditioners regularly, and change air filters.
- Recycle and buy recycled goods, which save production energy. Avoid disposables, packaging, incineration, and throwaways. Buy in bulk.
- Look into living off-the-grid with solar, wind, or hydropower.
- From your legislators demand new cars that get 100+ miles per gallon, stronger mandatory investments by utilities in energy conservation and renewables, and an end to nuclear power—it is not efficient.

Industrial/Political Solutions

What's Being Done

- The Asian Development Bank is investing directly in energy efficiency projects.
- Organizations like the International Institute for Energy Conservation and the Rocky Mountain Institute are helping utilities, governments, and citizens in dozens of countries obtain energy conservation and efficiency strategies, programs, and equipment.
- Research is being done on energy efficiency in air transportation, buildings and industry, and electric and hybrid vehicles.
- The National Energy Policy Act of 1992 gives utilities tax exclusions for utility-paid conservation subsidies for residential, commercial, and industrial customers, and sets efficiency standards for light bulbs

and electric motors. It also sets energy efficiency standards for washers, dryers, and dishwashers.

- A 1991 U.S. executive order mandates a 20% improvement in energy efficiency (over 1985 levels) of federal buildings and industrial facilities by the year 2000.
- The U.S. Environmental Protection Agency (EPA) has a Green Lights program that provides technical support for efficient workplace and home lighting, with a goal of reducing electric consumption of U.S. corporations by 50%; hundreds of corporations and a number of states are involved.
- The U.S. Department of Energy gives low-income grants in a Weatherization Assistance Program and Institutional Conservation Program.
- Super-efficient water heaters, office equipment, air conditioners, refrigerators, and industrial motors, and energy systems that automatically turn off lights in empty rooms or turn down thermostats at night, have been built.
- Many states require utility regulators to take environmental costs into account when deciding whether to approve new power plants, and they can approve only the least expensive, and thus most efficient, option presented.
- Electric utility plants must purchase energy developed by small renewable producers, and must do so at a rate equivalent to avoided costs (for new/replaced energy generation). This requirement began with passage of the Public Utilities Regulatory Policies Act (PURPA) in 1978.
- Some state and local governments are revoking utility franchises or municipalizing them to avoid building new power plants, which cost more than energy conservation and efficiency.
- Electric utilities can recover investments in energy conservation in most states; therefore they have an incentive to help customers use energy efficiently. Some utilities are giving grants to industry to increase energy efficiency, charging sliding scales depending on building efficiency, creating futures markets in saved electricity,

doing multiple energy audits for businesses, and giving customers contracts for saving or not increasing electricity usage. Utility customers can often get loans and rebates from their utility company for buying efficient appliances and weatherizing their homes. Some utilities have given away efficient light bulbs, insulation, and low-flow shower heads to customers.

- Electric utilities in the Midwest and Northwest and on the East Coast are studying and developing wind energy.
- In many states utilities are required to spend a percentage of revenues on efficiency.
- Electricity is being traded and auctioned to utilities in a number of states; this practice rewards the most efficient producers.
- A coalition of electric utilities formed a Super Efficient Refrigerator program; the Whirlpool units, which should be in stores in 1994, use no chlorofluorocarbons (CFCs) and consume 25% to 50% less energy than current models.

What Needs to Be Done

- Mandate higher fuel efficiency for all vehicles; give strong incentives for electric or high-mileage (100 to 300 miles per gallon) ultralight vehicles.
- Include the environmental costs of pollution in fossil fuel taxes.
- Mandate energy efficiency in all appliances; establish incentives like rebates for consumers and manufacturers.
- Create more incentives to use mass transit.
- End fossil fuel subsidies.
- Delete provisions in the North American Free Trade Agreement (NAFTA) that have a bias toward fossil fuel use, consumption, and trade.
- Improve state, local, and federal education of the public on energy efficiency.
- Mandate residential, commercial, industrial, and governmental retrofits for energy conservation.
- Mandate energy efficiency and conservation in all public utilities.

- Increase investment in renewable energy.
- End nuclear power subsidies, and phase out nuclear power plants.
- Decentralize and diversify electricity generation; support off-the-grid electricity generation.
- Focus international lending aid from the World Bank and other institutions on energy efficiency programs and diversified renewables, instead of on large power projects like hydroelectric dams.

Also See

- Fossil fuels
- Renewable energy

Environmental disasters

Definition

Large-scale disruptions of the environment caused by natural processes and cycles, whose frequency is sometimes increased by human activities.

Environmental Impact

Flooding: Flooding is a natural occurrence in rivers that often rebuilds coastal estuaries and floodplains with nutrient-rich silt deposits when a river overflows its banks. When this is prevented the river dumps the excess silt into the ocean; the coastal estuary erodes, and floodplain soil eventually loses its nutrients. River and land flooding is increased when soil is degraded through overgrazing or when deforestation or wetland destruction occurs. Then rainwater is not held by plants and absorbed into the soil but instead runs off the surface, carrying soil away in a rapid process of erosion. When the nutrients and topsoil are carried away, plants cannot grow and the soil goes through desertification. Aquatic fish and plants may die as a result of excessive nutrients and sediments in the water.

Volcanic eruptions: Eruptions spew sulfur into the air, adding to acid rain problems, which inhibit plant growth and acidify lakes. The sulfur particles can stay aloft from days to several years and form seeds for clouds, blocking sunlight and cooling the planet. The particles can reach the upper atmosphere and attract chlorine and ozone chemical reactions on their surfaces, thus temporarily accelerating ozone depletion. Local areas can be buried under ash and lava. Researchers have found that wildlife and plants sometimes can recover very quickly; in other places local wildlife is starved out because the ground is covered too heavily. Volcanic eruptions can also cause landslides and floods.

Hurricanes, cyclones, and tornados: Hurricanes, cyclones, and tornados can partially destroy coastal estuaries, rip out trees, and kill birds and other wildlife. Often these events are part of a natural cycle, and the land and species can recover. But where wildlife habitat is already severely limited, such disasters can damage it further and actually threaten coastal species survival.

Drought: In areas where drought is a seasonal occurrence, plants and animals have evolved to adapt to the changes of moisture and dryness. But when drought occurs in areas where humans have caused deforestation or soil degradation, local plants and animals often cannot survive or recover quickly. Then drought further accelerates desertification, baking impoverished soils hard so that when it does rain the water runs off and erodes the topsoil; drought can prevent vegetation from reclaiming the land for decades or longer.

Earthquakes: Earthquakes cause local environmental damage from which wildlife can often recover; earthquakes also can generate landslides and tsunamis—large ocean waves that can travel quickly from one side of the ocean to another. Chemicals or released radiation from facilities destroyed by an earthquake can threaten local plant and wildlife populations and contaminate the air, soil, or water.

Fires: Fires are a natural part of temperate ecosystems; fires help rejuvenate an aging ecosystem by building up soil nutrients, destroying tree-killing rusts and fungi, and allowing seeds and other plants to quickly invade an area that they could not enter before because of thick growth. Some seeds do not germinate unless exposed to forest fire heat. When humans practice fire prevention and do not allow forest fires to occur, a forest may become too dry and too old; then an inevitable fire may cover a much larger area than if there had been no control. Most temperate animals and plants can recover from a fire, but if their habitat is already restricted a fire places more pressure on them, especially if they run out of room to run from the fire owing to surrounding human development. In rain forests fires can destroy areas that may not recover; animals are threatened with extinction, and soil will quickly lose its nutrients or erode. Burning forests emit air pollution, greenhouse gases, and particulates.

Landslides: Landslides can carry tons of topsoil down a hill or mountainside. They often occur in an area already stripped of vegetation and thus prevent plants from regrowing.

Major Sources

Flooding: Floods are caused by too much rain, loss of wetlands, the diking of rivers, dams, soil degradation, or deforestation—especially in tropical or heavy rainfall areas. Local floods can be triggered by volcanic eruptions and earthquakes.

Volcanic eruptions: Volcanic eruptions are caused by geologic events; they are prevalent in the Pacific Rim. Up to 80% of eruptions are thought to occur on the ocean floor in some areas.

Hurricanes, cyclones, and tornados: Hurricanes, cyclones, and tornados are caused by natural, often seasonal, weather patterns. El Niño (see glossary), and perhaps global warming, may have some influence. These events are prevalent in coastal and equatorial areas.

Drought: Droughts are caused by seasonal natural weather patterns and possibly El Niño. Local droughts are exacerbated by deforestation, dam projects, or soil degradation. They are more common in semiarid and arid regions.

Earthquakes: Earthquakes are caused by geologic disturbances; they are most common in Pacific Rim countries. Earthquakes are also influenced by large dam reservoirs because of the massive pressure exerted by the water behind the dam upon the underlying and adjacent rock.

Fires: Fires can be started by lightning, earthquakes, or humans; they are exacerbated by drought and human fire control. Major fires have occurred in recent decades in forests in Brazil and California and in Yellowstone National Park.

Landslides: Landslides are caused by such geologic disturbances as earthquakes, volcanic eruptions, and heavy rains, as well as by defor-estation, mining, and soil degradation; they are most common in rain forests and hilly or mountainous areas. Landslides are often a natural part of shoreline erosion, which can be accelerated by wetland or coral reef destruction. Landslides are thought to be frequent on the ocean floor as well.

State of the Earth

Flooding throughout the United States is increasing as a result of the loss of more than 50% of total wetlands and the diking and damming of rivers. The great Midwest flood of 1993 (the heavy rains were believed to be due to El Niño) began in Minnesota (40% of its wetlands had been plowed under), continued to Iowa and Missouri (both of which had plowed under 90% of their wetlands), and ran south to Arkansas (which had lost 70% of its wetlands), and finally to Louisiana (which had lost 45% of its wetlands). All the way down the river, river banks were diked and dammed, and few wetlands were present to hold excess water. More than 15,000 square miles were flooded, thousands of people were left without drinking water, and the floods picked up pesticides and manure on flooded farmland that polluted downstream drinking water.

Major landslides and erosion due to deforestation have occurred in Borneo, Brazil, Indonesia, Malaysia, Nepal, and Thailand. In 1988, deforestation in Bangladesh, India, and Nepal resulted in flooding in much of Bangladesh; in 1993 the same countries experienced flooding that killed thousands, affected some 20 million people, and made home-less 6 to 7 million. Floods kill 1,500 people yearly in India. In 1993, 72 people and 15,000 livestock were killed in flash floods in northern China. In Haiti most farming is done on mountain slopes; the country loses up to 15,000 acres yearly to erosion and landslides. Landslides have occurred in every country in the world, as well as on the ocean floor.

In June 1991, Mount Pinatubo in the Philippines erupted and spewed 15 to 30 million tons of sulfur dioxide into the upper atmosphere; the particulates caused several years of cooler weather globally and disrupted weather patterns. Earthquakes and a typhoon accompanied Pinatubo's eruptions. The August 12, 1991, eruption of the Hudson Volcano in the Chilean Andes covered about 25,000 square miles with ash up to 6 inches deep. Overhunted wildlife such as the rhea (an ostrich relative) and guanaco (a llama relative) are expected to be further stressed. The May 18, 1980, eruption of Mount Saint Helens was equivalent to a 10-megaton nuclear explosion; it leveled everything within 15 miles, with a 200-square-mile "blast zone," creating the largest landslide ever recorded. Accompanied by an earthquake and 200-mile-per-hour winds, hot avalanches burned through forests in an area 12 miles wide and 5 miles long, dumping 300 feet of rock and ash into Spirit Lake. Fourteen years later the lake and surrounding areas seem to be rapidly rejuvenating; trees are growing back, but complete reforestation may take more than 150 years.

On August 24, 1992, Hurricane Andrew ripped through the heart of the Everglades. Mangroves were stripped bare, and 90% of the trees in some areas were knocked down, but ecologists expect the mangroves to recover fully. One-third of the coral reefs at Biscayne National Park were damaged; 50,000 homes were destroyed. Worldwide, from 1980 to 1985, windstorms killed more than 30,000, injured more than 35,000, affected 28 million people, and made more than 8 million homeless.

California had to ration water during the nearly 6-year drought that ended in 1991–1992; fires were frequent and burned housing and vegetation alike. Half of the water used in California is for grazing land and such water-intensive crops as rice, cotton, and alfalfa—this use contributed to drinking water shortages. In arid and semiarid regions, agricultural water uses and dammed rivers often leave people without drinking water. El Niño of 1982 was the worst in a century and was believed to have been the main cause of rare typhoons in French Polynesia, mud slides and floods in the United States due to heavy rains along the West Coast, and severe drought leading to brushfires and dust storms in Australia, India, and parts of Africa.

More than 3 dozen countries have a high probability of experiencing earthquakes; more than 110,000 people died in earthquakes from 1980 to 1990; more than 50,000 died in Iran in a 1990 earthquake; nearly 10,000 died in a 1993 earthquake in India. In 1989 a California earthquake killed more than 50 people and disrupted San Francisco and large portions of the state; a 1994 earthquake in Los Angeles also killed more than 50 people. A 1960 earthquake in Chile generated tsunamis that traveled to Japan and killed nearly 150 people. In 1952, Japan had an earthquake that killed more than 8,000 people. In July 1993, another earthquake left more than 100 dead in Japan and sent tsunamis that destroyed homes as far away as South Korea.

The 1988 fire in Yellowstone National Park charred 1/2 million acres. The fire is expected to result in a more diverse environment for wildlife, since fire does not burn in a clean sweep through a forest, but in a mosaic, leaving some spots less disturbed than others. Large wildlife in the area increased owing to more available grass, and lodgepole pines seeded 2 to 3 days after the fire; the pines will take 100 years to grow to the height of the burned pines. A fire started illegally in a forest preserve in Brazil threatened the existence of the golden lion tamarin, which is endangered and already limited in numbers and habitat range; researchers there will use techniques learned in Yellowstone to try to prevent further fires. Up to 50 million acres have been burned in the past decade in the Amazon basin, mostly in Brazil, for land and agricultural expansion; much of this land is in varying stages of degradation.

Human Impact

Flooding: Floods cause home loss, disease outbreaks, and mosquito and pest outbreaks. The water can also sweep over farmland, ruin crops, and pick up manure or pesticides, which can pollute downstream drinking water. Flooding in deforested areas can erode soil so it is lost for timber or food crop use. Soggy soils due to flooding increase aflatoxins (carcinogenic mold) in corn crops, root rot in soybeans, and vomitoxin (a disease that makes crops like wheat unusable).

Volcanic eruptions: Eruptions can cause burns or kill people outright by suffocating or burying them; hot lava can start towns on fire.

Hurricanes, cyclones, and tornados: Hurricanes, cyclones, and tornadoes can threaten developed areas, destroy buildings and crops, and pollute water resources.

Drought: Drought can deplete food resources and force evacuation of an area, which places stress on the other areas refugees travel to.

Earthquakes: Earthquakes threaten buildings, which can collapse; disrupt water and food supplies; threaten outbreaks of disease; and generate tsunamis, which can kill coastal inhabitants. Chemical companies, nuclear reactors, or other facilities destroyed in an earthquake could threaten local or nearby populations with dangerous chemicals or radiation.

Fires: Fires can threaten developed areas and tourist areas and destroy the biodiversity of a rain forest. Medicinal plants can be lost to extinction.

Landslides: Landslides can bury towns or houses and erode soil so the land cannot grow timber or food crops.

Reduced health: Any of the disasters mentioned can cause post-traumatic stress syndrome and susceptibility to other illnesses due to shock, weakness, or poor nutrition and care.

Costs: Costs are extremely high to rebuild areas devastated by earthquakes, fires, hurricanes, or other natural disasters. Moreover, most developing countries and their citizens cannot afford to build earthquake- or high-wind-resistant homes and buildings and therefore continue to suffer the highest casualties.

Individual Solutions

- When buying a house, find out if the region ever experiences flooding; avoid living on a floodplain or a coastal area with heavy erosion.
- If building a home in a high potential risk area for heavy winds, make sure the builder is certified and inspected for construction to withstand high winds.
- Practice emergency procedures with your family if you live in a high-wind or earthquake zone.
- Exercise care with fire in any forest.

- Eat less red meat, which is a major cause of deforestation, and thus localized flooding.
- Oppose large dam projects.
- Recycle paper products. Avoid disposable goods; use cloth or durable goods.
- Buy organic foods which preserves soils, wetlands, and wildlife habitat and thus lessens flooding problems.
- Don't buy such tropical hardwoods as teak, rosewood, mahogany, ramin, lauan, or maranti. Instead use oak, cherry, birch, pine, or maple.

Industrial/Political Solutions

What's Being Done

- The United Nations declared the 1990s the International Decade of Natural Disaster Reduction to stimulate efforts to improve disaster relief worldwide.
- Numerous efforts such as biotechnology research, extractive and biosphere preserves, debt-for-nature swap programs, and social justice and indigenous people's programs are being used to protect rain forests.
- Sustainable farming techniques, such as agroforestry, are being used in developing countries to prevent soil erosion.
- Paper products are being recycled worldwide in most industrial countries.
- Ongoing research is being conducted on early detection of earthquakes and volcanic eruptions.
- Earthquake-resistant construction of buildings is being done in high-risk areas.
- Authorities are setting managed fires in wildlife areas to allow natural processes to occur and to prevent too much deadwood and dry timber buildup.
- The U.S. Army Corps of Engineers and other agencies have been directed to rebuild wetlands for flood control instead of building more

levees, in response to the massive failure of more than 1,000 human-created levees during the 1993 great Midwest flood.
- The Conservation Reserve Program of the U.S. Farm Bill pays farmers to keep marginal wetlands covered in grass or trees.

What Needs to Be Done
- Reauthorize the Conservation Reserve Program in the 1995 U.S. Farm Bill.
- Reclaim wetlands, and institute a no-net-loss policy for wetlands; end excessive diking of rivers and major dam projects.
- Require international lending institutions like the World Bank to evaluate all lending projects for environmental impact and to fund energy efficiency programs instead of large power projects like hydropower energy.
- End deforestation; end clear-cutting and logging in old-growth forests; end timber subsidies.
- Reduce meat consumption; limit the amount of land devoted to cattle grazing, especially in arid and semiarid areas.
- Initiate government debt-for-nature swaps; industrial countries can save enormous tropical forest acreage in developing countries by swapping the owed national debts.
- Help developing countries meet their energy needs with renewable energy and energy efficiency.
- Pass strict building codes for construction of homes in high-wind and earthquake zones.

Also See
- Deforestation
- Soil degradation

Exotic species

A nonnative living organism (plant, animal, virus, bacterium, or insect) that has evolved in one location and travels or is introduced on purpose or by accident to an area where it does not naturally occur. This transfer can occur between countries, or between different regions of the same country.

With the onset of modern transportation, the large numbers of people traveling, and increases in shipments of different materials worldwide, thousands of exotic species such as pets, birds, seeds, viruses, and plants have been introduced into many countries. Many industrial countries now recognize the dangers some exotic species can pose and have severely restricted what species can or cannot enter or leave their countries. A number of developing countries as yet have no regulations over exotics.

Environmental Impact

Species introduced to a new area usually either die off because of a hostile environment or grow enormously because of the absence of predators and the presence of favorable conditions such as plentiful food sources. If successful, exotics can cause:

Extinctions: By competing with native species for the same food, by killing all the native species, or by destroying some part of the native species' habitat, exotic species can drive out the native species.

Ecosystem imbalances: If one native species becomes threatened, it may cause a chain reaction and result in other dependent species also being threatened.

Major Sources

Exotics have been introduced by:

Settlers, or those in contact with indigenous tribes: Early settlers and explorers often transmitted infectious diseases and introduced exotic plants and animals from their home countries.

Commercial ventures: People can bring an exotic animal or plant into a country to breed or raise for profit. The animal or plant then escapes or is intentionally let loose into the native area. This category includes pets such as birds, dogs, cats, reptiles, mammals, and fish. Illegal wildlife trade has also been responsible for some exotic introductions.

Tourists: Tourists may bring back or accidentally introduce plants, seeds, and insects.

Agricultural practices: Farmers sometimes use one insect as a biological control to kill off another insect that might be threatening crop production; some people import seeds for food crops or hobbies.

Sport: Nonnative species may be introduced for hunting and fishing.

Accidental: Exotics have "hitchhiked" on boats, cars, planes, in packaged goods, or among other traded goods between countries.

Scientific study: Insects, plants, or animals are brought into the country for research.

State of the Earth

Fire ants have spread across the southern United States, in some areas building up to 400 huge mounds per acre. Killer bees, imported from Africa to South America, are now in the southern United States. They occasionally attack people, and health officials are responding to a dozen calls a day. It remains to be seen what will happen to native honey bee populations. Tiger mosquitoes from Asia, introduced to the United States in 1986, are aggressive and can carry encephalitis (a fatal brain disease) and dengue fever.

Plants and animals in the United States are also threatened. The African walking catfish, which can walk between ponds and streams, competes with native fish in the southern United States. The gypsy moth from France kills trees. Asian fungus eradicated the American chestnut. European pine shoot beetles attack pines; European beech scale threatens beech; European purple loosestrife plants invade wetlands; and bee mites from Europe and Asia threaten honey bees. Domestic cats from Europe kill thousands of native wildlife species. Wild boars in the Rockies, oriental bittersweet, the tamarisk tree, carp, dandelions, and burros are other exotics causing problems in the United States.

In 1875 a measles epidemic in Fiji, brought by Europeans, killed between 20,000 and 40,000 people. Disease transmission, and resulting deaths, also occurred when the Pilgrims met Native Americans, when Spaniards brought smallpox to natives in South America, when Europeans brought measles and tuberculosis to islanders in the South Pacific, and most recently when city dwellers met Amazonian tribes.

More than 1/2 of all U.S. insect enemies of plants have been brought over from foreign countries on plants. Mediterranean fruit flies (medflies) caused millions of dollars worth of damage to California crops in the 1980s and early 1990s. The European corn borer, Japanese beetle, cotton boll weevil, codling moth, Argentine ant, and garden nematode are all introduced species that now are major concerns to crop growers. Green mites and mealybugs from Brazil and Colombia were brought to Africa, where they devastated crops like cassava.

Over 4,500 exotics are thriving in the United States, and the Office of Technology Assessment estimates that just 15 of them could cause up to $134 billion in losses over the next 50 years. Asian milfoil, a native plant of northern China, came to the United States in the 1940s and has rapidly spread across 37 states and 3 Canadian provinces, clogging lakes and waterways. Another introduced exotic, the milfoil weevil, is now being raised to fight milfoil. Kudzu was imported from Southeast Asia to control erosion in the southern United States; it grew so prolifically that it now covers fields and trees throughout the South; some people sell it for profit to the Japanese, who prize its roots and have a shortage. The water hyacinth, from Brazil, clogs waterways in Asia,

Africa, and the southern United States. The melaleuca tree from Australia has taken over 1.5 million acres of the Everglades; researchers are studying the impact of using an Australian weevil to eat the trees.

The Nile perch was introduced to Lake Victoria, in Africa, which had some 400 species of cichlids, one of the most diverse populations of such fish in the world; more than 1/2 are now extinct, eaten by the perch. Introduced European starlings are crowding out the indigenous Eastern bluebird in the United States. It is feared that 110 aggressive Nile crocodiles, recently imported into Brazil, could wipe out native amphibians, crowd out alligator-like caimans, attack humans, and forever upset the continent's ecological balance if they accidentally escape.

Australia and New Zealand demonstrate some of the worst effects of exotics. The European gray rabbit was introduced to Australia for sport. The rabbit population exploded—the species had no predators in Australia—and now competes with sheep, wallabies, and kangaroos. Predators like mongooses were brought in to eat the rabbits—instead they decimated native birds. Myxomatosis, a viral disease, was brought in to control the millions of rabbits. It worked for a while, but the rabbits soon began developing immunity to it; Australians may use a genetically engineered myxomatosis next.

The brushtail possum was imported by New Zealand from Australia for commercial fur trade. The possums now number about 70 million. They are destroying New Zealand's most beautiful trees, eat fruit and vegetables, and carry bovine tuberculosis, which threatens sheep herds. Huge poisonous Asian cane toads were introduced in Australia to control cane beetles—which fly and are rarely on the ground with the toads. The toads have run wild and eat small mammals, snakes, rodents, insects, and precious tropical plant seeds on the ground. Cane toads also have few native predators. House cats in Victoria, Australia, kill 13 million small animals yearly, including members of 67 bird species.

Wild pigs, introduced for hunting in many islands in the South Pacific, uproot precious native tropical plants. The Indian mongoose was introduced in Fiji, Hawaii, and Jamaica to control rats, but instead it decimates native bird populations. Cats, dogs, and goats threaten native species on the Galápagos Islands. Fleas, lice, and tapeworms that were accidentally spread have shifted hosts; for example, the sheep tapeworm now infests American deer.

An exotic jellylike organism is devastating anchovy fisheries in the Black Sea. It is suspected that the zebra mussel came from the Black and Caspian Seas in the ballast tanks of cargo ships. They first were introduced in the Great Lakes and now have spread to the Saint Croix River in the Midwestern United States. They are prolific breeders. Adults filter a quart of water a day; they cover intake pipes and valves on utility water equipment, up to 1,000 mussels per square meter, and can attach themselves to the bottoms of recreation vehicles. They crowd out native mussels and fish species such as walleye and lake trout. Ironically, new exotics, two species of goby fish, have now entered the Great Lakes the same way the zebra mussel did, and it is believed they eat zebra mussels.

Some exotics can be used as beneficial forms of biological control. The Netherlands is using the zebra mussel for water-quality management, to clean up lakes and filter water; the mussels filter algae, nutrients, pollutants, and phosphorus. A South American wasp was introduced in El Salvador to eat the eggs of pests and eliminated the need for 10 pesticide applications every year. A wasp has also been used to control mealybug on cassava in 25 African countries.

Human Impact

Disease: Infectious diseases have decimated native tribes when introduced by explorers. The Asian tiger mosquito carries dengue, yellow fever, and encephalitis and was recently introduced into the United States.

Food: Exotic insects have wiped out major crops in many countries. Conversely, many food crops in many countries are nonnative. And biological controls, such as nonnative wasps, have been used in a number of countries to successfully control crop pests.

Physical threats: Killer bees attack humans if the humans come too close. Fire ants can attack unwary people. Cane toads emit a secretion poisonous to humans.

Nuisances: Boaters in some areas must clean zebra mussels or milfoil off their boats. The Asian cockroach is attracted by light; at night it flies toward well-lit activity areas, like barbecues.

Recreational habitat: Milfoil prevents swimmers, fishermen, and boaters from utilizing lakeshores.

Sporting: Game fish or animals may be decimated by exotics. Conversely, some exotic game fish and birds (such as pheasant) have increased sport hunting.

Aesthetics: Native birds or other animals may be wiped out by exotics. Conversely, exotic plants and pets are desired by many.

Economic losses: Monies are spent to control an exotic, and monies are lost when a species is threatened by an exotic.

Individual Solutions

- If you are traveling from one country to another, do not bring a live plant or animal across the border; even if a country has no regulations, it is often unwise to transport such items.
- If you wish to bring an exotic species home with you, check with the U.S. Fish and Wildlife Service first.
- Do not purchase any exotic animals or plants, or products made from them, unless they are commercially raised in the United States; check with the seller, the U.S. Environmental Protection Agency, the Humane Society, or the U.S. Fish and Wildlife Service. Exotic commercial and illegal trade is threatening numerous species with extinction in many countries.
- Wipe your prop and boat clean when leaving a lake; milfoil is spread from lake to lake by pieces as small as 1 inch on a boat prop.

Industrial/Political Solutions

What's Being Done
- Many countries have stiff regulations regarding importing or exporting species.

- Programs exist in most industrial countries to control exotics once they arrive, to prevent further spread, and to minimize the damage they do to native plants and animals. One method that has been used with small animals and insects is to capture or breed male members of the species and make them infertile. The males are then released into the environment with the hopes that they will mate with the females, which will then have no offspring, thereby reducing the population enough so that it will not survive. This approach was taken with the medfly in California and the sea lamprey in the Great Lakes.

What Needs to Be Done

- Push all countries to regulate exotics.
- Mandate environmental impact studies before the introduction of any exotic.

Also See

- Ecotourism
- Extinction

E

Extinction

Permanent disappearance of a species of any living organism, plant or animal.

A species is defined as a group of organisms that can reproduce only among themselves. Different species of birds will usually not successfully reproduce together, even in zoos, and certainly not in the wild.

Even if all the individuals of one species are not killed, the remaining members of a species will go extinct if the genetic diversity, or gene pool, of their members is too small. A large gene pool helps the species adapt to environmental stress, problems, and disease.

Animal populations must be large enough so that their gene pool keeps them strong and enables them to survive adversity in their environment or pressures from human activities; this size of population is sometimes called a sustainable population. Under this definition, if only a few members of a species exist in the wild or in a zoo, the species is, for all practical purposes, extinct.

Extinction is a natural process that has been going on for millions of years. Human activities, however, have accelerated this process on a scale unprecedented since the dinosaurs went extinct. Most of the current extinctions are caused by pollutants and environmental damage that only humans can create. Most species that are going extinct have not even been identified or studied.

Biodiversity loss: Biodiversity refers to the number of different ecosystems, the number of different species, or the genetic diversity of individuals of one particular species in a given area. Genetic diversity among the individuals of a species means that there is a wide range of characteristics coded in the DNA of the individuals. For instance, if the DNA of a herd of elephants has many different characteristics for coloring, size, disease resistance, foot size, and other features, the herd has a rich genetic diversity. Productive areas, like tropical rain forests, coral reefs, and wetlands have many different types of animals and plants, and thus a rich species biodiversity. When one species (or many members of a single species or many species) is killed or taken away from a particular area, it lessens the area's biodiversity. An area has ecosystem diversity if it has, for example, a forest, grassland, and a lake, instead of just one of the three.

Food chain: If a particular species goes extinct, the ecological web and food chain is changed forever. If the extinct species was a plant or animal that other animals eat, those animals will now have to eat other plants or animals. One extinction may cause other extinctions. For instance, if eucalyptus trees were to go extinct, so would koala bears, since they feed primarily on eucalyptus leaves. The krill population of the Antarctic is the major source of food for many populations of fish, animals, birds, and baleen whales. If krill populations are overfished to the point of exhaustion, all the populations of mammals, birds, and fish that are dependent on krill as a food source could plummet, and even die off.

Population explosions: When one species goes extinct, other more undesirable species may increase dramatically or even take its place. When a marine fish is overfished, for example, the population may plummet and fail to recover. Populations of smaller, less desirable game fish may explode and take over the ecological niche of the larger fish. Or if a predator like a coyote is killed off, rabbit populations may increase dramatically since there is no predator to keep them in check. Rabbits may eat the grassland until it is so overgrazed that the soil is not held by plant roots and erodes with rain. The soil becomes hardened so nothing can grow, and the rabbit population also plummets.

Genetic erosion: When a local population goes extinct it may reduce the genetic richness or viability of the whole population; this process, referred to as genetic erosion, is a result of biodiversity loss. An example would be the wiping out of more than 100 major populations of salmon and steelhead on the west coast of North America. The loss of these populations means the genes of these populations have also been lost. Thus the western Pacific salmon population, as a whole, is less strong, less hardy, and more prone to further extinctions. Another example of genetic erosion would be the loss of many natural varieties of corn and rice when farmers cultivate only a few hybrid strains.

Ecosystem damage: No one really understands what a massive loss of species will do to the complex interactions of plants and animals in tropical rain forests, temperate forests, or the oceans. But it is suspected that if too many species are lost, ecosystems as a whole could break down in ways as yet not even understood.

Major Sources

Extinctions have been caused, accelerated, or threatened by:

Habitat loss: Forests, soils, wetlands, prairies, rivers, streams, lakes, seas, coral reefs, and caves have been destroyed, polluted, or fragmented beyond livable use. Wetlands and reefs are major areas for breeding and reproduction and are nurseries for numerous species. The destruction of such areas has been caused by expansion of agriculture, the raising of cattle and livestock, construction, fuelwood needs, aquaculture, timber industry, road building, urban enroachment, mining, locks and dams, acid rain, air pollution, and smog. Habitat loss is the largest cause of current extinctions, especially in the tropics.

Population increases: As the human population increases, it puts more pressure on existing resources by cutting down forests or mining for raw materials. It also increases manufacturing and resultant wastes, solid waste, and pollution from disposal methods such as incineration.

Overfishing and overhunting become a problem. Humans compete with wildlife populations for space, water, and food. In industrial countries increasing populations also place stress on wilderness resources through sport and recreation activities.

Agricultural practices: Farmers rely on only a few hybrid seeds and farm animals, while thousands of natural varieties go extinct. Since so much land is used to grow hybrids, little land is left for natural varieties to survive on their own.

Overfishing: Overfishing is caused by the use of driftnets, high-tech fishing fleets, and trawlers.

Overhunting: Overhunting is caused by whaling, hunting for food or sport, poaching, fur business, and predator control.

Wildlife trade: Animals or their body parts are taken from the wild legally or illegally and used for pets, zoos, product testing, medical testing, jewelry, ivory, aphrodisiacs, and folk medicine. Plants are sold commercially and for horticultural collections. These activities put pressure on species that already have low populations.

Keystone species loss: Keystone species are those whose activities other animals depend on for their own survival. For example, the sea otter eats sea urchins, which destroy kelp forests. If sea otters are killed off, kelp beds are destroyed by urchins and fish populations plummet, followed by falling populations of eagles and harbor seals.

Toxic chemicals: DDT, polychlorinated biphenyls (PCBs), pesticides, organic chemicals, and selenium, for example, often stress top predators.

Dams: Dams cause initial flooding for reservoirs, prevent silt and nutrient flow downstream, and prevent fish from swimming upstream to spawning grounds—as in the salmon fisheries.

Exotic species: Some introduced plants or animals are able to compete better than native species or simply eat native species or their prey.

Environmental disasters: Droughts, hurricanes, volcanic eruptions, and forest fires stress already reduced populations of plants and animals.

Ozone depletion: There is growing evidence that ozone depletion is causing and adding to the extinction of invertebrates and amphibians.

Multiple causes: When many species are placed under stress by a number of the above problems simultaneously, their survival and protection become even more difficult.

Natural causes: Extinctions occur naturally as a result of competition between species, environmental changes, and natural catastrophes. One theory holds that the extinction of the dinosaurs occurred after the earth was struck by a massive asteroid, whose crater has been found in the Yucatán. It is believed the impact produced a cloud of dust and smoke that over time blocked the sun, killing plants, and eventually plant eaters. Another theory is that fatal viral and bacterial diseases were transmitted by migrating species of dinosaurs when the continents of Asia and North America were connected by a bridge of land.

State of the Earth

Scientists have identified fewer than 1 1/2 million species of life forms, but estimates of existing species range from 10 to 30 million—most are insects. Estimates range from 5,000 to 50,000 for the number of species going extinct each year—15 to 150 each day. This extinction rate is 100 to 1,000 times higher than would occur without human activities. Fifteen percent of all world species are currently threatened; conservative estimates predict the loss of 20% of all species in 3 decades. Many temperate birds spend winter in the rain forests and are also threatened; 3/4 of the world's 9,600 bird species are declining or threatened with extinction.

Rain forests have the greatest biodiversity of any ecosystem in the world; 1/2 of all rain forests have already been destroyed. Rain forests hold 50% to 90% of the world's species, 40% of the birds of prey, 60% of the vascular plant species, 90% of the primates, more than 95% of the world's insects, more than 95% of the terrestrial invertebrates, and 50% of the vertebrates. A small number of countries and regions hold 60% to 70% of the world's biodiversity; countries with megadiversity (high biodiversity) are Australia, Brazil, Central Africa, China, Colombia, Costa Rica, Ecuador, India, Indonesia, Madagascar, Mexico, Peru, and Zaire. Brazil, Colombia, Indonesia, and Mexico have some of the highest biodiversity.

In the marine environment coral reefs have the highest biodiversity, followed by mangroves and coastal wetlands; coral reefs have higher productivity per square meter than tropical rain forests because the ecosystem is more "dense." But coral reefs are considered prolific if they contain only several thousand species of fish; the Great Barrier Reef has fewer than 10,000 species of fish, mollusks, crustaceans, turtles, marine mammals, and nesting birds. The Great Barrier Reef has one of the best ecosystem protection programs in the world; it is also under continually increasing pressure from fisheries and tourism. Five percent to 10% of all coral reefs have already been destroyed; 50% could be gone in 2 to 4 decades.

Depending on the area of the world, 75% to 90% of all commercial marine fish caught are species that depend on coastal estuaries for reproduction, food, migration, or nurseries; most of the rest depend on coral reefs. The United States has developed 50% of its coastal wetlands, as have tropical areas around the world; Italy has developed nearly 100%. Coral reefs, which are very sensitive to temperature, light, and water nutrient changes, support 1/3 of all fish species. Coral bleaching is being observed with more frequency in the Atlantic and Pacific Oceans. In some tropical countries coral is being mined for construction.

More than 1/2 the world's monkeys and lemurs are threatened with extinction, as are 40% of the world's turtles. Most wild cats and bears are in decline. Thirty percent of U.S. fish stocks have been in decline since the late 1970s; in the waters around Europe more than 100 species of fish are overfished. Most whales are still threatened with extinction (except perhaps the minke and gray). Fifty percent of Australia's mammals are threatened with extinction—more than 7 million kangaroos and wallabies were killed in 1992. Frogs worldwide are showing major declines. One-fourth of all tropical plants may be gone in 30 years. Three thousand of the 25,000 plant varieties in the United States are threatened—700 are near extinction.

Islands often suffer the highest extinction rates because native species have nowhere to go if they are threatened, if an exotic species is introduced, or if their habitat changes. Hawaii has lost 10% of its native plant species; 40% are threatened; 50% of its native birds have gone extinct. Madagascar is one of the most biologically diverse and rich islands in the world; 80% of its forests and vegetation have already been cut down.

No more than several dozen plants and animals provide 80% of all human caloric intake. Three-quarters of all rice today comes from one strain of rice; Indonesia has lost 1,500 varieties of rice in the past several decades. Since 1900, 85% of the known apple varieties and several thousand varieties of pears have become extinct. Almost all major U.S. crops are nonnative species. In 1846 the great Irish Potato Famine was caused by widespread use of a single strain of potato with no resistance to a potato blight. Some indigenous peoples still use diverse varieties of crops: the Ifugao people in the Philippines use more than 200 varieties of sweet potato; Andean farmers use more than 1,000 varieties of potato.

Worldwide, about 400 native species of livestock animals—pigs, cattle, goats, sheep, horses, donkeys, and buffalo—are in danger of going extinct as other livestock is introduced in their place. The native varieties are hardier and acclimate better to local conditions. Half of all Europe's livestock breeds are already extinct; a United Nations survey of 5 livestock species in Europe showed that of 700 breeds, 1/3 are near extinction. Of an estimated 4,000 breeds worldwide, 1/4 could soon be endangered; 600 breeds are already extinct. The United States has 46 endangered breeds.

A number of whale populations may have already dipped below a sustainable number. For some whales, even 5,000 individuals may be too few to maintain their population in the wild. The Atlantic northern right whale, with an estimated population of 300 to 350, has not increased its numbers over the past 50 years. Researchers believe that its numbers are being kept in check by deaths caused by discarded fishing nets and boat collisions, which are also believed to be causing increased humpback whale deaths. Iceland and Norway have threatened to begin whaling again, despite the International Whaling Commission's ban; Japan has been whaling limited numbers continually, under the guise of "scientific whaling." More than one-half dozen countries in the past decade have taken whales illegally through catches or pirating ventures.

Poaching is a serious problem worldwide for thousands of species, including elephants (though the poaching problem was reduced by 80% when ivory trade was banned). Poaching may bring to extinction hundreds of species such as the African rhinoceros, tigers, pandas, leopards, and black bears (Asian demand for claws, paws, and gall bladders places a value of $10,000 on American black bears). U.S. illegal wildlife trade grosses $5 billion a year. Millions of wild birds are captured and sold as pets yearly; 1 to 3 times as many die in the capture and transport process. Many zoos, worldwide, have high mortality rates for the animals they stock.

More than 90% of U.S. old-growth forests have already been cut down; 40 million acres of tropical forests are destroyed annually. Destroying 90% of a habitat means losing at least 50% of the species. Less than 5% of the world's land area is designated as preserves and parks—the majority of that is deserts and tundra. Most parks and preserves in the world have little or no protection, many have leased timber and mining rights, and many are subject to excessive tourism. Most healthy ecosystems are found in areas where indigenous people still have control.

One of the most serious pollution problems in 16 U.S. western states is selenium, a natural poison in soils that leaches out because of poor irrigation practices. Selenium is more toxic than arsenic, bioaccumulates, and kills young birds and fish. It is teratogenic and is believed responsible for killing up to 10,000 waterfowl a year. Some experts believe selenium could end waterfowl hunting in 20 years. The U.S. Fish and Wildlife Service, Justice Department, Department of the Interior, Bureau of Reclamation, and Geologic Survey have failed to pursue changes in human-created evaporation ponds and other contributing irrigation practices because of pressure from corporate farm and ranching interests; agents investigating the problem in the past have been transferred to other departments or told to keep quiet.

The disappearance of songbirds and butterflies, both indicator species, in an area is a sign that an ecosystem is deteriorating and has already suffered significant damage. Many species of both songbirds and butterflies are in serious decline and face threats of extinction in the next decade if their habitats are not preserved.

Human Impact

Food: Marine fish are threatened with extinction from overfishing, wetland and coastal development, coral reef destruction, and ozone depletion. Hybrid seed crops sometimes suffer large losses since they are not as hardy as natural varieties and often require more chemical input. Thus losing natural seed varieties and using only hybrids has led to sporadic food shortages. Food nutritional quality is also lost as dependence on chemical farming with hybrids increases.

Medicinal value: Tropical forest plants that could provide cures for diseases are going extinct. Coral reefs also hold many untapped medicinal possibilities.

Economic: Loss of forests, medicinal plants, and animals results in lost income for fishermen, indigenous peoples, and others who use these resources for their livelihood or daily needs.

Aesthetic: Extinction means a loss of heritage both for future generations and for our own.

Other threats: The causes of extinctions often threaten human health. Toxic chemicals are often just as toxic to humans as to wildlife. Exotic introductions like the African honey bee compete with native bee populations but also attack humans. Dams that cause fish or plants downstream to die off often force large populations of indigenous peoples to leave the area. Soil degradation ruins wildlife habitat and makes soil unfit for agriculture. Polluted water kills fish and is dangerous to drink. In almost all cases the causes of extinctions are harmful to humans, in the short or long term.

E

Individual Solutions

- Don't buy exotic pets, products, or supplies (ivory, tropical birds, tropical fish, coral), unless they are raised commercially and legally for pet use; ask where the pets or products originated; if you are not sure call the U.S. Fish and Wildlife Service, the U.S. Environmental Protection Agency, or the Humane Society.
- Don't buy sea turtle products anywhere; avoid butterfly and insect curios.
- Don't buy products using exotic rain forest wood, such as ebony, mahogany, and teak.
- If you eat frog legs, find out how they are produced; frog populations in many places are under stress and are unmanaged.
- Consult your local zoo and find out what it is doing to preserve animals in the wild.
- Don't use pesticides or herbicides.
- Don't pour anything toxic down the drain or sewer or onto the land; contact local officials or the EPA for hazardous collection sites and information.
- Be careful in your recreational activities: off-road vehicles trample animals and plants; boat propellers threaten the manatee in Florida.
- Don't take souvenirs from beaches, forests, coral reefs, or deserts: desert cacti populations and petrified forests are being decimated by tourists.
- If you are concerned about the ethical treatment of animals, contact People for the Ethical Treatment of Animals (PETA) or the Humane Society.
- Practice ecotourism; contact the Ecotourism Society or Cultural Survival.
- Practice energy conservation, which lessens pollution and greenhouse gases.
- Eat less red meat; cattle are a major cause of tropical forest destruction and desertification in the western United States, and a source of methane—a greenhouse gas.

- Buy organic food and local natural varieties of food to support those species and sustainable agriculture.
- Garden with natural seed varieties instead of hybrids. This will support organizations dedicated to preserving these seeds, which will help ensure the survival of natural seed varieties.

Industrial/Political Solutions

What's Being Done

- Driftnets were banned globally in 1993.
- A world ban on whaling is in effect.
- Most countries have agreed to ban chlorofluorocarbons (CFCs) by the mid-1990s.
- Zoos, botanical gardens, and other organizations are working to study, protect, and reestablish threatened species in their native habitats.
- Seed and pollen banks in a number of places in the world are growing natural varieties of plants and then storing their seeds and pollen to preserve the varieties for future generations. The Seed Savers Exchange (SSE) on Heritage Farm in Decorah, Iowa, a network reaching 14 countries, stocks seed varieties that are no longer available commercially. The U.S. Department of Agriculture runs the National Seed Storage Laboratory in Fort Collins, Colorado; it has 240,000 samples of seeds that are available free to public and private breeders.
- Several major organizations worldwide, such as the American Minor Breeds Conservancy and the Rare Breeds Survival Trust in the United Kingdom, are devoted to preserving endangered livestock breeds.
- Zoos and other organizations are collecting sperm and ova of threatened or near extinct animals and storing them in genetic banks to preserve the possibility of future regeneration of the species.
- The Convention on International Trade in Endangered Species of Wild Fauna and Flora (CITES) is the biggest international agreement to control or prohibit global wildlife trade, especially with respect to endangered species; more than 100 countries participate. Countries

file annual reports. Sometimes international trade bans or boycotts are used to pressure a country to protect endangered species.

- The Convention on Biological Diversity of the 1992 Earth Summit in Rio de Janeiro calls for national strategies for conservation and bio-diversity and for the creation of a list of species to be preserved. It also gives nations the right to regulate access to or charge payment for use of genetic resources within their territories. Signed by all industrial countries and many developing countries, the agreement went into effect December 29, 1993. The United States is spending $179 million in 1994 for a National Biological Survey, created October 1, 1993, to map all national resources such as wetlands, wildlife habitats, existing wildlife, and plant species.

- A fully protected ecosystem approach to maintaining biodiversity is being attempted; in other words, some countries are declaring that rain forests or old-growth forests are threatened and generating efforts to save them, which in turn saves all the species living in either area. This is a relatively recent approach and supported by the EPA, the Bureau of Land Management, the Fish and Wildlife Service, and conservation groups worldwide.

- Habitat programs like wildlife parks, refuges, preserves, and urban wildlife habitat efforts have been established; the United States has the National Wildlife Refuge System. Mining has been prohibited in Antarctica for 50 years—there are efforts to make it a world park. British Columbia recently set aside 2.3 million acres in the Tatshen-shini watershed as a provincial park; it is the largest in the world. World Heritage Sites seek to preserve internationally esteemed wilderness areas.

- Debt-for-nature swap programs trade money for land conservation with countries that wish to lower their debts by agreeing to put aside a piece of land for conservation purposes. This approach has been used in a number of countries, including Bolivia, Brazil, Costa Rica, Dominican Republic, Ecuador, Guatemala, Jamaica, Madagascar, the Philippines, Poland, and Zambia.

- The EPA, World Bank, and United Nations have shifted their focus from heavy pesticide use in farming to integrated pest management.
- Efforts are being made to help organize indigenous peoples to protect their homes in the remaining forests. Worldwide, indigenous people are organizing to fight for their homelands.
- Some corporations are paying fees and royalties to countries with tropical rain forests for the right to research possible medicines from tropical plants.
- Genetics is being used more and more to track animals being poached, to identify species and hybrids of species, and to manage both captive and wild populations of animals.
- The U.S. National Environmental Policy Act of 1969 requires federal agencies to assess in advance the environmental impact of any major federal project or action; the courts recently extended this mandate to cover projects in Antarctica (scientific research expeditions of the National Science Foundation had been dumping garbage), and thus worldwide.
- The U.S. Wild Bird Conservation Act of 1992 regulates the import, sale, and transport of wild birds, so species will not be endangered in the wild and captured birds will be treated humanely.
- Boycotts have been called to halt the use of rain forest beef in hamburgers and rain forest timber that has been procured through clearcutting.
- The U.S. Endangered Species Act of 1973 identifies threatened species and makes it illegal to traffic in them or harm them or their habitat. Noted successes are the American alligator, red wolf, whooping crane, and bald eagle. Many other species have not been so lucky and have perished while on the endangered list, or even before they were put on it. Reauthorization occurs in 1993–1994.
- In the United States farmers are paid to keep highly erodable land as grassland or forest to protect and conserve wetlands. An agreement among 65 countries calls for protection of at least one major wetland in each country.

- Oil tankers and barges larger than 5,000 tons and using U.S. ports will be required to have double-walled hulls by 2010; smaller barges and tankers have until 2015.

What Needs to Be Done

- Stop industrial discharges into waterways, work to lessen agricultural and urban runoff, end raw sewage discharges, and end coastal drilling: all threaten coral reefs and coastal wetlands.
- End irrigation practices that are leaching selenium into waterways.
- Place strict restrictions on the Department of Agriculture's Federal Animal Damage Control Program, whose agents killed 2.2 million wild animals in 1992, including foxes, coyotes, and wolves.
- End federal practices such as excessive hunting on wildlife refuges, military bombing exercises on refuges, and other practices that fail to sustain the refuges for wildlife.
- Protect indigenous peoples who live in wilderness areas; they are often the best stewards of the land.
- Establish world agreements to reduce greenhouse gases.
- Fund and emphasize energy conservation, energy efficiency, and renewable energy instead of use of fossil fuels.
- Make family planning services available in all countries.
- End wildlife trade except where absolutely essential; ensure especially that the pet industry and medical researchers do not obtain protected species. Governments getting aid from environmental groups need to ensure that poaching does not continue in their parks. Countries need to boycott goods and services to countries, such as Taiwan, that are allowing illegal wildlife trade to continue; in 1994 the United States placed a ban on all wildlife imports from Taiwan for its illegal trafficking of endangered tiger and rhinceros parts.
- Set worldwide sustainable fishing harvesting limits.
- Make worldwide agreements to protect endangered migratory species that cross international boundaries.
- Strengthen the U.S. Endangered Species Act with more funding,

wider coverage, and a broad ecosystem approach to wildlife preservation.

- Plan wildlife habitats so corridors exist between wildlife parks; this design ensures that if one area is stressed, animals moving out of the parks have habitat to which they can go and are still protected.
- Restrict introductions of exotic species. A number of countries do not regulate exotics.
- Offer funding to developing countries to provide adequate protection to their wildlife preserves.
- Use sustainable agriculture and grow different natural varieties of a food crop to maintain genetic diversity and lessen risk of crop losses.
- Establish a worldwide emphasis on sustainable development, which includes preserving biodiversity, forests, wetlands, and all other resources so they can be passed down to future generations intact.
- Mandate a quicker worldwide ban on CFCs and all ozone depleting chemicals, including HCFCs and HFCs.

Also See

- Overfishing
- Population pressures
- Sustainable development

Fossil fuels

Fuels that have been formed over millions of years through the compression of dead organic plant and animal matter in the earth's crust.

Though some fossil fuel reserves in the earth's crust are very large, they are considered finite energy sources, because eventually they will be used up and cannot be replaced. This feature is partly what distinguishes them from renewable energy.

Fossil fuels are primarily hydrocarbons (molecules of hydrogen and carbon). Examples of fossil fuels are coal (a solid), natural gas (a gas that is primarily methane), and oil (a liquid also called petroleum).

A number of other fuels can be made from fossil fuels. Methanol, a 3:1:1 ratio of hydrogen, oxygen, and carbon, can be made from coal or natural gas; methanol is also made from biomass. Propane is a byproduct of oil production and natural gas processing. Gasoline is made from oil.

Most fossil fuels are used in automobiles and power plants. Ten percent of the petroleum (oil) and natural gas is used by the petrochemical industry to make chemicals. Hydrogen from natural gas is used to make fertilizers.

Environmental Impact

Air pollution: Coal and oil burn easily and give off carbon dioxide, nitrogen oxides, carbon monoxide, water vapor, heat, and a variety of minerals and impurities such as dust, fly ash oil, fly ash coal, particulates, soot, smoke, sulfur dioxide, heavy metals, radon, benzene, toxics, and hydrogen sulfide. This pollution constitutes the major portion of acid rain, greenhouse gases, and smog. Acid rain, air pollution, and smog harm plants and animals. In the case of acid rain, fish populations have been destroyed, as well as entire lakes and forests. Heavy metals, toxic to plants and animals, bioaccumulate in the food chain.

Mining: Coal mining (strip mines and underground tunnels) destroys natural habitats and causes erosion and leaching of acidic particles and heavy metals into nearby soils and water. Oil and natural gas production destroys natural habitats; pipeline leaks, drilling wastes, and oil spills kill wildlife, ruin beaches, and pollute waterways and oceans. Drilling for oil or natural gas often produces heavy metals, toxics, and radioactive wastes. These toxics can enter the food chain and pollute soils and water; they can cause diseases and death in animals. Discarded used fuels cause many of the same problems.

Ozone depletion: Burning coal produces sulfur dioxide, which changes to sulfates like sulfuric acid in the atmosphere. Sulfates are particulates that can increase ozone destruction by increasing the chemical reactions between chlorine and ozone. Ozone blocks ultraviolet rays from the sun; thus thinning ozone can lead to damaged plants. Ultraviolet also damages plankton in Antarctic waters and may affect the whole food chain there. Sulfates, however, may also block incoming ultraviolet, and the net effect is unclear for local areas.

Natural gas: When burned, natural gas produces no soot, unlike coal and oil; fewer hydrocarbons, carbon dioxide, and carbon monoxide than gasoline; and almost no sulfur dioxide. It yields, however, more nitrogen oxides, a major factor in acid rain. Leaking natural gas in production, transportation, and use also increases the greenhouse effect, since the main component of natural gas is methane, a significant greenhouse gas.

Petrochemicals: Chemicals made from fossil fuels are called petrochemicals. They are often toxic, persist in the environment, and can enter the food chain and harm the breeding of birds and other animals. Predators bioaccumulate and biomagnify these toxic chemicals and are especially susceptible to their harmful effects.

Fresh water and soil: Fossil fuel power plants use billions of gallons of water yearly and contribute to thermal pollution. Residual coal ash, which can be high in radon, heavy metals, and trace contaminants, is put in landfills.

Major Sources

Mining: Coal can be found deep underground or close to the surface, in which case strip-mining is used. Coal beds can be a thin film only inches thick, several feet thick, or a layer up to 50 feet thick. Oil and natural gas reserves can also be found in shallow reservoirs or very deep beneath the earth's surface. Strip mining recovers 85% to 90% of the coal in a site; 85% of a natural gas reservoir can be recovered. Only 35% of U.S. crude oil reservoirs are usually recovered; the cost of recovering the remaining oil is currently too great for this to be practicable.

Sensitive areas: Coal, oil, and natural gas reserves are huge, and more reserves are found yearly with newer, more sophisticated technologies. These reserves, however, are often in ecologically sensitive areas. Examples are the oil reserves in Alaska's Arctic National Wildlife Refuge, Antarctica, and Siberia or reserves on the continental shelves of the oceans.

Difficult deposits: Sometimes fossil fuel reserves are in areas that are either too deep or too difficult to reach or in deposits, such as shale, that are difficult to mine. New technologies are being developed to recover deposits such as these, but the expense may be great enough to encourage renewable energy use instead.

State of the Earth

Estimated fossil fuel reserves vary widely but are probably at least 30 to 50 years for oil, 60 to 120 for natural gas, and 130 to 200 for coal. What makes these reserves difficult to estimate is the increasing use of fossil fuels by developing nations, the uncertain size of some reserves, and the uncertainty of our technological and economic ability to recover all known reserves. Environmental factors and the resultant increasing costs are also playing a stronger role in determining whether some reserves will be used.

Oil reserves are concentrated in the Middle East and the former Soviet Union, but oil exporters include Indonesia, Nigeria, and Venezuela; the United States has 3% of world oil reserves. Transportation (mainly cars) is responsible for 2/3 of all oil used in the United States. Worldwide, transportation accounts for 1/3 of all oil use; the rest is used mainly in heating and manufacturing.

According to the U.S. Environmental Protection Agency (EPA), 179 million tons of oily wastes are buried at refinery sites yearly; millions of gallons of water contaminated by the drilling processes are dumped yearly, untreated, into waterways and oceans; about 1 billion gallons of oil pollute the oceans each year. In the past ten years several hundred oil spills have released an average of 1 million gallons of oil each month; one gallon of gasoline (refined oil) can contaminate up to 1/4 million gallons of drinking water.

China is now the largest producer of coal worldwide and relies heavily on coal for its energy needs. The United States and the former Soviet Union have equally large coal reserves totaling 30 to 40% of world reserves; China (with about 11% of world reserves), Australia, Germany, India, Poland, and South Africa also have significant reserves. Coal is used mainly for heating and electricity; in the United States 87% of coal is used in electric utilities. U.S. utilities are planning to open 60 new power plants between 2001 and 2011: 28 for coal, 18 for gas, and 8 for lignite; the number could climb to 150 plants. A typical utility power plant may burn 10,000 tons of coal each day; with even 5% waste (a low estimate), that leaves 500 tons of solid waste to dispose of each day. The coal industry generates 90 million tons of waste annually, most of which goes into landfills.

Carbon dioxide represents 40% to 50% of human-created greenhouse gases, and 70% of it is produced from burned or used gas, coal, and oil. Lifeless lakes and rivers (caused by acid rain, which is largely a result of burning coal) have been documented in past decades in nearly every country in the world; motor vehicles such as cars, buses, and trucks burning oil, gas, or diesel are responsible for more than 50% of the smog-producing chemicals.

Iran has 15% of world reserves of natural gas; the former Soviet Union, 40%; and the United States, 4%. Natural gas can be readily produced in Canada and the United States and may undergo rapid growth in the next decades. It is already used in limited fleets of cars and buses in several countries; natural gas provides 95% of the power for cooking, boiling water, and heating for 3 of the 4 anchors in the Mall of America, currently the largest mall in the world.

Since 1900 use of fossil fuels has increased 4 times as fast as the world population. Fossil fuels generate 3/4 of the world's electricity. The United States meets its energy needs, roughly, with 40% oil, 23% coal, 25% natural gas, 6% to 8% nuclear energy, and 8% to 10% renewable energy (hydroelectric, biomass, wind, solar, and geothermal). The United States meets its electricity needs, roughly, with 55% coal, 20% nuclear energy, 10% hydroelectric power, 12% natural gas, and 3% oil-derived electricity. Globally, commercial energy sources are 40% oil, 27% coal, 22% natural gas, and 6% nuclear. Biomass (mainly fuelwood in developing countries) provides 15% of the total world energy needs. Global electricity production is derived from coal (40%), hydroelectric power (20%), nuclear energy (15%), natural gas (13%), and oil (12%).

Coal gives off 95% more carbon (and thus greenhouse gases) than natural gas, and oil releases about 40% more carbon than natural gas; however, natural gas has only 17% of the energy content gasoline has, so more must be burned. As of 1991 the United States contributed about 25% of the worldwide carbon emissions from fossil fuels; the former Soviet Union contributed about 20%; China, 10%; Japan, 5%; and Germany, less than 5%. Other countries contributed lesser amounts. Brazil and other countries without oil reserves are using biomass energy and energy conservation to provide their energy needs.

Human Impact

Health: Burning fossil fuels adds to acid rain, air pollution, and smog, which increase respiratory problems. Acid rain also leaches cancer-causing heavy metals into water and fish. Smog causes lung, respiratory, and other health problems. When methanol is burned, the toxic chemical formaldehyde is produced; formaldehyde can cause lung disorders, asthma, and cancer.

Food: Acid rain, air pollution, and smog from burned fossil fuels all damage food crops; mercury can render fish stocks toxic.

Natural gas: Natural gas is dangerously volatile as a liquid and can cause explosions and fire if not transported carefully.

Petrochemicals: Petreochemicals are some of the most toxic human-created chemicals and can cause a number of health problems, including neurological problems and cancer.

Ozone depletion: Use of fossil fuels contributes to ozone depletion and thus increases ultraviolet exposure, which increases skin cancer and cataracts.

Individual Solutions

- Support renewable energy and taxes on fossil fuels; look into solar and wind energy for your home energy needs.
- Purchase cotton products instead of synthetics, which are petroleum products.
- Minimize use of plastic, another petroleum product.
- Practice energy conservation to minimize use of fossil fuels; use ethanol or gasohol in your car.
- Support strict pollution standards for all fossil fuel use and high fuel efficiency for cars.
- Support an end to coastal oil drilling and a continued ban on oil exploration in Antarctica and in Alaska's Arctic National Wildlife Refuge.
- Write or call your legislators for an end to fossil fuel subsidies and more support for renewable energy.

Industrial/Political Solutions

What's Being Done

- The Antarctic Treaty of 1961 (updated in 1991) prevents any mineral extraction in Antarctica for at least 50 years, until 2041.
- Many countries, including a number of European countries, Canada, Japan, and the United States, are reducing carbon dioxide emissions by the year 2000.
- An international scientific team is calling for a 10-year moratorium on oil and gas exploration in Siberia in response to pollution problems.
- The National Energy Policy Act of 1992 offers variable incentives to renewables in the form of investment and production tax credits.
- The U.S. Clean Air Act of 1990 is mandating tougher air emission standards for sulfur and nitrogen oxides and requiring use of cleaner fuel and fewer air emissions in automobiles.
- The environmental impact of oil drilling is being taken into account during approval of drilling in coastal areas, Alaska, the Antarctic, and the Arctic.
- The U.S. Surface Mining Control and Reclamation Act of 1977 requires coal mining companies to restore strip-mined areas to their initial conditions as nearly as possible.
- The U.S. Coal Mine Health and Safety Act of 1969 reduced risks to miners by requiring better ventilation and safer tunnels; it was amended in 1977 by the Mine Safety and Health Act.
- Electric utility plants must purchase energy developed by small renewable producers, and must do so at a rate equivalent to avoided cost (of new or replaced generation). This measure began with passage of the Public Utilities Regulatory Policies Act (PURPA) in 1978.
- An increased federal gas tax began in 1993.
- In Minnesota and a few other states, utility regulators must include environmental costs when deciding whether to approve new power plants, and they can only approve the least expensive option presented.

F

- California is requiring zero-emission cars to be sold by the mid-1990s and 200,000 to be sold by 2003. Ten other states have said they will follow California's lead.

What Needs to Be Done
- Tax fossil fuels to pay for cleanup costs of pollutants.
- End all fossil fuel subsidies.
- Increase investment in renewable energy; set stronger incentives for off-the-grid renewable energy use in the commercial, residential, and industrial sectors.
- Mandate higher fuel efficiency and lower tailpipe emissions in automobiles; phase in zero-emission cars, like electric cars.
- Support mass transit and light rail.
- Factor into the costs of electric utilities the environmental damage and health problems related to fossil fuel use.
- Mandate scrubbers for sulfur dioxide and nitrogen oxides on coal plants; end emissions trading and focus on pollution prevention.
- Set world air emission limits for fossil fuel pollutants.

Also See

- Coal
- Mining
- Oil

Freshwater degradation

Definition

Fresh water that is either polluted or used up faster than it can replenish itself.

Fresh water is inland water that has a low concentration of minerals, salts, and dissolved solids. Fresh water is either surface water (held in streams, rivers, lakes, ponds, and springs) or groundwater (water that percolates downward through the ground and is held in caverns or rock formations as underground lakes called aquifers).

A special type of aquifer is an artesian system, where impermeable rock surrounds the water and often holds it under pressure. The major source of all fresh water is ocean water, which evaporates and later falls over land as rain, snow, or other precipitation. Thus, if managed properly, fresh water is a renewable resource.

Environmental Impact

Water pollution: Sediments, salts, pathogens, suspended matter or particles, and chemicals can kill fish and other aquatic life and inhibit plant growth. If the pollutants are fertilizers, manure, sewage, or other sources of phosphates, nitrogen, or nutrients, abnormal algae growth can be stimulated in a process called eutrophication, which eventually uses up the oxygen in the water, killing fish and plants. Water used for cooling by industry is heated, and when released into coastal waters it affects coral reefs and plant and fish growth. In general, freshwater pollution from rivers is a major factor in ocean degradation. Rivers can cleanse themselves of pollutants more quickly than lakes, ponds, and aquifers—which can take decades to hundreds of years to cleanse.

Toxics: Toxic pollution can be taken up by plants or fish and enter the food chain. Higher predators may eat the fish, and thus the toxins may bioaccumulate and biomagnify in birds and mammals. As water resources become used up, the toxics in them become more concentrated; this is especially true in lakes and aquifers.

Water resource depletion: Depletion of water resources occurs when rivers, aquifers, lakes, streams, and ponds are used up faster than they can replenish themselves through rainfall, snow, springs, or other moisture. In the case of aquifers, which hold groundwater, loss occurs when the use exceeds the rate at which water percolates down through the ground to the aquifer. As lakes and rivers become depleted through use and drought, fish populations and other wildlife populations that use these water resources decline. Forests and vegetation downstream from dams die when the water tables become so low that the trees cannot send their roots down to them.

Major Sources

Freshwater pollution can occur through:

Agriculture: Pesticides, chemical fertilizers, sediments, and manure can be washed off farmland by rain into rivers and lakes. Irrigation can lead to salt-ladened soils through salinization; runoff from these soils can raise salt levels in rivers and lakes.

Industrial wastewater: Industrial wastewater often contains organic chemicals, heavy metals, wastes, manufacturing by-products, and heated or warm water and is frequently dumped directly into sewage systems, rivers, and streams.

Sewage: Sewage can contain toxic chemicals, heavy metals, phosphates, and bacteria. While sewage is often treated in industrial countries, it is often left untreated in developing countries and dumped directly into rivers and streams. Leaking septic tanks are also a source of sewage pollution.

Urban runoff: Pesticides, chemical fertilizers, road salts, and oil and chemical residues on roads are all washed by rain into sewers, lakes, and rivers.

Air pollution: All particulates, chemicals, and gases eventually are precipitated out over water or soil. If they do not bind with the soil, the contaminants can be washed into streams, rivers, and lakes. Or they can percolate down with water to aquifers. Acid rain changes the pH of water and leaches heavy metals into the water from nearby soils.

Oil or petroleum chemicals: Contamination by petrochemicals can result from accidents of inland river barges and small tankers, leaks in underground storage tanks, oil or gasoline that is sent down the sewer system or onto soil where it is washed into water sources, or yachting boat engine leaks.

Hazardous waste: Landfills, mining wastes, oil drilling, buried hazardous wastes, nuclear waste, and other land-based waste all have the potential to be leached by rain runoff into rivers, streams, lakes, and aquifers.

Construction sediments: Soil erosion due to urban construction, or large dam and stream channeling projects, can send sediments and construction chemicals into lakes, rivers, or oceans.

Logging runoff: Tree farms produce pesticide, fertilizer, and sediment runoff; natural forest logging and deforestation produce sediment runoff.

Lead: Old lead piping can leach lead directly into drinking water.

Chlorine and fluoride: Chlorine enters water as a disinfectant in drinking water and as a pollutant; fluorides are added for dental health and also appear as a result of pollution.

Freshwater usage, or scarcity, is increased by:

Irrigation of farmland: Irrigation is a major use of fresh water and a concern in all countries, especially in arid and semiarid regions.

Industrial uses: Water is used in manufacturing, cooling, and chemical reactions.

Public usage: Public usage includes landscaping use and home use.

Draining of wetlands: Wetlands serve as sinks to hold water that percolates down to aquifers. If wetlands are eliminated, less water is held than can reach aquifers. Wetlands are also important natural purifiers of polluted water, especially from nonpoint sources. As more wetlands are lost, less water is cleansed naturally.

Urban pavement, roads, culverts, cities: Urban areas expose less soil to rainfall, which then runs into sewers and rivers instead of soaking into groundwater supplies.

Building and construction around lakes: Construction around lakes diverts runoff water away from lakes so they do not refill.

Hydroelectric facilities and dams: Dams reduce water reaching downstream areas.

Mining: Mining uses water to mix with ores or minerals as slurry, to wash down mining operations, to assist in production, and to drill.

Livestock: In livestock production water is used to grow grain and feed crops and for daily watering.

Drought: Drought can seriously strain already overused water resources.

State of the Earth

Less than 3% of the earth's water is fresh water (97% is seawater), and roughly 80% of all fresh water on earth is locked into ice caps and glaciers. Of the remaining 20%, which is the world's available fresh water (less than 1% of all water on the planet), 97% is held underground as groundwater in pools and lakes called aquifers; of the remaining 3% (only 0.6% of all fresh water), 57% is held in rivers and lakes, 33% in soil, 7% in the atmosphere, and 3% in living organisms.

Worldwide, agriculture accounts for more than 65% of all freshwater use, industry 25%, and domestic use 6% to 10%; 70% of the water used for irrigation never reaches the crops. The United States has the highest freshwater usage in the world (over 340 billion gallons a day); industry accounts for 45% of freshwater use; agriculture, 40%; and domestic uses, more than 10%. Industrial water use is also high in Europe and increasing worldwide.

Roughly 1/2 of Americans use groundwater for domestic needs. Two-thirds of groundwater use is for irrigation. The Ogallala Aquifer, the nation's largest, stretching from Texas to the Dakotas, may be depleted within decades; groundwater overuse is lowering water tables past critical levels in Mexico, the western United States, parts of Asia, and arid and semiarid areas of Africa. Aquifers can take years, even centuries, to establish themselves, and most are considered nonrenewable in the short term. On a normal human time scale, groundwater contamination can often be considered permanent—nearly impossible to clean up. Rivers and lakes can take anywhere from years to decades to purify themselves; rivers with locks and dams on them take even longer.

Globally, more than 90% of all sewage in developing countries is untreated; Mexico sends 30 million gallons of raw sewage into California daily, via the Tijuana River. Industrial countries have more sewage treatment than developing countries, but many still do not treat all their sewage. Europe treats 70% of its sewage; the former Soviet Union treats 30%. Canada, Japan, and the United States have significant percentages of sewage that is not treated or is released raw when sewers overflow. Sewage treatment generally removes only 1/3 to 2/3 of the toxics, 1/3 of the phosphorus, and 1/2 of the nitrogen. Israel reuses 67% of total sewage for irrigation; the United States reuses 2.4%.

More than 1/3 of the people in developing countries do not have access to an adequate supply of clean water. In the Central African Republic 92% of the people lack such access; in Ethiopia, 82%; Bolivia, 54%; Pakistan, 50%; Vietnam, 50%; India, 25%; Mexico, 22%; South Korea, 22%; and Egypt, 14%. In the United States, 1/5 of the more than 7,100 industrial and public waste treatment facilities have serious or chronic violations of their discharge permits; 300,000 drinking water violations occurred during 1991–1992. It is estimated that more than 1/2 of all illnesses in India are due to drinking water pollution caused by sewage. Diarrheal diseases resulting from contaminated water kill 4 million children yearly; 700,000 cases of cholera, transmitted through contaminated water, were recorded in 1992 and 1993 in Latin America and the Caribbean, resulting in 6,400 deaths; cholera killed 1,000 people in India in 1993; 80% of all diseases in developing countries are due to contaminated water.

F

The U.S. Environmental Protection Agency (EPA) has estimated that of 2 million underground storage tanks, used for oil, gas, and other fuels or chemicals, 20% may be leaking. In California 1/5 of the major drinking wells are polluted; nationwide, 10% of wells tested do not meet Safe Drinking Water Act standards. Public drinking water is tested for only a small fraction of all possible contaminants; 1/5 of Americans drink untreated or untested water. In Russia, 75% of freshwater lakes and rivers are dangerous to drink from, and 50% of tap water has microbes, worms, chemicals, or other contaminants; 30% of groundwater is polluted by fertilizers, pesticides, or industrial wastes.

In 1990 the EPA estimated that industry legally dumped 7 million gallons of oil, 90,000 pounds of lead, 900 pounds of mercury, and 2,000 pounds of PCBs into the Great Lakes, which hold 1/5 of the world's surface freshwater supply. About 250 million gallons of toxic industrial waste is dumped into U.S. rivers yearly; Almost 20% of all U.S. industrial, municipal, and federal facilities repeatedly violate the Clean Water Act; agricultural runoff in the United States pollutes 100,00 river miles. About 1,300 bodies of water in the United States are so polluted that authorities have had to limit public consumption of fish from them.

Rivers worldwide are becoming more polluted. The 7 major rivers flowing through China's cities are choked with sewage and industrial waste; Russia does not have a single large river that is not polluted by industrial waste and agricultural runoff. The Rhine is the worst in Europe; the Dnieper is the worst in Ukraine; the Minnesota River is one of the worst in the U.S. Midwest. In 1990, 28,000 miles of rivers contained hazardous levels of toxins, and a 1990 EPA inventory showed 1/3 of assessed rivers, 1/2 of estuaries, and 1/2 of lakes did not meet fishing and swimming standards. Some of the worst pollution occurs when toxics settle into river and lake sediments and are routinely "resuspended" by storms, currents, or other natural causes. In addition, salinization of rivers is occurring in agricultural areas around the world, including the San Joaquin Valley in California, the Murray River in Australia, and rivers in Asia and the Middle East.

One-third of U.S. shellfish harvest areas are closed because of pollution. Sixty-five percent to 70% of mussel and crayfish species are rare or extinct (mussels are an indicator species of water quality). One-third of North American freshwater fish species (more than 350 species) are imperiled.

World fertilizer use has increased 5-fold in the past 30 years. Farming areas have the highest concentration of nitrates, which are used in fertilizers, in wells and groundwater. Europe and North America have the most serious nitrate problems, but such problems are beginning to increase in developing countries. Human-created wetlands are being used to filter the sediments, phosphorus, and nitrates produced by agriculture from water. More than 200 human-made wetlands are now in operation in the United States, filtering polluted water before it reaches freshwater sources, such as rivers and lakes.

Until the 1960s the Aral Sea in the former Soviet Union was the fourth largest inland sea in the world. It is drying up because the two main rivers that feed it have been diverted to irrigate arid land. Fish populations have plummeted. Also, yearly windstorms sweep up more than 40 million tons of salt, dried pesticide, and other chemical residues from the dried areas of the Aral Sea, depositing them up to hundreds of miles away in salt rain. The rain ruins farming soils and causes widespread serious illnesses. Lake Baikal in Siberia holds 1/5 of all surface fresh water and is under a number of threats from industrial and agricultural river inflow, coal utility pollution, and direct discharges from pulp mills.

Water shortages are serious problems in East and West Africa, the north China plain, the southwestern United States, and the Middle East; 25 nations are experiencing water shortages affecting 1/2 the world's population. Water shortages are exacerbated in some regions, since 40% of the world's people depend on river systems shared by 2 or more countries. It is projected that by 2010 Israel and some other Middle Eastern countries will have only enough water for drinking and industry; currently 2/3 of available water in the Middle East and Africa goes to agriculture. Accelerated pollution makes fresh water more scarce.

More than 7,500 desalinization plants in at least 120 countries are operating to obtain drinking water from ocean or brackish waters; they are also being used to obtain pure water from groundwater and rivers that are too polluted for municipal and industrial needs. About 60% of global desalinization takes place in the Persian Gulf; Saudi Arabia has more than 30% of the world's desalinization capacity; Kuwait and some other Gulf states are almost totally dependent on desalinization plants.

Human Impact

Infectious diseases: Water fouled by raw sewage can spread diseases such as diarrhea, hepatitis, typhoid, cholera, dysentery, schistosomiasis, elephantiasis, parasites, flukes, yellow fever, malaria, and gastrointestinal illnesses.

Drinking water shortages: As a result of poor management practices, water shortages are occurring in a number of countries and can lead to dehydration, weakness, increased susceptibility to other diseases, and death.

Food: Waterborne toxics that bioaccumulate in animals and fish can be ingested with food. Mercury in fish is reducing the number of fish that can be eaten. Sediments, toxics, drought, and water usage can all result in plummeting fish and shellfish populations, and thus lessen food supplies. Forests are also harmed by water shortages, and the resulting wildlife loss can further affect local food supplies. Water shortages can create food losses if crops cannot be irrigated. Drought and changing weather patterns, together with poor water management, have been responsible for increasing famines in a number of countries.

Toxics: Many cases of cancer can be directly linked to the water supply. Drinking wells all over the United States have been tested and have often been found to contain toxic chemicals such as pesticides, herbicides, organic chemicals, and nitrates. Many of these chemicals can cause kidney and liver problems, immune problems, neurological and reproductive problems, and cancer leading to death. Other contaminants found in public water systems include lead, PCBs, bacteria, and viruses.

Chlorine: Chlorine can combine with organic chemicals, such as industrial organics or decaying vegetation in water supplies, to form trihalomethanes (THMs), which cause cancer. Chlorinated drinking water has been shown to increase the risk of rectal and bladder cancer in communities where it has been studied.

Nitrates: Bacteria that live in humans can change nitrates to nitrites, which can reduce the oxygen-carrying capacity of blood; in babies this is referred to as blue-baby syndrome. Babies and young children are especially susceptible, since their bodies convert nitrates to nitrites faster than those of adults, and because their hemoglobin combines faster with nitrites. Nitrites pose an additional risk, for in the body they can form nitrosamines, which are suspected of causing stomach cancer.

Land subsidence: Massive groundwater use has resulted in land subsidence, or sinking, in major urban areas such as Beijing, Houston, Osaka, Tokyo, and Venice.

Individual Solutions

- Do not dump toxics or hazardous wastes down the drain, into sewers, or onto soils; consult the EPA or local officials for hazardous waste pickup sites.
- Buy and use nontoxic household products; consult local environmental groups or the EPA for a list.
- Use biodegradable soaps and nonphosphate detergents.
- Practice energy conservation, since this lessens fossil fuel use and the resultant air pollution and acid rain.
- Conserve water by using low-flow faucets and shower heads. Turn off taps when not using water. Fix plumbing leaks. Take quick showers. Minimize sink, faucet, and shower use in all activities. Use a small toilet tank of 1 to 1 1/2 gallons. Use front-loading instead of top-loading washing machines (which use more water). Use the dishwasher and washing machine only when full. Water lawns sparingly in the morning and evening. Sweep your driveway and sidewalk instead of using water.

- Eat less meat, especially red meat, which demands a great deal of water to produce.
- Drive less, use mass transit, bicycle, walk; these practices lessen air and acid rain pollutants.
- Buy organic food, which generally means less toxics will be applied to the crops, soils, and thus water.
- Find out the source of your water; have your water checked for bacteria, lead, radon, and nitrates (call your state health department or a private water testing company). If you're concerned, buy a home water filtration unit.
- Do not use pesticides, herbicides, or artificial fertilizer; consult local landscaping companies for natural alternatives. Compost your clippings.
- Make sure your septic system does not leak.

Industrial/Political Solutions

What's Being Done

- Water conservation and efficiency programs, especially in agriculture, are being pursued worldwide. Some cities and electric and water utility companies are giving away low-flow faucet heads to save water resources and cut energy use for heating water; long-term planning for water needs is being implemented by city and state officials.
- Some dam projects are being rejected by local communities and governments because of water usage needs below the planned dam reservoir.
- The International Mussel Watch Project assesses mussel populations to track chemical contaminants.
- The Earth Island Institute and other agencies are developing programs to protect Lake Baikal and classify it as a World Heritage Site; a Canadian-U.S. agency is proposing cleaning up the Great Lakes by phasing out all industrial uses of chlorine.
- Mexico, the United States (Seattle), and a number of other countries are using sewage for farm fertilizer (it needs to be treated; otherwise it could be toxic). Human-created marshes have been constructed to

purify sewage, which can then be used as soil and fertilizer, while supporting a wetland wildlife habitat. In addition, large sewage ponds and lakes have been created; after sewage solids settle out, in about 1 month, fish are introduced and raised. Ponds have been built in China, Germany, India, Israel, Vietnam, and elsewhere.

- The National Energy Policy Act of 1992 set standards for water efficiency for toilets, urinals, shower heads, and faucets manufactured after January 1994, which will reduce this type of water use by more than 50% and save related electricity to pump, heat, clean, and treat the water.
- The U.S. Clean Water Act of 1972 was designed to protect water resources from pollution, especially industrial discharges and municipal sewage. It was later broadened to target a small number of toxic chemicals, nonpoint source pollution such as agricultural runoff, acid rain problems, and lake cleanup efforts. It comes up for renewal in 1993–1994.
- The U.S. Safe Drinking Water Act was passed in 1974. It established maximum contaminant levels (MCLs) in drinking water for a small number of chemicals, including some pesticides; MCLs for known carcinogens are set at 0. Groundwater protection came in later amendments. In 1986 lead was banned from any new public water construction. The act comes up for renewal in 1993–1994. Methods for purifying drinking water include distillation, gas exchange, coagulation, flocculation, sedimentation, filtration, adsorption, ion exchange, and disinfection.
- The U.S. Clean Air Act of 1990 limits acid rain pollutants and some auto emissions.
- Desalinization projects and plants are converting ocean water to usable drinking water.
- In November 1992 the EPA enacted rule 503, the Sewage Sludge Use and Disposal Rule, which set standards for pathogens and 10 heavy metals in sewage applied to land, incinerated, or destined for surface disposal; paper mill sludge (often containing dioxins) must be treated separately.

- The EPA is supporting integrated pest management instead of heavy pesticide use.
- Farmers are controlling nonpoint source runoff by regulating pesticide and livestock chemical fertilizer usage, and livestock manure rainwater runoff.
- Use of gray water is legal in some states; this practice entails reclaiming used domestic water (except from the toilet) for uses like irrigation, cooling, or toilet use. Some water treatment plants are reclaiming used water and recycling it for nonpotable needs. Florida reclaims 26% of its water; Arizona, 20% to 30%; and California, 8%.
- More communities are taking chlorine out of their drinking water; disinfectant substitutions include ozone, ultraviolet light, and other safer alternatives.
- Once-through air conditioning, in which billions of gallons of pristine aquifer water are used to cool air conditioners, is being limited in some areas in the United States.
- Subsurface drip irrigation is being studied and used for urban landscapes and agriculture, such as vineyards; it reduces runoff and water loss. Israel pioneered drip irrigation and saline water agriculture.

What Needs to Be Done
- Strengthen the U.S. Clean Water Act to include more chemicals, more prevention measures, and strict enforcement of its provisions. Focus on pollution prevention instead of cleanup strategies, especially nonpoint source pollution and zero discharge from industrial sources.
- Strengthen the U.S. Safe Drinking Water Act to also include more chemicals, more prevention measures, and strict enforcement of its provisions.
- Legalize gray water use in all states.
- Offer more money to farmers for equipment and training to use manure as a fertilizer and lessen runoff problems. Enforce feedlot manure regulations so runoff is prevented. Support sustainable organic agriculture.

- Mandate water conservation and efficiency in commercial, industrial, and government facilities; mandate installation of retrofitted low-flow faucet heads in all such facilities as well as in residential buildings.
- Mandate long-term health testing of all chemicals applied to the land, soil, or water; limit lawn care chemicals and fertilizers.
- End the injection of toxics underground.
- Ban overflow dumping of raw sewage; help developing countries develop sewage treatment systems. Improve sewage treatment systems to better purify wastes.
- End all nonessential use of aquifer water, such as for once-through air-conditioning cooling.
- Initiate stricter regulation of underground storage tank maintenance.
- Ban single-hull barges on rivers to prevent oil and hazardous waste spills.
- Reclaim more wetlands; adopt a no-net-loss wetlands management program.
- Regulate construction and buildings around lakefronts to maximize water replenishment.
- Explore desalinization technology, which converts tropical water to electricity and uses the water's heat to provide the required energy for desalinization.
- Ensure that the North American Free Trade Agreement (NAFTA) does not supersede tribal, local, state, or federal laws.
- Limit human population growth, which is a factor in water pollution and shortages worldwide.
- Limit population growth and development in desert or arid areas, where it can only be continued by unsustainable water practices.
- Limit and slow boat and barge traffic on rivers, both of which can increase shoreline erosion and water turbidity.

Also See

- Ocean degradation
- Wetlands degradation

Geothermal energy

Definition

Heat contained beneath the earth's surface, usually held by hot molten rock called magma, which has been formed by the partial melting of the earth's mantle.

The magma heats up nearby water or creates steam, which can be brought to the surface through drilling. Sometimes the molten rock heats up nearby rock that does not melt (called a dry-rock system). In such a case cold pressurized water can be sent down to the hot rocks, where it becomes heated or vaporizes into steam and then is brought back up. Average temperatures in geothermal operations are 250° to 400°F.

The water or steam that is brought to the surface can be used as heat energy, but often it is used to power turbines and produce electricity. Heated water or brine used to produce electricity or heat is called hydrothermal energy. Geothermal power cannot be easily transported and thus is best used locally; up to 90% of the heat energy can be lost if not used in the immediate area. Geothermal operations often recover methane, or natural gas, which can then be used as fuel for heating or other purposes.

In closed-loop geothermal designs, the liquid or steam is not exposed to the air when it is brought up, and it can be sent back down when its heat is removed. Open-loop systems release the water or steam above ground when the heat is removed.

Environmental Impact

Open-loop geothermal systems: Steam and brackish water below ground can pick up toxic heavy metals, sulfur, minerals, salts, radon, and noxious fumes and gases. If the water or steam is released above ground, it can cause air pollution, water pollution, and hazardous waste, which must be disposed of. Scrubbers can filter out some of the pollutants, but this results in a toxic sludge. Geothermal plants may release carbon dioxide, but only 5% of the amount released by a coal or oil power plant. Hazardous waste and sludge contaminate the soil wherever they are buried, and they may contaminate groundwater; these toxics are still far less in quantity than those created during fossil fuel production.

Closed-loop geothermal systems: In closed-loop systems the steam or water is returned to the drilling reservoir, causing less pollution and fewer hazardous waste problems. Minerals the water has picked up, however, may have to be removed. This design is characteristic of dry-rock systems, which are still being researched.

Sensitive areas: Sometimes geothermal energy is accessible only in an ecologically sensitive area. Building plants in such areas could pollute and damage the local ecology.

Water use: Most geothermal plants need large amounts of water; local ecologies may be threatened if streams or rivers are diverted, dammed, or siphoned for the water.

Major Sources

Worldwide: A layer of molten rock called magma lies some 15 to 30 miles beneath the surface crust of the whole world. This molten rock usually comes closest to the surface in areas where there are volcanoes and earthquake fault lines. For every 100 feet down into the earth, the temperature increases by about 1°C, or 30° per kilometer; this rate can double in hotter areas. Because this heat resource is so large, it is considered renewable. A power plant may use up the local available heat in a given area, however, and thus, in this sense, geothermal energy can be depleted. Technology might be able to bring deep geothermal energy to the surface easily and cheaply in 1 to 2 decades.

Easiest access: Geothermal energy can be reached most easily in surface hot springs, geysers, and molten rock that pours forth from volcanoes.

The best areas: The best places to capture geothermal energy are along the Pacific Rim, including the western coastlines of North, Central, and South America, as well as in Hawaii, New Zealand, Papua New Guinea, Japan, and eastern Siberia. Of course, as technology improves, geothermal energy may become accessible anywhere in the world.

State of the Earth

More than 30 countries, including Australia, Bulgaria, Canada, Denmark, El Salvador, Greece, Hungary, Italy, Mexico, Norway, and the former Yugoslavia, now use geothermal energy; 40 countries will use it by 2000. Tibet has the highest availability of geothermal energy in China, with more than 600 discovered fields. In Iceland more than 60% of all homes are heated by geothermal energy; Japan also has extensive geothermal development. El Salvador and the Philippines met 20% of their electricity needs with geothermal energy in 1990. Another two dozen countries have good, but as yet unexplored, potential for developing geothermal energy.

A geothermal plant has been operating on the active volcanic island of Hawaii since mid-1993. It is supplying the island with 25 megawatts of electricity, almost 15% of total usage. Thus far it has not appeared to damage the lowland forest within which it is sited—the brine is reinjected into the ground, and the only problem (since rectified) was initial escaped emissions of hydrogen sulfide. There are plans to study other islands, including Maui, for future geothermal energy.

Geothermal energy is available along Africa's Great Rift Valley, especially in Ethiopia and Kenya. One plant supplies 10% of Kenya's power needs. Both Ethiopia and Kenya plan further development.

There are hundreds of locations in Italy, New Zealand, the United

States, and other countries where space and water heating is accomplished by piping fresh water through sealed pipes underground, into warmer rock or soils, and back to the surface. This heat is used for greenhouses, fish ponds, soil heating, and industry; geothermal energy is also used to power paper mills in New Zealand.

Geothermal fluids and brines are being studied for commercial use. The solutions can hold boric acid, potassium salts, silica, lithium, cesium, rubidium, silver, and gold, and these elements are worth more than the energy produced from the heat. In New Mexico dry-rock geothermal energy is being researched. Shafts are drilled 12,000 feet deep to rock at 240°C; cold water sent down under pressure has returned at 200°C at a rate of 100 gallons per minute—hot enough to create electricity. Researchers believe that once dry-rock geothermal energy is perfected, it could supply all U.S. electricity needs for 1,000 years.

The geothermal energy held under the continental United States to a depth of 6 miles is thought to be equal to the energy of burning trillions of tons of coal. The Geysers, in northern California, the largest geothermal development in the world, has been operating since 1960; it supplies the bulk of electricity to the San Francisco Bay Area and 6.5% of California's electricity. Recent studies, however, indicate the available geothermal energy in northern California may be only 1/2 of previous estimates; since 1988 The Geysers facility has reduced its energy generation.

Human Impact

Health: Some geothermal plants emit toxic air pollution and create hazardous waste. If toxic sludge is buried, it can contaminate soil and underground water supplies. Heavy metals can be released into the air or water supplies; they affect the body by damaging the brain and the nervous, glandular, and organ systems. Most heavy metals are highly toxic in even microgram amounts and can cause cancer. The amount of toxics released in geothermal operations, however, is still far less than that produced by fossil fuel development. Closed-loop designs will further eliminate many of these hazards.

G

Methane: Methane is a by-product fuel that can be used for heating and energy needs; drilling releases may also contribute to global warming.

Hydrogen sulfide (H_2S): Hydrogen sulfide, sometimes released from geothermal plants, releases a "rotten egg" smell; it also can produce nausea, headaches, shortness of breath, sleep problems, and eye and throat irritation.

Individual Solutions

- Look into energy independence for your home with solar or wind energy.
- Practice energy conservation.
- Write and call your legislators to press for stronger investments in renewables and an end to fossil fuel subsidies.

Industrial/Political Solutions

What's Being Done

- Researchers are attempting to locate strategic areas for geothermal plants. They are also studying how to drill deeper into the earth and reach deeper veins of heat, including the magma chamber.
- The United States offers a 10% tax credit for geothermal investment.
- The Public Utilites Regulatory Policies Act (PURPA) of 1978 requires public utilities to purchase electricity from small producers at rates equivalent to avoided cost of replacement or new generation.

What Needs to Be Done

- Mandate that geothermal plants use scrubbers to detoxify any air emissions.
- Mandate that hazardous wastes be destroyed or detoxified, not buried.
- Require environmental impact statements for any geothermal plant sited in a sensitive area.

Also See

- Renewable energy

Global warming

Definition

The rise in average temperature of the earth's surface atmosphere over a long period of time.

In the past the earth's atmosphere has undergone many periods of warming and cooling. Usually these changes have taken place over thousands of years, allowing plant and animal populations to adjust to the changes. If the current predictions of increased global warming are accurate, the earth's average temperature would rise higher than it has in the past 125,000 years. The change would occur relatively quickly, which means many ecosystems, plants, and animals would not have time to adjust. Extinctions would be the most likely result.

At this point global warming estimates are just predictions. The earth's temperature has risen 1°F in the past hundred years, but no one knows for sure how fast it will continue to rise. This uncertainty is due to the complexity of the earth's atmosphere and ecosytems and the fact that we still don't understand many of the interrelationships between atmospheric gases, living systems, and environmental sinks and how our own activities are affecting all three.

Temperatures on earth have been relatively stable for a long time. Experts believe this stablility is due to the greenhouse effect. The greenhouse effect is a natural process in which atmospheric gases trap solar heat reflected from the earth and warm up the planet. Without this process the planet would be much colder. The greenhouse effect allows life to flourish.

Humans are adding large amounts of greenhouse gases to the

atmosphere. Most scientists agree that adding greenhouse gases to the atmosphere might cause global warming. And most experts agree that global warming is occurring now, even if they are not sure how long, or how fast, it will continue. If we wait for absolute proof that global warming is a serious problem that humans are causing, and it does occur, it may be too late to reverse the trend, for atmospheric gas composition does not change quickly.

Environmental Impact

If global warming continues, expected changes are:

Melting of polar ice: A melting of polar ice would cause sea levels to rise. Some ocean islands would be covered, as would coastal areas. Coastal estuaries and wetlands would be lost, disrupting breeding for birds, ocean animals, and fish, as well as nutrient losses. Coastal erosion would be intensified. Other ecosystems affected would be coral reefs, mangroves, Arctic tundra, and polar landmasses and seas.

Rainfall changes: The increased heat would evaporate more moisture into the air. Thus some areas would experience more rainfall and flooding, while others would experience drought. Some lakes and rivers would dry up. Some animals might have difficulty reproducing, and plant growth would be affected.

Erratic weather: Weather patterns would be more erratic and unpredictable owing to the climatic changes.

Pest increases: Pests might migrate into temperate latitudes, competing with or destroying plants and animals that had no exposure to them before such a change.

Extinctions: It is predicted that the net effect would be extinctions of many species of plants and animals that would be unable to either adjust or migrate to handle these changes. Loss of biodiversity would weaken the stability of natural ecosystems.

Some theories of past global temperature changes are that they were caused by:

- Solar cycles.
- Meteor impacts or volcanic eruptions.
- Continental drift, which created the mountains and resultant ocean currents.
- Carbon dioxide fluctuations due to coral reef growth and decay.
- Changes in atmospheric gas composition.
- Disruptions of the greenhouse effect.

State of the Earth

Many scientists believe humans are increasing the earth's temperature on the basis of the following evidence:

Greenhouse gases: Humans are adding large amounts of heat-trapping greenhouse gases to the atmosphere (carbon dioxide, chlorofluorocarbons, methane, nitrogen oxides).

Deforestation: Trees hold carbon, so destroying them releases more carbon and thus more greenhouse gases. Also, trees take in carbon dioxide for photosynthesis—they act as a sink. When more trees are destroyed, less carbon dioxide is taken out of the atmosphere.

Temperature increase: The earth's temperature has risen 1°F over the past 100 years.

Record yearly temperatures: The past two decades have held the seven hottest years on record.

Some variables whose influence on the greenhouse effect is uncertain are:

Clouds: Clouds may reflect sunlight during the day (keeping days cool) and hold in the heat of the earth at night (keeping nights warm). Some believe observed increased cloudcover is due to increased evaporation from increased heat; others feel the increased cloudcover is caused by pollution particulates in the air. The effects of clouds vary with their height, size, and moisture content.

Ozone depletion: Ozone thinning destroys phytoplankton, which is a carbon sink; less phytoplankton would mean less carbon dioxide removal from the air and thus more greenhouse gases and more heat. Ozone also traps heat—less ozone means less heat is trapped, and may result in a cooling effect. The net effect of ozone on global warming is unclear.

Particulates: Burning fossil fuels releases airborne particulates, which reflect the sun's heat and may have a net local cooling effect. The June 1991 eruption of Mount Pinatubo emitted massive amounts of particulates, which are believed to be responsible for a temporary cooling of the earth. They also increase ozone-destroying reactions between chlorine and ozone.

Feedback loops: Feedback loops are effects of global warming that can in turn accelerate global warming. For example, as a result of increased heat:

- Plant decomposition might speed up and release carbon faster.
- Trees might grow more in polar regions (forests could shift north as much as 125 miles) and thus absorb more heat and increase temperatures further.
- Oceans might warm, thus absorbing less carbon dioxide (the chief greenhouse gas) and maybe even releasing carbon dioxide.
- Warmer temperatures may release methane from frozen marine sediments and tundras and thus add to the greenhouse gases.
- The melting of polar snow and ice would reveal darker surfaces on the earth (like dirt), which absorb more heat and thus hasten warming.
- More water might be evaporated from the oceans (water vapor is a greenhouse gas) and thus help retain more heat in the atmosphere.

Or:

- Polar ice might not melt as fast as expected.
- The oceans might absorb extra greenhouse heat, or carbon dioxide, and slow warming.

- The oceans might not heat up and thus might not release more water vapor.
- Poorly understood biological processes might absorb more carbon dioxide and slow warming.

Human Impact

If global warming continues it will affect:

Freshwater sources: Coastal and island areas will experience increased seawater flooding, leading to freshwater shortages. Drought in other inland areas may prevent groundwater drinking supplies from being replenished.

Food: Farming areas will change owing to weather changes. Erratic and severe weather may result in crop wipeouts, topsoil losses, and food shortages. Migrating pests may also affect crop production.

Home loss: Millions of people along coastlines and on low islands will be displaced because of flooding.

Diseases: Some pests will migrate to new areas, bringing new diseases with them; in particular, tropical diseases might migrate to temperate zones.

Hydroelectric power production: Rainfall patterns will change and may affect river direction, requiring new dam and hydroelectric power stations to be constructed.

Individual Solutions

- Practice energy conservation.
- Drive less; support mass transit; buy fuel-efficient cars.
- Recycle and use recycled products.
- Plant trees; they remove carbon from the air.
- Investigate solar and wind energy for your home needs.
- Demand higher fuel efficiency for all cars and production of zero-emission vehicles.
- Ask your legislators to call for the United States to lead the world in more stringent, mandatory carbon dioxide emission reductions and an

immediate phaseout of chlorofluorocarbons (CFCs), including planned replacements of hydrochlorofluorocarbons (HCFCs) and hydrofluorocarbons (HFCs); both are greenhouse gases.

Industrial/Political Solutions

What's Being Done

- Several thousand scientists worldwide have called for limiting greenhouse gases.
- CFCs will be phased out worldwide by 2000; the United States plans to phase them out by 1995. Planned alternatives are HCFCs and HFCs.
- Europe, Japan, and the United States are reducing carbon dioxide emissions by the year 2000. The U.S. program, the Climate Change Action Plan, is voluntary; it is aimed at reducing greenhouse gas levels to 1990 levels, which would eliminate 110 million tons of carbon dioxide from current levels. It will be accomplished by energy efficiency and conservation.
- Thirty-five island nations at the Earth Summit (see Glossary) demanded reduction of carbon dioxide emissions; they fear global warming may result in rising ocean water.
- Japan has launched a major research project to learn how to strip carbon dioxide out of industrial emissions and recycle the gas. It also is training 10,000 people in developing countries over the next 10 years in energy-efficiency technology.
- The 1990 U.S. Clean Air Act is mandating reductions of nitrogen oxide emissions.
- CFCs are being recycled and recaptured from appliances and cars.

What Needs to Be Done

- Sign a world agreement to limit greenhouse gases in all countries; industrial countries need to help developing countries accomplish this.
- Work to end worldwide deforestation; increase reforestation.

- End CFC production and use worldwide immediately, including HCFCs and HFCs.
- End subsidies of fossil fuels; tax fossil fuels to reflect costs of environmental damage and health effects.
- Phase in requirements for zero automobile tailpipe emissions and zero carbon dioxide emissions at utility plants.
- Establish stronger incentives for energy conservation and off-the-grid home, commercial, and industrial solar and wind energy use.
- Help developing countries use energy-efficiency standards in all sectors of energy use.

Also See

- Energy conservation
- Greenhouse effect

Greenhouse effect

Definition

A natural process that occurs when heat from the sun is reflected off the earth's surface and then held in the atmosphere, instead of escaping into space.

Gases that float up from the surface of the earth are responsible for partially blocking and trapping the heat. These gases are called greenhouse gases.

The sun sends visible light to the earth, some of which is reflected by clouds and particulates back into space and some of which can penetrate the atmosphere. Once it strikes the earth, light is reflected as heat (long-wave, or infrared, radiation); some of this heat is trapped or held by the greenhouse gases.

Scientists believe that long ago the greenhouse effect helped earth's air become warm enough to support life and that it continues to do so today. Without the greenhouse effect, the average surface temperature would be about 0°F (60° cooler)—too cool to support most life. It is believed that the greenhouse effect has helped to stabilize the temperatures of the earth over a long period of time.

More greenhouse gases in the atmosphere will trap more heat in the air around the planet. There is concern that humans are influencing the temperature of the planet by releasing large amounts of greenhouse gases, the most significant being carbon dioxide, methane, nitrogen oxides, and chlorofluorocarbons (CFCs). CFCs are created by humans, but the other gases occur naturally. Human activities, however, are releasing more of the naturally occurring gases than natural processes and cycles normally would. The ozone layer in the upper atmosphere, water vapor, and smog also trap heat around the planet and contribute to the greenhouse effect.

Environmental Impact

Global warming: It is clear that the greenhouse effect traps heat around the earth. It is still unclear, with the effects of human activities, how much heat will be produced, how fast, and how major the resulting problems might be.

Greenhouse gases are also responsible for a number of other serious environmental problems independent of the greenhouse effect:

Nitrogen oxides: Produced from the burning of fossil fuels, nitrogen oxides contribute to acid rain, air pollution, smog, and water pollution. These problems inhibit plant growth; destroy lakes, fish, and animal populations; and leach heavy metals into waterways. Smog also holds in heat, like greenhouse gases. Nitrogen oxides also play a role in ozone depletion.

CFCs: CFCs accelerate ozone depletion. The thinning of the ozone layer can lead to plankton destruction in the Antarctic, which could affect the whole ocean food chain. The increased ultraviolet radiation that reaches earth through the depleted layer of ozone inhibits plant growth.

Methane: Methane also plays a complex role (still under research) in the stability of the ozone layer.

Major Sources

Greenhouse gas contributors and their relative impact are thought to be:

Carbon dioxide (CO_2); 40% to 55%: Two-thirds of global emissions of carbon dioxide are produced from burned or used gas, coal, and oil; biomass burning (burning of savannas, agricultural wastes, fuelwood, and forest fires) and other deforestation contribute the remaining 1/3. Emissions of carbon dioxide are expected to double in 50 years at current growth rates; there are some efforts to reduce emissions. Natural processes in the carbon cycle are forest fires, volcanoes, organic decomposition, and breathing of animals and humans. CO_2 is recycled naturally (see carbon cycle in Glossary) through the atmosphere every 4 years; excessive levels of CO_2 may last 200 to 500 years in the atmosphere.

Methane (CH_4); 20% to 25%: Half or more of methane comes from human activities. It is produced by the belching of cattle, bacteria in rice fields, burning of biomass, landfills and sewage, and natural gas leaks during production, transportation, and use. Methane production could rise dramatically with increased natural gas production. Natural sources are wild animals, wetlands, and termites (though dramatic increases in termite populations are due to deforestation). Methane lasts 8 to 12 years in the atmosphere.

CFCs; 15% to 20%: All CFCs are human-created. They are used in air conditioners, solvents, insulation, fire extinguishers, aerosols, and foam. Most CFCs are being phased out, but some replacements, hydrochlorofluorocarbons (HCFCs) and hydrofluorocarbons (HFCs), are also greenhouse gases. Some evidence shows that ozone depletion caused by CFCs might be cooling the upper atmosphere, possibly offsetting the greenhouse effects of CFCs. CFCs remain in the atmosphere from 6 to 400 years, with the majority lasting more than 100 years.

Nitrogen oxides (NO_x); 5% to 10%: Half or more of nitrogen oxides result from human activities. They are produced by the burning of fossil fuels in automobiles and by utilities (50%), factories, deforestation, global biomass burning (more than 20%—includes deforestation due to burning), aircraft engines, and the breakdown of chemical fertilizers. Some efforts at reductions have been made, and they are expected to increase. Natural sources of nitrogen oxides are biological processes in soils. Nitrogen oxides last 120 to 150 years in the atmosphere.

Smog and water vapor; 6%: Smog is human-created and remains in the atmosphere less than 1 year.

Nongreenhouse gases that affect the earth's atmosphere and temperature are:

Ozone (O_3—stratospheric): Ozone is produced in the upper atmosphere with the combination of sunlight and oxygen. Since ozone holds heat around the earth, a thinned layer of ozone will hold less heat. Ozone is in a constant state of formation and destruction.

Carbon monoxide (CO): Half of carbon monoxide emissions are human-created, resulting from land clearing (50%) and from incomplete combustion (car exhaust, incinerators, fuelwood use, and elsewhere) and oxidation of human-created hydrocarbons (50%). Natural sources are plants, wildfires, oceans, and oxidation of natural hydrocarbons. Carbon monoxide has a lifetime of 2 to 4 months in the atmosphere. It is not a greenhouse gas, but in the atmosphere it is converted to CO_2 and thus adds to CO_2 levels. More important, CO reacts with and reduces the quantity of hydroxyl radicals, thus increasing methane levels.

Hydroxyl radicals (OH): Hydroxyl radicals occur naturally in the atmosphere. They combine with methane (CH_4) and carbon monoxide (CO); as CO and CH_4 increase, OH decreases, leading to further increases in CO and CH_4 and thus to higher greenhouse gas levels.

State of the Earth

Seven of the 10 hottest years of recorded average surface temperatures were 1980, 1981, 1983, 1987, 1988, 1989, and 1990; 1990 was the warmest on record. The earth's surface temperature has risen 1°F over the past 100 years; in the past several thousand years the earth's temperature has never varied more than 1°F. Ice cores from Greenland, however, show that rapid climate changes in the past occurred as fast as every few years. All greenhouse gases are still increasing their concentrations in the atmosphere: 60% from energy production, 20% from CFCs, 10% from deforestation, and 10% from agriculture.

In 1991 the National Academy of Sciences concluded there was clear evidence of global warming and recommended immediate reductions of greenhouse gases. They and others have predicted temperature increases of 3° to 9°F in the next century at current rates of greenhouse gas emissions, with sea levels rising from 3 to 25 feet. It is estimated that sea levels have risen 4 to 8 inches in the past 100 years.

About 700 billion tons of carbon dioxide naturally cycles through the atmosphere each year in the carbon cycle. Plants remove 100 billion tons of CO_2 every year from the atmosphere, which is 15% of total world CO_2. Carbon dioxide makes up only about 0.03% of the total atmosphere. Soils hold up to 1.5 trillion metric tons of CO_2, some of which is lost through deforestation. In warmer climates soils emit more CO_2 than they absorb; over the past 10 years tundra soils have also been found to be emitting CO_2 faster than they are absorbing it.

Humans generate 24 billion tons of carbon dioxide each year; only 1/2 of this amount is absorbed in natural processes (the oceans take up 50% of the absorbed CO_2). Thus carbon dioxide is increasing in the atmosphere and is about 25% higher than 100 years ago. In the past decade Brazil was the fourth largest carbon emitter in the world, releasing several hundred million tons of carbon each year from deforestation.

Industrial nations create 65% to 80% of the greenhouse gases; developing nations will produce 45% of carbon emissions by 2050, because of desires for quick economic growth. China already emits 10% of the world carbon as a result of coal burning. The United States holds 5% of the world population, uses about 25% of the world's oil, and releases about 25% of the world's nitrogen oxide, carbon dioxide, CFCs, and halons.

Methane is 30 times more efficient in trapping heat than carbon dioxide; nitrogen oxides are up to 200 times more efficient; CFCs are up to 15,000 times more efficient. Carbon dioxide is still the biggest factor in the greenhouse effect, because so much more of it has been released into the atmosphere. It is estimated, however, that a 1% leakage of methane from large increases of natural gas production and distribution could offset carbon dioxide reductions; natural gas leakage in the former Soviet Union is estimated to be as high as 10%.

One gallon of burned gasoline gives off 20 pounds of carbon dioxide. A 5-mile-per-gallon improvement in auto mileage would cut CO_2 emissions by several hundred billion pounds a year.

Deforestation is a major contributor to greenhouse gases, especially carbon dioxide. Trees store carbon, and when burned or destroyed they release it. Trees also remove carbon dioxide from the air for photosynthesis and thus act as a sink for carbon dioxide. Tree planting programs have been started worldwide, but they are not keeping pace with the rate of deforestation.

A recent study found that greenhouse gases could be cut by 40% at little or no cost with improved energy efficiency standards for cars, appliances, and buildings; a moderate reforestation program; and more mass transit. Worldwide, population increases are responsible for more deforestation, fossil fuel use, cars, and rice consumption, and accompanying methane production; all lead to more greenhouse gases.

Human Impact

It is unclear how much and how fast the greenhouse effect will increase global warming. But the production of greenhouse gases is causing:

Acid rain, air pollution, smog, and water pollution: Acid rain, air pollution, smog, and water pollution are created by fossil fuel mining, processing, and use. They aggravate lung, skin, blood, and respiratory problems and can cause kidney and liver problems, immune system weaknesses, and cancer. They also destroy fish and timber stocks.

Ozone thinning: Ozone thinning, due to CFCs, will increase skin cancer and cataracts. It threatens the whole ocean food chain.

Individual Solutions

- Practice energy conservation; investigate solar and wind energy for your home.
- Drive less; support mass transit; demand higher fuel efficiency for all cars.
- Make sure that your air conditioners are not leaking and that Freon (a CFC) is recycled.
- Recycle; use durable, nondisposable products.
- Plant trees; they remove carbon from the air.
- Support reduced nitrogen oxide tailpipe emissions on cars and at utility power plants.

G

- Support an immediate phaseout of CFCs and the planned substitutes HCFCs and HFCs; all are greenhouse gases, and HCFCs are ozone depleters. Safer alternatives are already available.
- Write your legislators to call for the United States to lead the world in more stringent, mandatory carbon dioxide emission reductions.

Industrial/Political Solutions

What's Being Done

- At the 1992 Earth Summit, 153 countries signed a document agreeing to reduce greenhouse gases to 1990 levels; unfortunately no deadlines or target goals were set, mainly at the request of the United States. Thirty-six countries, including the United States, have since ratified it; fourteen more need to endorse it to make it binding.
- Japan has launched a major 20-year research project to attempt to strip carbon dioxide out of industrial emissions and then either dispose of or recycle it. Japan is also training 10,000 people in developing countries in energy efficiency technology over the next 10 years.
- CFCs will be phased out worldwide by 2000; the United States plans to phase them out of production by 1995. Planned alternatives are HCFCs and HFCs.
- CFCs are being recycled and recaptured from appliances and cars.
- Canada, Europe, and Japan plan to reduce carbon dioxide emissions to 1990 levels by the year 2000. The United States has created the voluntary Climate Change Action Plan to reduce CO_2 emissions by 110 million tons to 1990 levels (1.5 billion tons) by 2000. A National Energy Strategy is projected to cut greenhouse emissions 7% to 11% by 2000.
- Thousands of scientists have called for globally limiting the production of greenhouse gases.
- Thirty-five island nations expressed demands at the Earth Summit for a reduction of carbon dioxide emissions; they fear global warming could lead to rising ocean water.
- The International Energy Agency (researchers from Canada, Europe,

Japan, and the United States) is researching ways to capture and dispose of utility CO_2.

- The 1990 U.S. Clean Air Act is mandating reductions of nitrous oxide emissions.
- The U.S. Department of Energy has a $4.6 billion Clean Coal Technology Program to reduce carbon dioxide emissions.
- New energy efficiency requirements will slow carbon dioxide emissions slightly.
- In Hawaii and Western Samoa, insurance companies have refused coverage to owners of beach property less than 5 meters (16 1/2 feet) above sea level; they may extend this policy to Florida and the Gulf Coast.
- Companies are ending use of CFCs as refrigerants and cleaning solvents.
- Halon (a form of CFC) is being phased out of fire extinguisher use.

What Needs to Be Done

- Develop a world agreement to limit all greenhouse gases in industrial countries; help developing countries do the same.
- Work to end worldwide deforestation; increase reforestation.
- End CFC production and use worldwide immediately, including the intended alternatives, HCFCs and HFCs.
- End subsidies for fossil fuels; tax fossil fuels to reflect the cost of environmental damage and health effects.
- Increase incentives for off-the-grid commercial, industrial, and residential use of renewable energy.
- Mandate that utilities show yearly increases for energy conservation and solar or wind energy investments.
- To reduce heat retention, design urban areas with more vegetation, heat-reflective colors, and building patterns that allow better circulation and wind flow.
- Strengthen the U.S. Clean Air Act to significantly reduce automobile tailpipe emissions and phase in zero-carbon-dioxide emissions at utility plants.

G

- Phase in zero-emission automobiles, like electric cars; as an intermediate step, mandate a fuel efficiency level of 50 miles per gallon.
- Help developing countries use energy efficiency standards in all sectors of energy use.
- Make family planning services available worldwide; one of the biggest reasons for increased fossil fuel use and deforestation is increased population.

Also See

- Energy conservation
- Global warming
- Methane (CH_4)

Hazardous waste

Solid, gaseous, or liquid waste, or a combination thereof, that can present a serious threat to human health or the environment if it is not stored, treated, transported, or managed properly.

Hazardous waste is usually highly flammable, corrosive, reactive (explosive or unstable), or toxic. It can possess one or all of these characteristics. It can be generated from industry, homes, agriculture, or the environment. There are as many possible types of hazardous wastes as there are possible combinations of hazardous and toxic chemicals: millions.

Organic compounds, inorganic compounds, and radioactive waste make up three major classes of hazardous waste. Organic compounds are made primarily of carbon, hydrogen, and water; one example is dioxin. Inorganic compounds do not contain carbon; mercury is an example of an inorganic heavy metal. Hazardous waste is usually created by humans, but it can also be a naturally occurring substance, such as mercury or uranium, that has been mined and released in large quantities.

Environmental Impact

Plants and animals: Hazardous wastes in soils, fresh water, or oceans can kill fish and aquatic organisms or be taken up by plants that are eaten by animals. Such wastes can thus contaminate the food chain. Predators are most susceptible to the harm of any contaminants, since toxics often biomagnify in predators. Hazardous wastes can cause liver and kidney damage, behavioral and neurological changes and damage, reproductive problems, cancer, and death in young and adult animals. Often hazardous wastes do not biodegrade and can take years or even decades to be cleansed from the environment; ponds, lakes, bays, soils, and other areas with low circulation can take much longer. Thus hazardous wastes can pose a threat to living organisms for a long time in a given area.

Soils: Often hazardous wastes end up in soils or sediments and do not biodegrade. These wastes can "migrate" in soils over large distances, carried by water. Hazardous waste landfills and impoundments leak; waste drums and containers rust, break, and leak; vitrified wastes in cement or other materials break down over time; and hazardous waste deep wells or pits might have unseen cracks and fissures. If the waste is highly radioactive, as in nuclear waste, there is no guarantee that any method of disposal will be safe for more than 10,000 years. No one can guarantee that a geologic disturbance, such as an earthquake, will not occur and break open a storage area of hazardous waste. Chemicals in soils can affect plant growth or be incorporated into plants and affect insects or animals eating the plants.

Groundwater: Buried or dumped hazardous wastes (or toxic ash from incinerated hazardous or municipal waste) can leak into soils and percolate down to groundwater aquifers. There it can impregnate the surrounding rock and slowly leach out into the aquifer's water, making it unusable and often impossible to clean up.

Transported wastes: Air emissions of hazardous waste can often be carried on the wind virtually anywhere until the waste is deposited in rain, snow, or other moisture. Incineration of hazardous and municipal waste releases toxic fumes.

Major Sources

Industrial sources: Industrial hazardous wastes include a wide range of chemicals, toxics, metals, sludges, and hazardous mixtures. Industry is the largest multiple source of all hazardous waste. Ninety-five percent of industrial hazardous wastes are disposed of on the site where they are produced. Industrial wastes are landfilled, buried in deep pits or shafts, placed in surface impoundments (covered and lined artificial ponds), encased in drums or similar containers, mixed and hardened with cement, incinerated, or vitrified in silica or a similar type of material. There is now more emphasis on detoxification and reduction.

Chemical industry: The chemical industry is the largest industrial source of hazardous wastes, which include metals, organic chemicals, inorganic chemicals, acids, and hazardous mixtures. Fifty percent are emitted into the air, and 50% are injected into deep wells.

The military, U.S. Department of Defense, and U.S. Department of Energy: Military hazardous wastes come from weapons projects, research, and nuclear naval ships; they include radioactive materials, chemicals, pesticides, mixtures, and used weapons. The military is probably the largest single producer of hazardous wastes. Wastes are stored, buried, or incinerated.

Nuclear waste: Nuclear waste is produced by nuclear power plants, medical labs, hospitals, and research. It is stored in pools, buried, dumped, or landfilled.

Medical waste: Hospitals, clinics, labs, and research facilities generate medical waste, including toxic chemicals, radioactive material, heavy metals, contaminated syringes, and other infectious materials. The waste is incinerated, autoclaved, or dumped.

Municipal solid waste: Municipal solid waste includes household hazardous waste and toxics such as batteries, plastics, household cleansers, resins, synthetic chemicals, and paint. They are landfilled, incinerated, or recycled.

Fossil fuels: Hazardous wastes generated by fossil fuels are drilling and production by-products such as organic chemicals, radioactive materials, and heavy metals. They are dumped or buried.

Transportation: Hazardous transportation by-products include spent motor oil, radiator fluid, and batteries containing organic chemicals, heavy metals, acids, and toxics. They are dumped or recycled.

Sewage sludge: Sewage sludge can contain heavy metals, household toxics, and other chemicals (anything thrown down the drain). This waste is treated, incinerated, recycled, or dumped.

Mining by-products: Mining produces hazardous wastes such as radon and uranium. They are left exposed or buried.

Dredge spoil: Dredge spoil can contain heavy metals, PCBs, and other chemicals. It is dumped, buried, or treated.

State of the Earth

In the United States 60% to 80% of the hazardous waste goes into injection wells, surface impoundments, and landfills; as much as 20% goes into streams and rivers. It is estimated that 90% of landfills and impoundments leak.

Americans, the largest producers of hazardous waste per capita, generate 275 million tons each year, which is more than 1 ton for each person; more than 3 million tons of infectious and medical waste is generated every year; more than 5 million tons of hazardous waste are burned yearly in 350 hazardous waste incinerators, industrial furnaces, and cement kilns. There are about 200,000 hazardous waste generators in the United States. China produces 400 million tons of industrial waste, which is dumped on the land or in rivers. Europe generates 25 million tons of hazardous waste, and the European Community has 600,000 contaminated land sites, of which 154,000 require cleanup; 1.5 billion tons of hazardous wastes have piled up around Russia.

The U.S. Defense Department estimated that in 1992 the United States had about 11,000 active hazardous waste sites at more than 1,800 military installations in all 50 states and outside the U.S., with 80,000 tons of hazardous waste produced in 1991 and 500,000 tons released into the air and water. The U.S. Environmental Protection Agency (EPA) estimates that bringing military bases into compliance with environmental laws will cost $400 billion to $100 trillion. Land mines and other unexploded firing range ordnance are strewn over 100,000 acres of government land.

Chemical and petrochemical companies produce more than 70% of all hazardous waste in industrial countries. The United States has about 1,500 hazardous waste landfills, up to 15,000 landfill sites that contain hazardous waste, and several times that number of polluted impoundments and industrial waste landfills. The EPA lists 34,000 hazardous waste sites, but fewer than 1,300 are scheduled for cleanup under Superfund; possibly 75,000 sites in the United States need cleanup. The average cleanup of a Superfund site costs $25 million and takes 7 to 10 years; the EPA has already spent $7.5 billion to remediate only 163 of 1,204 sites. Estimates are as high as $500 billion to clean up Superfund sites and the nuclear weapons network of hazardous wastes created by the Departments of Defense and Energy.

More than 190 million tons of radioactive uranium mining waste (with high levels of radium, and thus radon) have been heaped on southwest U.S. deserts. One of the worst hazardous waste sites is 50,000 acres along 140 miles of the Clark Fork River in Montana; mining, milling, and smelting left substantial heavy metal pollution. At Superfund sites the top 10 chemicals are lead, arsenic, mercury, vinyl chloride, benzene, cadmium, PCBs, chloroform, benzo(a)pyrene, and trichloroethylene (TCE).

Studies have shown that most hazardous waste has been dumped in areas that are economically poor, have minorities, or are in developing countries. Studies by the U.S. government in 1980, and the United Church of Christ in 1987, concluded that for the areas studied in the United States, 3 out of 5 toxic waste sites were located in poor African-American and Hispanic-American communities, in states where these groups constituted only 20% of the total state population. Canada is currently the largest dumping ground for U.S. hazardous waste—close to 1/2 million tons a year. In 1991 U.S. corporations exported 200 million pounds of plastic waste, more than 60% going to Asia.

Millions of tons of hazardous waste are shipped internationally each year. In 1990 more than 100 countries agreed to the Basel Treaty, the first step to regulate international transportation of hazardous waste. It requires notification of intent of international hazardous waste transport and prior informed consent (PIC) by receiving parties; it also allows hazardous waste transport to continue. In response, a coalition of countries in Africa in 1991 adopted more stringent hazardous waste transportation laws in the Bamako Convention; the Lomé IV Convention also banned hazardous waste export among more than 80 countries in Africa, the Caribbean, Europe, and the Pacific. As of March 1994, industrialized countries agreed to an immediate ban of exporting hazardous wastes to developing countries for incineration or burial; export of hazardous wastes for "recycling" will be banned as of December 31, 1997.

The Emergency Planning and Community Right-to-Know Act resulted after a number of serious crises involving hazardous waste in the United States. An example was the illegal dumping of hazardous waste in Love Canal, New York, without the knowledge of the residents or community; many people developed cancer and other diseases. A similar situation occurred in Bhopal, India, when a Union Carbide company allowed 30 tons of methyl isocyanate gas to escape into the air of a slum area without informing the residents; more than 2,500 died, and nearly 20,000 were disabled. Thousands more are expected to die from related diseases. Now U.S. workers and area residents have the right to know of toxic and hazardous substances, wastes, and emissions, and accidents in which local companies are involved.

Human Impact

Health: Hazardous wastes can cause cancer, organ diseases, birth defects, miscarriages, blood diseases, hyperactivity, asthma, and allergies. Mercury and other heavy metals can harm the kidneys and central nervous system. Hazardous wastes have been emitted into the environment over the years in many places and many forms. It is impossible to document what all these chemicals have done, or are doing, except in specific cases where known pollutants resulted in specific health problems. Most hazardous waste has not been tested for long-term health effects. In the United States nearly 30% of the population will contract cancer in their lifetime.

Synergy: Incineration or mixing of hazardous waste allows many diverse toxic chemicals to unite synergistically to form other dangerous toxics that threaten public health.

Worker safety: Workers in the chemical and other industries have higher risks of exposure to hazardous and toxic substances than the general public. For instance, uranium miners have a 5 times greater chance than the rest of the population of getting lung cancer.

Individual Solutions

- Buy nontoxic household products; minimize plastic use; recycle; buy in bulk. If uncertain about a certain chemical or product, check with environmental groups or the EPA for a list of nontoxic alternatives.
- Have your water checked by a water testing company; if you have concerns, use a water filter.
- Oppose nuclear power and incineration.
- Never dispose of toxics down the drain, into your garbage, or onto the soil; call your state officials for hazardous waste disposal sites.
- Demand reduction of toxic use for industry, zero discharge of toxics, and large, prohibitive cleanup fees for polluting industries.
- It is now possible for citizens to know what pollutants local industry emits, thanks to the Emergency Planning and Community Right-to-Know Act. Private citizens also have a right to know of any stored toxic chemicals or hazardous wastes in their area or of any accidental releases. Industries must also publish yearly Toxic Release Inventories (TRIs) on more than 300 chemicals. If you are concerned about a local industry, contact the EPA or state environmental officials to request data on its chemicals and wastes.
- Report to the EPA if you think your company, or another company, is illegally dumping wastes, emissions, or other toxic or chemical substances; this behavior is punishable by fines and prison sentences.

Industrial/Political Solutions

What's Being Done

- Industries are recapturing and recycling hazardous wastes.
- Household hazardous wastes are being collected.
- Research is being conducted on dozens of methods to detoxify hazardous wastes. Some examples are bioremediation (using bacteria to detoxify waste); plasma arc torch furnaces (they burn hotter, leaving

only hydrogen—a potential fuel—and metal pellets); plasma arc torches (smaller, mobile versions of the furnace); ultraviolet (concentrating ultraviolet with solar collectors on liquid waste containing a catalyst that breaks down organic wastes into methanol); reverse-burn gasification; chemical detoxification; filtration; and settlement of wastes.

- The U.S. Resource Conservation and Recovery Act (1976; renewal is slated for 1993–1994) and the Hazardous and Solid Waste Amendments (1984) define the EPA's enforced management of industrial hazardous waste production, its safe transport, and disposal. The EPA has concluded that the best way to manage hazardous waste is to not create it.

- Under the 1986 U.S. Superfund Amendments and Reauthorization Act, more than $15 billion is being used by the federal government to clean up abandoned industrial hazardous waste sites and respond to emergency spills and leaks during toxic or hazardous waste treatment, storage, or disposal. This legislation began after Hooker Chemical Company irresponsibly dumped thousands of barrels of toxics at Love Canal, New York. Reauthorization occurs in 1994; the EPA is looking at revamping Superfund to make it more effective.

- The 1990 U.S. Clean Air Act Amendments require the EPA and the Occupational Safeth and Health Administration (OSHA) to develop regulations for chemical safety management.

- The EPA's Bioremediation Field Initiative, which began in 1990, collects data from 150 sites around the country where bioremediation is being used or considered. A number of Superfund sites are using bioremediation because it is cheaper than incineration and can be done on site.

- Toxic use reduction programs in the United States and a number of industrial countries focus on incentives and mandated reduction of toxic use at production sites by replacing toxics with safe alternatives in manufacturing.

- A 1994 environmental justice order requires government agencies to develop strategies to end environmental racism (see Glossary) with regard to pollution.

- Many states have developed their own superfund programs.
- Medical waste disposal has been delegated to state authority.
- Some companies have employee incentives to reduce pollution. For more than 15 years the 3M company of Minnesota has used a program called Pollution Prevention Pays (3P), in which employees have been encouraged to find ways to reduce pollution. A total of 3,000 3P suggestions have saved the company $500 million to date and have eliminated more than 1 billion pounds of waste. In 1991 3M pledged to reduce air emissions by 70% by mid-1993 and all emissions by 90% by 2000. This task will mainly involve eliminating hydrocarbon-based solvents.
- An 18-month freeze on building new hazardous waste incinerators ends in 1995, while operating incinerators and industrial furnaces are coming under tougher permitting controls for dioxins and heavy metals.
- Researchers are seeking methods and locations for long-term storage of high-level nuclear waste. There are 2 commercial sites for low-level nuclear wastes. Most nuclear waste is stored on site.
- More corporate executives and company owners are being prosecuted and given prison sentences for illegally dumping hazardous wastes.
- Fire departments and community agencies are aware of toxics used on company premises for emergency purposes.

What Needs to Be Done

- Phase out all hazardous waste production, and detoxify or recycle any hazardous waste that is generated.
- Phase out incineration and landfilling of hazardous waste; autoclave or use other safer alternatives for medical waste.
- Phase out nuclear power.
- Immediately ban all transport and sale of hazardous wastes for "recycling" from industrial to developing countries.
- Help developing countries with their debt. The reason developing

countries accept international shipments of hazardous waste is to help pay off their debt.

- Enact international laws overseeing hazardous waste production and disposal in all countries.
- Prevent manufacturers from putting toxics into household products.
- Tax hazardous waste production as an incentive to reduce it and pay for pollution cleanup costs; make hazardous waste polluters pay for all cleanup costs.
- Set mandatory executive prison sentences for illegal hazardous waste dumping.
- Adopt uniform product labeling that is clear for consumers.
- Require double-walled hulls on inland barges carrying hazardous wastes.
- Ensure that the North American Free Trade Agreement (NAFTA) does not override local, state, tribal, or federal laws.
- Strengthen Superfund with more research on safe treatment and disposal of hazardous wastes and aggressive recovery of money from polluters.

Also See

- Heavy metals
- Nuclear energy
- Toxic chemicals
- Warfare effects on the environment

Heavy metals

Metallic elements with a high atomic weight, or high density (gravity of 4 grams per cubed centimeter and above), that are literally heavy.

Examples are aluminum, arsenic, beryllium, cadmium, chromium, copper, lead, mercury, selenium, silver, and zinc. Some metals, such as iron, manganese, copper, and zinc, are essential trace elements in the human diet. Others, such as mercury, lead, and cadmium, however, are very toxic even in minute amounts.

Metals cannot be destroyed, but some can be converted to other compounds (for example, mercury can be converted to methylmercury). When burned, metals can be carried by the wind over long distances.

Environmental Impact

Environmental buildup: Heavy metals do not biodegrade, so whatever is released into the environment builds up year after year, becoming more and more concentrated in the soils, food chains (of fish, birds, predators, shellfish, and vegetation), and water.

Animals and plants: In living organisms heavy metals bioaccumulate in greater and greater concentrations. In mammals and birds heavy metals can cause reproductive problems, neurological problems, kidney disease, cancer, and the death of young and adults. Fish are affected by metal buildups and can develop lesions and tumors; aluminum can invade their gills and kill them. Heavy metal exposure can also kill other aquatic organisms. Heavy metals can affect soil chemistry and biology and thus affect tree and plant growth. Heavy metals such as zinc, copper, and nickel are taken up by plants in acidic soils; lead usually binds with plant roots and soil molecules.

Acid rain: Acid rain can cause heavy metals to become more active in soils and to leach out of the soils and into the waterways.

Major Sources

Mining: All heavy metals are mined out of the earth.

Industrial uses: Heavy metals are used in many industrial processes worldwide.

Products: Heavy metals are used in a variety of products: lead in batteries and plastic, copper in electronics and alloys, cadmium in solders and batteries, and mercury in paper production and pesticides. Metals are also used as additives in industrial chemicals such as in photographic materials, leather, plastic, refining fertilizers, agricultural pesticides and herbicides, and pharmaceuticals.

Processing and incineration: Heavy metals are released back to the air, soil, and water through smelting, incineration of solid waste items (batteries, electronic equipment, medical equipment, household goods), burning of leaded gas, or landfilling of incinerator ash or other garbage. Incineration is one of the largest sources of mercury emissions into the environment.

Fossil fuels: Heavy metals are present in small amounts in fuels like coal and oil and are released when they are burned. The small amounts add up to large emissions because of the amount of fuel burned worldwide. Leaded gasoline is the largest contributor to lead in the environment, and a number of developing countries still use leaded gas. Lead is the most prevalent heavy metal waste in the environment. Mercury is rapidly becoming more prevalent than it used to be.

Sewage: Sewage often has a high content of heavy metals such as cadmium.

Tobacco smoke: Tobacco smoke can contain arsenic, cadmium, selenium, and lead.

Natural processes: Erosion and volcanic activity release heavy metals from rocks and ores. This amount, however, is small in comparison with human-released heavy metals.

State of the Earth

Data on mercury, lead, and cadmium being released into the environment are not thorough, and for many other heavy metals data are almost nonexistent. Since 1900 more than 200 million pounds of mercury have been used in the United States. More than 40 million tons of lead are produced each year worldwide. Most heavy metals are difficult to recycle because they are used in complex mixtures.

High concentrations of heavy metals have been found in the North Sea. Lakes across the United States show high mercury levels; very small amounts of mercury can render a whole lake contaminated and toxic. Heavy metal pollution in U.S. waterways has increased since the Clean Water Act of 1970. Lead has been found in Arctic air, water, fish, mammals, and seabirds.

Urban areas have high levels of lead. Tap water in a number of countries has been tested and found to contain high concentrations of lead and heavy metals. Poland's soil has some of the highest lead contamination ever measured. Worldwide, sewage is often a toxic waste owing to the presence of heavy metals, especially cadmium.

Flashlight, camera, watch, car, and accessory batteries contain mercury, cadmium, lead, lithium, manganese dioxide, nickel, silver, zinc, and sulfuric acid, all of which are toxic materials. Worldwide, millions of disposable batteries are discarded each year. Incineration of small batteries is one of the major sources of mercury in the environment. Incinerators have been legally permitted to release, in air emissions, hundreds to thousands of pounds of a number of heavy metals yearly; incinerator ash also contains high levels of heavy metals.

Aluminum, the most common metal in the earth's crust, is used in kitchenware, aircraft, aluminum foil, cars, and cans. Workers in industry are most exposed. Aluminum can be held in the lungs and is believed to be a neurotoxin. Antimony, a skin irritant and carcinogen, is used in pigments and with lead and other metals as an alloy in semi-conductors and thermoelectric products. Beryllium is used in light alloys in the space and aircraft industry and is a carcinogen. Barium, used in pesticides and coatings, causes paralysis. Cadmium is found in phosphate fertilizers, sewage, and tobacco leaves and is used in batteries, alloys, and nuclear power plants; it causes cramps and diarrhea and affects the lungs, liver, and kidneys. Chromium is used in steel, plating, and pigments and is a carcinogen. Cobalt is used in pigments in ceramics and glass and in alloys in the electrical, car, and aircraft industries; it causes lung disease and is a possible carcinogen. Copper is widely used in alloys and can cause lung damage. Excess iron can cause liver damage. Nickel is used in batteries and is a carcinogen.

Arsenic occurs widely in the environment, often with other metals, and has been used in animal poisons, insecticides, paints, and coatings. In the past, spraying cotton fields with arsenic wiped out bee populations and sickened cows. Coal-burning plants release large quantities of arsenic, and some 20,000 tons are imported yearly for manufacturing. The U.S. Environmental Protection Agency (EPA) estimated in 1992 that there were more than 1,300 sites of arsenic and mercury contamination alone and that to "clean up" these sites would require decontaminating more than 150,000 tons of contaminated soils.

Human Impact

Health: Heavy metals can be ingested with food, inhaled from the air, or absorbed through the skin. They tend to accumulate in the lungs, liver, kidneys, and brain and over time can build up in the body; the greater the concentration, the greater the health problems. Most heavy metals are highly toxic in even microgram amounts. They can cause cancer and damage the brain, nervous system, glandular system, organ system, and reproductive system; they also cause kidney disease, heart disease, increased blood pressure, brittle bones, skin diseases, and lung damage. Workers in industry often risk the highest exposure to metals, though mercury, cadmium, and lead are now very common in the general environment.

Children: Heavy metals can cause the same health problems to unborn children. Young children are much more susceptible than adults to all the above health hazards since their nervous and organ systems are still developing and thus will incorporate any heavy metals in their environment at a much more rapid pace. Learning disabilities and retardation are common symptoms for children with lead exposure.

Individual Solutions

- Recycle appliances, fluorescent lights, electronic equipment, and any other equipment that may contain heavy metals; call the EPA if you are unsure.
- Avoid disposable batteries or products that use them; use rechargeable, electric, or hand-powered products.
- Take batteries and toxics to hazardous waste collection sites. Call the EPA or a local environmental group for information.
- Support toxic use reduction in industry and an end to heavy metal use wherever possible.
- Oppose incineration.
- Demand more support for renewable energy and an end to fossil fuel subsidies.
- Test your water for lead contamination; if you have old piping, run your water until it is cold.
- Have professionals remove old lead piping or leaded paint.
- When buying an older house, first have it inspected for leaded paint and lead piping.

Industrial/Political Solutions

What's Being Done

- There has been a global move away from lead and cadmium use in any products.
- Europe is phasing out mercury in most consumer batteries.
- Leaded gas was phased out in Japan, the United States, and much of Europe.

- Some European countries are limiting heavy metal use.
- A number of European countries have lower pollutant emission standards for incinerators than the United States; the EPA toughened heavy metal emission regulations for hazardous waste incinerators in 1993.
- Household hazardous waste is being collected in the United States and in a number of industrial countries.
- Mexico City has introduced unleaded gasoline; all new cars made there must use it.
- The U.S. Department of Energy is studying reducing mercury and lead from coal plants.
- By July 1, 1994, Illinois, Maine, New Jersey, New York, and another dozen states will not allow sales of products such as inks, dyes, pigments, adhesives, or stabilizers that intentionally have heavy metals (lead, cadmium, mercury, chromium) added; unintentional heavy metal content of over 100 parts per million will be banned July 1, 1996.
- A number of states have set deadlines of 1996 for mercury-free batteries.
- The U.S. Food and Drug Administration (FDA) does not allow fish with more than 1 part per million of mercury to be marketed commercially.
- Mercury has been eliminated from indoor paints; labeling is required for outdoor paints.
- Lead use in new drinking water systems has been banned.
- Incentives have been established to discourage the mining of virgin lead.
- Car batteries are being recycled; stores that sell them in the United States are required to collect the old batteries.
- Fluorescent bulbs and high-intensity lamps, considered hazardous waste for their mercury content, are being recycled.

What Needs to Be Done

- Fund toxic use reduction strategies; phase out industrial use of heavy metals, except where they can be fully recycled; phase in zero-discharge heavy metal emission limits for industry and utilities.
- Phase out incineration; treat all incinerator ash as hazardous waste; ban its use in any construction material. Mandate that all incinerators separate out batteries and other heavy metal–containing items before burning waste.
- Phase out heavy metals from products wherever possible, especially in household products.
- End fossil fuel subsidies and support renewable energy, and energy conservation and efficiency.
- Develop world policies to eliminate leaded gas from all markets.
- Separate sewage sludge from industrial wastes, which often contain heavy metals.
- Tax heavy metal use.
- Initiate a worldwide ban on lead use in water pipes and other areas where it is a direct source of environmental pollution.
- Require quicker overhauls of water supply systems with lead piping.

Also See

- Incineration
- Lead (Pb)
- Mercury (Hg)
- Mining

Hydroelectric energy

Definition

Renewable energy created when water passes over turbines to produce electricity.

This process is usually controlled by building a dam across a river or stream and controlling the amount and rate of water flow over the turbines. The more water that passes over the turbines and the farther it drops, the more energy is produced. Thus the larger the river, the bigger the dam, and the more energy is produced; conversely, hydropower can be greatly reduced in times of drought and reduced river flow. Less common hydroelectric plants called run-of-river plants do not attempt to control the water flow and do not have reservoirs for water storage.

Pumped storage hydroelectric plants use two reservoirs: water is pumped from the lower to the higher reservoir in periods of slack demand to store the water, so that its energy can be recovered when demand is higher (with efficiency of 70% to 80%).

Dams force a large reservoir of water to build up behind them, and much less water travels down the river beyond the dam. The larger the dam, the larger the associated environmental problems.

Small dams, which can often be built with local materials and supply local electricity, sometimes are not as disruptive to the environment; small hydroelectric projects are often run-of-river plants that have minimal impact on river flow. Hydroelectric dam reservoirs have often been used to provide water for irrigation of crops and energy for mining.

H

Environmental Impact

Clean energy: Compared with coal plants, hydroelectric power is very clean and gives off few air emissions after initial dam construction.

Downstream effects: Fish populations downstream from a dam often cannot migrate upstream for spawning; a pathway for the fish can be made, but many current dams need retrofitting for this. Fish and wildlife populations below the dam, which are dependent on the water flow, can be lost. Water temperatures and quality below the dam can be changed by the dam reservoir, perhaps harming downstream animal populations. Rivers often carry silt, salts, and nutrients that are blocked by dams. Thus soils downstream from dams may become less fertile. Also, river basins below the dams begin to erode because their supply of silt is not being replenished.

Flooding: Reservoirs behind dams initially flood the local landscape, killing wildlife and, in a number of countries, forcing indigenous people to leave the area. Extreme flooding due to reservoir overflow in heavy rainfall has been known to happen in some countries.

Water loss: Water evaporation in reservoirs in hot climates is enormous. Lake Nasser in Egypt is created by the Aswan High Dam; it loses enough water to evaporation each year to irrigate 2 million acres of farmland. Aquatic weeds in reservoirs use 2 to 6 times the amount of water they would use in natural settings since nutrient levels, and thus growth rates, are much higher.

Greenhouse gases: Carbon dioxide and methane output from some reservoirs is high because of bacterial decomposition of flooded forests.

Salinization: Salt builds up in the dam reservoir, and if reservoir water is used to irrigate farm soils, it leads to salinization.

Nutrient loss: Introduced fish populations in some reservoirs may soon peak and then plummet when nutrients are used up.

Toxics: Bacteria in the reservoirs can release heavy metals, such as mercury, from the bottom rocks; the mercury is changed into methylmercury and taken up by fish. Decomposing plants in the reservoirs can make the water so acidic that fish and plant life cannot grow; this acidic water also affects fish and wildlife downstream.

Earthquakes: The weight of reservoir water increases chances of earthquake in the immediate area, resulting in dam breaks and flooding.

Sedimentation: Silt can build up behind a dam and fill the reservoir. This problem is usually more frequent in tropical latitudes where soils erode more easily and water silt levels are higher.

Weather: Large reservoirs can influence local climates by reducing cloud cover or increasing humidity.

Small dams: Small dams may have the same impact on local plants, fish, and animals in smaller ecosystems as large dams, depending on whether fish can swim upstream past the dam and how much water is allowed to flow through the dam.

Major Sources

Rivers: The appeal of hydropower is that almost any river or flow of water in the world can be dammed and used to create electricity. Large projects require huge amounts of startup money and cause greater environmental problems than small ones. Many countries are shifting their focus to smaller hydroelectric projects, but large projects are still being planned in a number of countries.

Mountainous regions: Mountain regions are best for generating hydropower because they offer a larger vertical drop for the water, which increases the power generated. In the United States, mountainous areas in the Northwest, West, and Southeast are most favorable, though hydropower is used in other areas as well. Hydropower is used in many countries in areas of varied terrain.

State of the Earth

A 1989 survey found that there were more than 36,000 large dams worldwide over 15 meters in height and nearly 8,000 more than 30 meters in height. Slightly more than 600 of all these dams were fitted for hydropower. About 150 large dams account for 40% of all hydroelectric generation.

Hydroelectric power now supplies nearly 100% of electricity needs in Paraguay; 90% in Brazil; 75% in Venezuela; 51% in Sweden; 36% in Argentina; 30% in China; 30% in the Philippines; 26% in India; 24% in Malaysia; and 7% to 10% in the United States. It meets about 20% of world electricity needs and is used in almost 100 countries. It is rapidly increasing in use in many countries but is not expanding much in the United States owing to environmental problems, increased demand for agricultural and drinking water, and efforts to preserve river fish stocks. The United States has 37 pumped storage hydroelectric projects, and there are about 300 worldwide, with 40% in Europe. Japan is building the world's first seawater pumped storage site; it will be completed by 1995.

In the United States, hydropower dams on the Columbia River have seriously depleted the sockeye salmon and trout populations, which have dropped from 16 million to 2.5 million. In the Northwest and New England, local groups are attempting to block hydropower and dismantle existing plants to restore native trout and salmon populations; two dams on the Elwha River are being removed to restore salmon runs. The Colorado River is one of the most dammed and diverted rivers in the world; by the time it reaches the Gulf of Mexico it is scarcely more than a trickle.

Large dams such as the Tucurui and Itaipù Binacional in Brazil and Paraguay and the Mahaveli in Sri Lanka have caused extensive environmental damage. Brazil's projects have caused worldwide controversy by flooding huge areas of the Amazon and displacing thousands of indigenous people. The Kariba Dam in Zambia displaced 57,000 people. In Ghana's Volta River basin, 70,000 people suffered permanent river blindness, and a million more were weakened by the disease. It is estimated that in the past decade the World Bank supported projects (through lending) that forcibly resettled nearly 1 million people.

Between 1950 and 1980 China constructed 6 million small dams. China has more than 80,000 small hydropower plants, which generate the energy equivalent to the output of 10 nuclear power plants; it plans to double this capacity during the next decade. The proposed Three Gorges Dam project on China's Yangtze River will flood 60,000 acres of farmland, cost $11 billion, take 18 years to build, and displace 1.2 million people.

Argentina, Canada, Chile, Malawi, Mexico, Paraguay, Thailand, and Vietnam also have large hydroelectric projects under construction or on the drawing boards. India is building the Sardar Sarovar Dam on the Narmada River; its reservoir will displace 100,000 people and cover 200,000 acres of land.

Small dams are in greater favor in a number of countries: France, Italy, Sweden, and the United States all have about 1,500 small hydropower projects. Japanese companies have created collapsible rubber dams that prevent silting and are used for small hydroelectric and irrigation projects. Nepal has microhydroelectric projects—2- to 3-person enterprises of less than 100 kilowatts, used to mill grain, expel oil from oilseed, and generate electricity for lighting.

Because of the Aswan High Dam, the Nile deposits only a few tons of rich organic sediment on shore banks annually, compared with more than 100 million tons annually before the dam was built. Artificial fertilizers must make up the nutrient loss. Water behind the Aswan High Dam is estimated to hold as much as 10% more salt than water entering the reservoir; this water is used for irrigation and has led to massive salinization problems and farm soil degradation. Egypt gets 28% of its energy from hydropower when there is strong river flow.

Large dams have led to more earthquakes in the immediate area of the reservoir; the Hoover Dam experienced 6,000 shocks in the first 10 years after filling its reservoir. Large dams in China, Greece, India, and Zimbabwe have increased major seismic activity in their regions; minor reservoir-related earthquakes have been experienced in Australia, France, Italy, Japan, New Zealand, the former Soviet Union, Spain, Turkey, and the former Yugoslavia.

Hydro-Quebec of Canada is planning the Great Whale hydroelectric project on James Bay that will flood millions of acres of wilderness, dislocate Cree and Inuit tribes, dam 11 rivers, and increase mercury pollution; Quebec currently gets 95% of its electricity from hydropower. Canada is exploring using hydroelectric energy to create hydrogen fuel through electrolysis and selling the hydrogen to Europe.

Human Impact

Health: Millions of people have been physically displaced by the reservoirs of large dams, and their health has been jeopardized by waterborne diseases from the reservoirs and irrigation projects. Examples of waterborne diseases are malaria, carried by the anopheles mosquito; schistosomiasis, carried by snails; filariasis, carried by a parasitic worm; onchocerciasis, or river blindness, carried by the "black-fly" simulium, which breeds in reduced flow downstream of dams; cholera; dysentery; and diarrhea. Pesticides used in the reservoirs and downstream rivers to control disease and vegetation can threaten both fish and wildlife populations, and thus the humans that eat them.

Water shortages: Dams may prevent fresh water from reaching downstream populations, depriving them of a water supply.

Food: Downstream fish and shellfish population declines affect sport fishing and food supplies. Downstream farmland may lose its productivity, which may also threaten food supplies. Fish that take up mercury in the reservoir pose a health threat when consumed as food.

Individual Solutions

- Oppose large dam construction projects.
- Support renewable energy alternatives, such as wind and solar.
- Demand that all dam projects take into account the effects on local upstream and downstream ecologies.
- Practice energy conservation.
- If you live near a hydropower project, find out when the permit is renewed; many are due for renewal in the coming few years. Get involved in the process, find out the effects on the river, and determine if you wish to support or oppose the permit renewal (see *Rivers at Risk: The Concerned Citizen's Guide to Hydropower,* by John D. Echeverria).

H

What's Being Done

- In the United States and worldwide there is a growing movement to demand that large dam projects meet other criteria besides just energy outputs (such as the defeated Two Forks Dam in Colorado).
- The World Bank, which has funded a number of large dam projects in developing countries, is beginning to consider the environmental impacts of such projects.
- Some dams are being retrofitted with fish "ladders" to make it possible for fish to swim upstream past the dams (some older dams cannot be retrofitted).
- Some small existing dams are being retrofitted with hydropower technology, and many others—some large—are being dismantled.

What Needs to Be Done

- Invest in energy conservation and other renewable energies; this approach could save enough energy to eliminate the need for many new proposed hydroelectric projects in most countries.
- Call for industrial countries to help developing countries create renewable energy options through joint technology development programs.
- Give worldwide support to indigenous peoples when their homes and livelihoods are threatened by proposed hydroelectric plans.
- Insist that any proposed hydroelectric plans meet criteria ensuring minimal human and environmental impact on upstream and downstream ecologies. Any suggested benefits for local populations should be adequately demonstrated.
- Demand that the World Bank fund only projects that are not only fair and equitable for all parties affected but also environmentally sound. Any proposed lending project plans should be made available to the people in the country proposing the plan.

Also See

- Indigenous peoples displacement
- Renewable energy

Incineration

Burning waste at a high temperature in a furnace.

Incinerators can be land based or operated at sea. They burn solid waste, hazardous waste, medical waste, and biomass energy materials. Waste-to-energy facilities are incinerators that use burned material to generate either heat or electricity.

Bottom ash composes 75% to 85% of the burned substances; the rest is fly ash—ash going up the stacks. Scrubbers are used to remove fly ash, toxics, and gases from the air stacks; this removed material is then added to the bottom ash or disposed of in ponds or wastewater. Bottom ash is landfilled and as of 1993 was not considered hazardous waste.

Fly ash, air emissions, and bottom ash are all toxic. Even if the materials to be burned are not toxic in themselves, when incinerated they often break down into toxic chemicals or combine with other chemicals to form toxic chemicals. This is especially true for materials like plastic, petrochemical materials, and even paper with oil-based ink.

Environmental Impact

Air pollution: Incinerators give off high air emissions of mercury, lead, other heavy metals, metal chlorides (from metals binding with chlorine in plastics), dioxins, furans, particulates, sulfur dioxide, carbon dioxide, nitrogen oxides, and other greenhouse gases; all add to acid rain, smog, ozone depletion, and the greenhouse effect. Plant growth is inhibited, and forests and lakes are acidified.

Plants and animals: The emitted toxics are carried by the wind and end up in surface waters and soils. They are taken up in water by plants and wildlife and bioaccumulate in the food chain. They can inhibit plant growth and in animals cause reproductive problems, birth defects, neurological problems, cancer, and death. Sea-based incinerators spread air emissions over the oceans, and again toxics can enter the food chain.

Landfills: Leftover ash, which contains heavy metals and other toxics, is landfilled. Most landfills leak rainwater runoff, called leachate, and thus can contaminate nearby soils and groundwater. As air pollution control equipment becomes more sophisticated in capturing airborne pollutants, the fly ash, and thus landfilled ash and leachate, becomes more toxic.

Raw materials: Incinerators put stress on raw resources, such as forests and minerals, because they burn them up instead of recycling them. This pressure causes further mining pollution, deforestation, and energy waste, because more raw resources must be harvested to replace the products and materials that are not being recycled. To be effective, incinerators must burn at a very high heat, which means more energy, and thus resources, is required to sustain the necessary temperature.

Major Sources

Incineration is used in industrial and developing countries around the world. Incinerator types include:

Old incinerator designs: Old designs simply burn unsorted solid waste, landfill the ash, and filter air emissions. About 3 dozen are in operation in the United States and even more in developing countries.

Mass-burn incinerators (waste-to-energy plants): Mass-burn incinerators burn unsorted solid waste; the heat is used to create steam or electricity. The ash is landfilled, and air emissions are filtered.

Controlled air (modular) incinerators (waste-to-energy plants): Controlled air incinerators burn both unsorted waste in a lower chamber and rising gases from the lower chamber in an upper chamber to destroy toxic chemicals. The heat is used to create steam or electricity. The ash is landfilled, and the remaining air emissions are filtered. This type often uses a smaller design and burns medical bioinfectious waste and radioactive waste.

Refuse-derived fuel incinerators (waste-to-energy plants): Refuse-derived fuel incinerators presort waste to separate high-burning wastes (which burn easily and yield high heat), often mixing them with another fuel (such as coal) for burning. The heat is used to create steam or electricity. Unburned waste is recycled or landfilled, ash is landfilled, and air emissions are filtered.

Plasma arc torch furnaces: Plasma arc torch furnaces generate heat as high as 10,000°F in a process called pyrolysis. They are in limited use and are under study as a way of incinerating toxic and medical waste. They generate no ash by-product and give off no toxic fumes. By-products are metal slag and a hydrogen-rich gas.

Transportable and mobile incinerators and plasma arc torches: Transportable incinerators are much smaller incinerators used for hazardous materials at contaminated sites. The plasma arc torch is a mobile version of the plasma arc torch furnace.

State of the Earth

For every 3 tons of incinerated solid waste, roughly 1 ton of toxic ash is generated. Nationally, 35% of all bottom ash of incinerators tests as hazardous, and 100% of the fly ash tests as toxic. Incinerator ash has been found with the same lead content that lead-based paint once had—about 5 pounds per ton. Most incinerator ash has been landfilled with municipal solid waste.

Often hazardous waste sites contain contaminated soils, which are transported to incinerators and burned. Other techniques for cleaning up hazardous soil on site are rapidly being developed.

The U.S. Environmental Protection Agency (EPA) estimates that 71% of lead and 88% of cadmium emissions come from garbage incinerator emissions, mainly from plastic packaging material. More than 1/4 of U.S. airborne mercury and much of the heavy metal contamination is from incinerators; up to 74,000 pounds of mercury are released annually. Incinerators have released, legally under air permits, as much as

700 pounds of mercury yearly and 15,000 to 20,000 pounds of heavy metals; it takes a small fraction of a pound of mercury to pollute a small lake and require fish advisories. In Minnesota more than 300 lakes (95% of 325 tested from 1989 to 1991) were found contaminated with mercury and required fish advisories; the list of mercury-contaminated lakes across the United States is growing as more lakes are tested.

Incineration air emission standards in the United States for chemicals like dioxin have been as much as 200 times more lenient than those in Sweden, which has state-of-the-art equipment. Ocean sediments in the North Sea were found to contain toxic chemicals whose source was a sea-based incineration program. The United States has a giant incinerator on Johnston Atoll, some 800 miles southwest of Hawaii, where it has incinerated nerve agents and other outdated weapons since 1990.

Waste-to-energy facilities are in use in many countries; the United States has about 150, which generate enough electricity for more than 1 million homes. They burn 100,000 tons of municipal solid waste daily and 30 million tons of trash yearly—15% of the nation's trash. One-third of the trash ends up as toxic ash. Incineration uses more energy than it produces. For example, burning 1 pound of paper gives off 500 Btus, but if recycled, it saves 2,000 Btus. Even the most efficient incinerator releases up to 7,000 pounds of toxic chemicals yearly.

Human Impact

Health: Particulates from incinerator air pollution can be very small, and a number of toxics, such as heavy metals and dioxins, can attach to them. Humans can breathe in these toxic particles, which cause cancer, birth defects, central nervous disorders, kidney and liver problems, endocrine problems, and respiratory illnesses. Leachate from ash landfills can add lead and other toxics to drinking water supplies. Lead causes cancer and neurological disorders; children are especially vulnerable.

Food: When food crops grow on soils contaminated by airborne toxics or leachates, some become toxic; when eaten, those toxics can cause the same health problems already mentioned. Mercury-contaminated lakes render fish inedible.

Raw materials: Incineration destroys materials, which results in increased raw material needs, related mining pollution, related health problems, and aesthetic concerns.

Individual Solutions

- Oppose incineration.
- Compost your food scraps and yard waste; experts believe 80% of solid waste could be composted.
- Recycle; buy recycled products; buy durable products you can reuse; reduce all garbage you cannot recycle.
- Buy products without (or with minimal) packaging, especially plastic and foam materials; avoid disposable goods.
- Buy groceries and other food in bulk so you can reuse containers.
- Avoid buying toxic products; consult the EPA or an environmental group for a list of safe household alternatives.

Industrial/Political Solutions

What's Being Done

- Recycling programs and source reduction programs are being pursued.
- Composting programs are being used and researched widely in the United States and much of Europe.
- Countries bordering the North Sea (Belgium, Denmark, England, Holland) banned sea incineration by 1994; the United States has already abandoned incineration at sea.
- Household hazardous waste is being collected in the United States and in many European and other industrial countries.
- Incinerator ash is landfilled; in May, 1994, the U.S. Supreme Court ruled that incinerator ash must be treated as hazardous waste when it doesn't meet federal standards, and then must be placed into hazardous waste landfills.

- Air emission standards for incinerators are becoming stricter in the United States.
- The United States has a current 18-month freeze on building new hazardous waste incinerators, while operating incinerators and industrial furnaces face tougher permit controls for dioxins and heavy metals. The freeze will end by 1995.
- The EPA is focusing on reducing hazardous waste.
- Some states are requiring presorting of garbage to remove toxics.
- Incinerator ash is being mined for metal content in a few new programs in the United States.

What Needs to Be Done

- Phase out incineration of solid or hazardous waste.
- Institute a world ban on incineration at sea.
- Strengthen recycling and source reduction incentives; charge people for their garbage by weight, unit, or volume.
- Mandate stringent pollution emission standards on existing incinerators, and set close-down deadlines for all incinerators that cannot meet these standards.
- Treat incinerator ash as hazardous waste.
- Separate any hazardous waste, like batteries, from municipal garbage before it is burned.
- Use autoclaving instead of incineration for medical waste; minimize the use of disposables and plastics in medical practices.
- Ban any use of incinerator ash in construction material of any kind (such as roads, buildings, and paved lots); virtually all incinerator ash is toxic.
- Set packaging limits; reduce plastic production.

Also See

- Hazardous waste
- Solid waste

Indigenous peoples displacement

The forcing of original peoples, usually tribes living in often undeveloped wilderness areas, from their land or from a part of its original range, either through physical force, destruction of their homeland, or depletion of its livable resources.

Worldwide, indigenous peoples are fighting to secure land and other rights where they are living, often in conflict with the governments that wish to displace them. In many cases governments have used the fact that indigenous people have no land deeds, or are nomadic, to refuse to acknowledge any claims they make about land rights or ownership. Indigenous peoples have banded together, however, creating local, regional, national, and international organizations to help each other achieve justice.

Indigenous people are inhabitants of an area before modern development, often living off the land by hunting, fishing, gathering fruits and nuts, and employing small agricultural methods. Often they are nomadic within a small or large range or shift their agriculture to different areas to allow the soil fertility to be replenished. One way of defining any distinct culture, including an indigenous culture, is by its unique spoken language.

In some cases today's indigenous peoples replaced other indigenous peoples at some time in the past. Indigenous people usually live in tribal units of varying sizes and interact with the environment in a sustainable manner, preserving food sources, fish supplies, trees and plants, and wildlife. They generally do not have formal education, but nevertheless they often are better caretakers of the land than "educated" people. Their ties to the land include spiritual and practical beliefs. The legacy they often wish to leave their children is the forest, the land, and its resources, intact.

Environmental Impact

Indigenous people are often good caretakers of the environment. When they are driven out, the pressures that drove them are frequently detrimental to the environment. The negative impacts include:

Deforestation: Most tribal peoples that live together live in rain forests or other wilderness areas that are intact; when they are removed, the forests are usually also destroyed. Trees hold soil, and when they are removed the soil can erode or become compacted. Soil erosion allows nutrients, silt, and heavy metals to wash into rivers, causing algae growth and toxicity and killing fish and plants. Loss of trees leads to increased evaporation of water supplies, as well as to the erosion of soil, which holds water and allows it to percolate down to groundwater supplies.

Extinction: Animals dependent on forest reserves die or leave the area. In some cases indigenous people may come into conflict with efforts to save endangered species. For instance, the Eskimo of Alaska are allowed to hunt bowhead whales as part of their continuing heritage. Only 3,000 to 5,000 bowheads are left, however, and hunting seems to be enough to keep the population from making a comeback. Thus the Eskimo are coming under increasing pressure to end their hunts.

Major Sources

Deforestation: Mining, logging, ranching, agriculture (food crops and palm and rubber plantations), oil drilling, road building, and urban development all have been responsible for driving indigenous people from their forests. This process is occurring in tropical rain forest areas throughout the world.

Population increases: Population increases have increased demand for raw materials and for agriculture and cattle ranching, all of which have put more pressures on governments to exploit forests and their tribal caretakers.

Dams: The reservoirs of large dams force indigenous people out when they first flood, and the lack of water, nutrients, and thus fish and fertile land downstream has sometimes forced indigenous people there from their land. This kind of displacement has occurred in Brazil, Canada, China, the United States, and numerous other countries.

Government land programs: Government programs sometimes give or sell land rights to nonindigenous people, without recognizing that the indigenous people, often nomadic, have any rights to the land. Other programs by governments, such as defining land ownership by ranching, have forced indigenous people to cut down their own forests to gain land titles. Such programs have been carried out in Australia, Brazil, the United States, and other countries.

Unfair agreements: In the past, unfair or misunderstood contracts have been arranged with indigenous people to force them from their home-land. In a few cases contracts have been fair monetarily, but the indigenous people have not understood the impact that the loss of land, such as through logging, would have on their people over time. These agreements have been made in Australia, Brazil, and the United States, and areas of Asia.

Tourism: Some countries have forced indigenous people off their land to use it as a tourist attraction. This has occurred in Central America.

Disease: Diseases brought in by urban contacts often require indigenous people to move in order to gain access to Western medical help, beginning a cycle of dependence. This process has occurred in Asia and North and South America.

Warfare: Warfare between factions in a country, or between two different countries, often dislocates indigenous people and is prevalent in Africa.

Cultural influences: Members of a tribe can be influenced by modern technology, wealth, or other goods and might wish to trade away their land rights for them. Tribes in South America and the United States have taken this action.

Overfishing: Along the coastlines of small tropical islands, countries such as Japan, Taiwan, and the United States have overfished shorelines inhabited by indigenous peoples, depleting their food supplies and thus forcing them to move elsewhere.

State of the Earth

Indigenous people are found in most countries and on every continent. Of 6,000 cultures worldwide, 4,000 to 5,000 are indigenous. These 300 to 600 million people make up 10% of the world. Fifty million live in the rain forests. Some indigenous tribes are rapidly going extinct, and so are the forests they take care of. In Canada and the United States up to 33% of coal reserves, 50% of uranium deposits, and 90% of freshwater reserves are on indigenous treaty lands.

The Coordinating Body of Indigenous Peoples' Organizations of the Amazon Basin represents more than 1.2 million indigenous people in the Amazon basin; the Tungavik Federation of Nunavut represents Inuit tribes in the Northwest Territories of Canada; the World Council of Indigenous Peoples is a global federation representing the cause of all indigenous peoples. The United Nations Working Group on Indigenous Populations has finished a document called the Universal Declaration on the Rights of Indigenous Peoples, which covers land and resource rights. Convention 169 of the International Labor Organization asked that the cultural integrity of indigenous peoples be respected.

Native Cree Indians have fought Hydro-Quebec of Canada to stop a dam project that will destroy their fishing, hunting, and trapping land. Brazil's dam projects have caused worldwide controversy by flooding huge areas of the Amazon and displacing native tribes in the process. The Kariba Dam in Zambia displaced more than 50,000 people; the proposed Three Gorges Project on China's Yangtze River will displace more than 1 million people.

In 1991, native Inuit tribes in Canada were given political domain (self-government and resource control) over 770,000 square miles of northernmost Canada. The Gwich'in people in northeastern Alaska and Canada were part of the opposition that stopped the proposed oil drilling in the Arctic National Wildlife Refuge; the Gwich'in believed the drilling would disrupt their caribou herds. Indigenous peoples such as the Udege in Siberia are working to stop logging in their homelands in favor of more sustainable practices. The Kayapó and some other indigenous tribes in Brazil and elsewhere have also used their land rights to sell mahogany, mineral rights, or other rights to their land, thereby contributing to deforestation and other problems.

Australia gave aboriginal peoples land rights long ago, but kept all mineral rights. Bolivia, Colombia, Ecuador, Greenland, New Zealand, Papua New Guinea, Sweden, and Venezuela have recognized indigenous land rights; parts of Africa, and many Asian countries such as Indonesia, Malaysia, and the Philippines, are refusing rights of indigenous peoples. Native Americans in Minnesota and Wisconsin have gone to court claiming that treaties signed in the 1800s give them rights to spear and net fish in certain regions and lakes; to date they have won. Their success has caused wide conflict with area fishermen and landowners, who want the rights revoked. Lummi, Muckleshoot, and Tulalip tribes in the northwestern United States regained salmon rights and now manage those fisheries jointly with state and federal fisheries.

Just as in U.S. history, rain forest ranchers and loggers elsewhere have displaced, killed, and scared off indigenous tribes worldwide. In the Amazon, many rubber tappers, Indians, and activists (such as world-renowned Chico Mendez) have been killed by ranchers and others who want their land. Recent mass violence, including murder and rape, perpetrated by governments and armies against indigenous people has occurred in Burma, East Timor, Guatemala, Malaysia, and Thailand, and in a number of countries it is still occurring. The Penan have been arrested for creating dozens of roadblocks over the past half dozen years in Sarawak, Malaysia, to block logging that is ruining their homeland forest; the Asmat of Indonesia are facing the same problems. Indigenous people in Bolivia and Ecuador have also risked violence for their recent protests.

The Kayapó and Yanomamo Indians (largest of the remaining Stone Age tribes) in Brazil have both suffered from introduced malaria, tuberculosis, and mercury poisoning from pan mining. In 1991 Brazil finally recognized the rights of Yanomamo Indians and gave them a reserve of more than 37,000 square miles of Amazon wilderness. Nearly 3/4 of the original 1 million Amazon tribal people have been killed or displaced. In 1993 more than 30 Yanomamo Indians were slaughtered by miners in Brazil; 2,000 have been killed since 1987. Thousands of miners have invaded the Yanomamo tribe's territory in search of gold, diamonds, tin, and other minerals. In 1993 Maoist Shining Path rebels murdered 55 Indian villagers in Peru's central jungle.

From 1945 to 1986 some countries used indigenous nations and tribal areas as nuclear test sites; China, France, the Soviet Union, the United Kingdom, and the United States detonated nuclear bombs on indigenous peoples of Kazakhstan, Polynesia, and elsewhere.

Nearly all of the next 5 billion people born on the planet will be born in developing countries, posing more stress on rain forests and the people who live in them.

Human Impact

Effects on displaced indigenous peoples include:

Poverty: Displaced indigenous peoples often have no skills, and no desire, to fit into modern society, and thus end up chronically poor.

Racism: Indigenous peoples are often treated as second-class citizens and are denied citizenship or other rights by the governments that have evicted them; this problem is still occurring in a number of countries.

Violence: Some indigenous peoples have been murdered or forcibly removed.

Loss of culture: The culture of indigenous peoples is tied inextricably to their homelands.

Health problems: Indigenous peoples are introduced to diseases and lose access to their healing plants.

Effects on people other than the indigenous tribes are:

Loss of medicinal knowledge: Tribal medicine practitioners have been described as "walking encyclopedias" for their plant knowledge; rain forests are acknowledged as living stores of healing medicines that are as yet untapped.

Loss of genetic material: Literally millions of possible biotechnology products could be lost with the rain forests.

Diminished ecosystem knowledge: Nonindigenous people understand little about the complex ecosystems of rain forests, or even what other types of knowledge might be available in a rain forest and from its inhabitants.

Loss of cultural understanding: Indigenous peoples' beliefs are just beginning to be explored for their value, especially in regard to human relationships with nature; once these cultures are destroyed, so is this access.

Reduced environmental quality: As go the indigenous peoples, often so goes the environment—timber, soil, water and air quality, climate stability, and species of animals.

Individual Solutions

- Practice ecotourism, which preserves wilderness habitat and supports local populations of people; contact the Ecotourism Society, Conservation International, or the Nature Conservancy for programs.
- Recycle paper and wood products. Buy reusable products, such as cloth towels instead of paper towels. Don't use disposable wood products, like chopsticks.
- Don't buy tropical hardwoods, such as teak, rosewood, mahogany, ramin, lauan, or maranti. Instead use oak, cherry, birch, pine, or maple.
- Eat less red meat; cattle production is a major reason for deforestation of indigenous people's homelands.
- Support the rights of indigenous peoples in your country; join a group that supports indigenous people worldwide, such as Cultural Survival.
- Write your legislators to support ending all aid to any country not granting rights to its indigenous peoples.
- Beware of companies and products claiming to benefit indigenous people. Some merchandising chains are helping indigenous people by selling their goods; others are not. Find out if the chains buy their

ingredients or products directly from indigenous people, which ensures that the funds directly benefit the local people.

Industrial/Political Solutions

What's Being Done

- December 10, 1992, began the United Nations International Year for the World's Indigenous People to draw attention to indigenous rights and to educate the public.
- Worldwide, indigenous people are organizing to fight for their homelands; nonindigenous groups are joining in the struggle by providing expertise, political clout, marketing, and organizing.
- Land rights have been given to indigenous peoples in many countries; there is growing pressure worldwide to do the same everywhere. Indigenous people in Honduras and Panama have used maps from their tribal surveyors to prove their land ownership and to secure occupancy and land rights.
- Extractive and biosphere reserves allow rubber, nuts, and other forest products to be sustainably harvested from rain forest land where other development is prohibited.
- Programs have been started to fund individuals to study with indigenous natives to learn medicinal and other uses of tropical rain forest plants and trees.
- Beginning in 1993, the U.S. State Department was mandated to report on the status of indigenous peoples' rights in countries receiving U.S. aid.
- Boycotts are discouraging the use of rain forest beef in hamburgers and rain forest hardwood timber.
- Los Angeles and Berkeley, California, have changed October 12 from Columbus Day to Indigenous Peoples' Day.

What Needs to Be Done

- Make family planning available to all families everywhere.
- Give indigenous peoples the land and resource rights to the forests

and other areas they live in. They must have freedom to organize politically; furthermore, their land rights must be protected.

- Help developing countries with debt relief and poverty; debts and poverty are a driving force in deforestation. The United States relieved substantial debt to Bolivia, Chile, and Jamaica in 1990.
- Pursue government debt-for-nature swaps; industrial countries can save enormous tropical forest acreage in developing countries by swapping the owed national debts.
- Mandate labeling of rain forest beef in all foods, pet foods, and other products.
- Institute a worldwide ban on rain forest exotic woods and on rain forest destruction; industrial countries like Japan and the United States should meet their wood needs with sustainable forestry.
- End mining in rain forests.
- End all international funding by the World Bank and other lending institutions for the building of dams, roads, mines, hydroelectric plants, and other projects that damage rain forests or evict indigenous people.
- Use international boycotts and pressure on countries that are ignoring the rights of their indigenous peoples.
- Promote a worldwide emphasis on sustainable development, which includes protecting the rights of indigenous peoples.
- Ensure that the North American Free Trade Agreement (NAFTA) recognizes the rights of indigenous landholders and does not supersede tribal, local, state, or federal laws.

Also See

- Deforestation
- Sustainable development

L

Landscaping problems

Definition

Problems caused by the methods and equipment used to maintain lawns, golf courses, public parks, recreational areas, and other private and public lands.

Often landscaping involves maintenance of grassy areas, which are used by birds and other animals as feeding grounds.

Environmental Impact

Pesticides: Pesticides (and herbicides) used on grass can threaten the health of small animals like birds and squirrels that use the grass. Pesticides also kill the microorganisms in the soil that are responsible for keeping the soil healthy by breaking down organic matter; as a result more fertilizer is necessary. Pesticides are washed off lawns into rivers, streams, lakes, or ponds and are taken up in the water by fish and wildlife, thus entering the food chain. These chemicals are toxic and over time can cause wildlife serious harm, such as neurological and behavioral problems, reproductive problems, cancer, and death.

Air pollution: Recent studies have shown that gas-powered lawn care equipment (such as mowers, blowers, chain saws, trimmers, and tillers) emit more air pollution for their size than cars. This air pollution is a component of smog. Smog and air pollution inhibit plant growth and can destroy forests. Lawn care equipment emits greenhouse gases, adding to concerns about global warming.

Fertilizer: Chemical grass fertilizers are washed into lakes, ponds, streams, and rivers. They add excess nutrients to the water and increase algae growth, which can eventually deplete the oxygen in the water in a process called eutrophication. Eventually fish can die, and natural aquatic plants can be crowded out. The mining and production of fertilizers adds to air pollution, water pollution, and soil degradation. Phosphate rock, mined for the fertilizer phosphate, releases radiactive radon gas.

Soil degradation: Discarded oil or other chemicals from lawn care equipment can add to soil pollution. Oil, gas, and other chemicals used in lawn care cause soil pollution both when they are mined and during production. Many of these chemicals are toxic and often include heavy metals, which can bioaccumulate in animals and cause diseases, cancer, and death.

Freshwater depletion: The amount of water used to water grasses is a serious factor in freshwater depletion. Water taken from rivers and lakes causes further stress to wildlife species dependent on those resources.

Major Sources

Industrial countries: Industrial countries are still the main sources of landscaping problems. Most of the equipment used in lawn care is not used in many developing countries. And even though pesticides and fertilizers are used in agriculture in developing countries, they are usually not used to maintain private and public lawns. This characteristic has been changing, however, in the past decade through rapid expansion of golf courses and other tourist attractions in some developing countries.

Urban areas: The highest concentration of landscaping problems exists in urban areas, where these problems also contribute to smog, air pollution, and water usage and pollution.

Runoff: Rainwater or watering runoff from grassy areas enters sewers, which often dump directly into rivers or oceans. Runoff from golf courses or other public or private lands that have had pesticides or fertilizers applied is especially a problem because of the large area they cover and the fact that these lands often are adjacent to rivers, lakes, and oceans. They can contribute to serious nonpoint pollution.

L

State of the Earth

There are nearly 100 million lawns in the United States, using more than 40 million pounds of toxic chemicals to preserve them. Acre for acre, lawns in the United States are covered with 4 to 8 times the chemicals and pesticides that farmland receives. Of the dozens of pesticides used on lawns 1/3 can cause cancer or birth defects; damage the nervous system, kidneys, or liver; or affect reproduction. Roughly 1/2 million gallons of water are needed each week to keep grass green in the United States. Yearly, the total amount of time spent mowing grass in the United States totals 6 to 8 billion hours.

Golf courses use up to 126 different pesticides. Each year 12 million pounds of pesticides are used on U.S. golf courses. Golf courses use an average of 18 pounds of pesticide per acre compared with 2.7 pounds per acre on farms. Only 20 states require clubs to post notices of pesticide use for players. Golf courses in New Mexico used up to 1 million gallons of water a day; the Palm Springs Country Clubhouse uses 430 million gallons of water a year. Golf courses take up about 6 million acres worldwide; there are 14,000 golf courses in the United States.

The U.S. Environmental Protection Agency (EPA) has done a study that found using a riding mower for 1 hour creates as much smog pollutant as driving a car for 20 miles; 1 hour of using a power lawn mower is equivalent to driving 50 miles; 1 hour of using a leaf blower is equivalent to driving 100 miles; 1 hour of using a chain saw is equivalent to driving 200 miles. The 80 million lawn mowers in the United States produce pollution equivalent to 3.5 million newer cars. Electric mowers are more than 70 times less polluting.

More than 90% of pesticides have not been tested for long-term health effects. In Iowa, Minnesota, and Wisconsin, nearly 1/3 of all wells tested are toxic with pesticides; in a survey of midwestern streams, 90% tested had herbicide residues. Pesticides have fouled drinking wells in more than 40 states in the United States; EPA estimates 10% of all drinking wells have pesticide residues. Dozens of pesticides now in use in the United States are carcinogenic in animals; the weedkiller 2, 4-D increases cancer in dogs.

Turf grasses are not native to America. "Weeds" are often native species that have evolved to survive in the local conditions. Natural, beautiful meadowlands could replace lawns; they would need no care, and they would support natural wildlife species. Apple orchards are one of the best ways to make meadowland fertile; they need little care and can survive even in poor soil conditions.

Human Impact

Pesticides: Pesticides can enter the food chain and water supplies and cause cancer, neurological problems, kidney and liver failure, heart problems, skin disease, and death.

Fertilizer: Fertilizer adds nitrates, which cause cancer, to water supplies. Fertilizers also ruin lakes and rivers by eutrophication. Mining for fertilizer increases radon releases and thus local background radiation levels.

Air pollution: Smog causes respiratory diseases, aggravates asthma, causes eye and skin irritations, and increases cardiovascular problems.

Water depletion: Water used on lawns, golf courses, and other nonessential areas can exacerbate drought conditions, potentially leading to crop failures. There is already a struggle for water between farmers and other community needs.

Individual Solutions

- Xeriscape—choose native grasses. They are hardier and require less water and care. Plant less grass; use rock gardens and low-growing flowering plants.
- If choosing nonnative grasses, choose varieties resistant to local pests and weather conditions.
- Aerate your lawn; air is taken naturally into your lawn by earthworms, which pesticides kill.

- Water sparingly, in the early morning; overwatering carries away nutrients.
- Use only natural organic fertilizers; have your soil tested for acidity and nutrient imbalances. Consult a natural organic lawn care company with questions or problems.
- Don't use pesticides or herbicides; pesticides on your shoes can dry indoors and enter the air you breathe. Never walk barefoot on chemically treated lawns.
- Contact the U.S. Department of Agriculture's Soil Conservation Service for a list of plants you can grow in your garden to repel pests.
- Compost your clippings, and use them as mulch on your lawn or in your garden.
- Cut your lawn less frequently; let grass grow longer (uses less water). Prevent too much thatch buildup; thatch is the dead grass on top of the soil, which needs to be raked or cut if it gets too thick. It can choke your grass and kill it.
- If you buy a mower, buy an electric or a reel (push) mower.
- Never dump oil and other residues on the soil or grass, in the sewer, or down the drain; contact local officials for hazardous waste collection sites.
- If you are a golfer, ask if your club posts notices when chemicals have been used; demand that your course use natural remedies that do not harm wildlife and soil organisms.
- Call or write your legislators to demand long-term health testing by independent firms of all lawn and landscaping chemicals and to demand that pesticides and herbicides be kept off our soils and out of our water systems until such testing is completed.

Industrial/Political Solutions

What's Being Done

- The EPA is supporting integrated pest management to minimize pesticide use in agriculture; this emphasis should be extended to landscaping.

- Many states have banned yard compost from garbage pickups, as well as burning of compost; a number of states have aggressive composting policies and large compost sites.
- In California pollution from nonvehicular engines must be cut 46% by 1995 and another 55% by 1999. Federal regulations will be developed and implemented by 1996.
- Watering bans have been put in place in some areas with water shortages; cities are restricting lawn care chemicals near waterways.
- Organic lawn care companies are providing safe alternative methods for keeping landscapes healthy.
- Some individuals have planted natural varieties of wildflowers and other plants in their lawns in urban areas; they have been challenged in court by their neighbors and have sometimes won and sometimes lost.

What Needs to Be Done

- Enforce mandatory testing of all pesticides and herbicides for long-term health effects; industry should pay for the studies to be done by independent organizations, and all chemicals should be withdrawn from the market until proven safe.
- Place limits on watering lawns, golf courses, and other public lands.
- Landscaping and lawn care equipment should be regulated for air pollution emissions.
- Increase water fees to encourage water conservation.
- Ban yard compost from garbage pickups; mandate that all states compost organic garbage.
- End artificial fertilizer usage on nonessential areas, especially urban landscaping areas.

Also See

- Pesticides
- Sustainable agriculture

L

Lead (Pb)

Definition

Silvery gray heavy metal that is indestructible and cannot be changed into something else.

Lead is a highly corrosion-resistant metal. This quality is why lead was widely used in pipes, solder, faucets, and plumbing until this use was banned in 1986.

Environmental Impact

Environmental buildup: Lead does not biodegrade and thus tends to accumulate in the environment. When burned, lead can be carried by the wind over long distances. Airborne lead accounts for most of the lead that is spread into the environment.

Plants and animals: In soil and water lead can be taken up by animals and thereby enter the food chain. Lead can bioaccumulate in animal tissues and cause reproductive problems, birth defects, kidney and liver diseases, cancer, and death of young and adult animals. In general, lead is not taken up by plants; it binds with the soil or with plant roots.

Major Sources

Mined: Lead has been in widespread use for thousands of years. It is mined, often with zinc and copper. Mining wastes can have high lead content.

Gasoline: Lead is still used in gasoline as an antiknock additive for cars in developing countries. Because of its use in car fuel, lead is a common atmospheric pollutant and is found worldwide in dust, oceans and lakes, and soils. It can fall out of the air in rain or soot particles and contaminate water and soil. It is especially common in urban soil, dust, and water.

Industry: Lead is used in paints, inks, alloys, lead-soldered food cans, ceramic glazes, lead-acid batteries (majority of current usage), building materials, and plastic production. It is produced during the smelting of zinc and the burning of coal and is a common component of sewage sludge and industrial waste water. Lead is also one of thousands of chemicals found in tobacco smoke.

Incineration: Lead is emitted during garbage incineration.

Water systems: Old piping and solders made extensive use of lead, which is now leaching into drinking water systems.

State of the Earth

The majority of all lead use is for lead storage batteries; 15% is used in gasoline additives; 20% is used in metal products such as solders and bearings; and 6% is used in ceramics and pigments. The United States is the highest consumer of lead, at well over 1 million tons a year, and the second-highest producer after Australia. Worldwide, 35% to 60% of lead is recycled in batteries and scrap metal.

Lead is toxic in microgram quantities (there are 454 million micrograms to a pound). The concentration of lead in some incinerator ash is so high that the ash is as dangerous as lead-based paint, which was banned in the United States in 1973. One large incinerator can legally put out thousands of pounds of lead annually in the air and in its ash. There are current plans to use incinerator ash in road aggregate and other construction, where it would eventually break down and end up in the water, soil, and air.

Though leaded gasoline is banned in the United States, it is still in use in many developing countries since it is cheaper than unleaded gasoline; its use is rising as the number of vehicles rises. China, Indonesia, Korea, Sri Lanka, and Thailand use leaded gas; China and Korea also release large amounts of lead through coal combustion and industrial pollution; Japan banned leaded gas in 1988, and incineration is its major source of lead emissions. In Mexico many children have lead levels in excess of World Health Organization standards; Mexico is now requiring all new cars made in Mexico to use unleaded gasoline. Areas where unleaded gas has been banned have seen large reductions in airborne lead.

Leaded paint and tap water with lead are the two greatest sources of lead in the home; it is estimated that 3 million tons of old leaded paint is in the homes of 57 million Americans, and 74% of all houses built before 1980 contain some lead paint. The U.S. Environmental Protection Agency (EPA) suspects that 40% of all medium-sized water systems (3,000 to 50,000 people) exceed safe lead levels (0.015 milligrams per liter) and considers lead the number one health threat to children. One in 5 Americans drink lead-rich tap water; 3 to 4 million children are at risk from lead in their water. Water sources cause 20% of all lead poisoning. In the United Kingdom more than 5% of tap water tested showed high levels of lead, which leached from the lead piping in homes tested. Citizens in a number of industrial countries have 500 to 1,000 times more lead in their bodies than early humans did. Lead has been in use since 2500 B.C.

In the environment, lead birdshot was responsible for large amounts of lead, which waterfowl swallowed while feeding. Officials noted the incidence of lead-related diseases and banned lead for this use. Lead has been used in sinkers for fishing, and some believe this use also endangers waterfowl. Lead from industrial emissions has been found in Antarctic waters; levels of lead in the ocean have risen since leaded gas use began.

Human Impact

Health: Lead can be breathed through the air (in old paint dust and air emissions), taken in through water (having leached from lead pipes, faucets, solder, and soil), or ingested in food (vegetables, fish, and meat). Workers in smelters and battery plants have high exposure. Lead causes serious health effects at both high and low exposure levels; there is no safe level of exposure for lead. It can accumulate in the body over months or years. Lead can cause headaches, dizziness, muscle and joint pain, hypertension, constipation, and abdominal cramping. Lead can also interfere with enzyme action and cause kidney damage, high blood pressure in adults, reproductive problems in women and men, and cancer. Lead can be absorbed into the bones, and if blood levels drop, the bones can release more lead into the bloodstream.

Children: Lead in dust, paint, and soils is the number one health hazard for children. A powerful neurotoxin, lead can cause brain damage, seizures, and neurological disorders such as learning disabilities and retardation even at trace levels. These risks are especially great for children, who absorb lead at 4 to 5 times the rate of adults (because their nervous systems are still developing). Thus pregnant women should avoid lead whenever possible.

Individual Solutions

- Take any old paints to hazardous collection sites; contact the EPA or local officials to find out where.
- Support recycling; oppose incineration.
- Recycle batteries. Avoid disposable battery products; buy electric or hand-powered products instead.
- Minimize the use of plastic.
- If you live in an older house, have your wall paint tested; lead can wear off as a dust and be inhaled. Have it professionally removed.
- Buy nonpetroleum-based paints.
- Test your water, and have professionals remove any old lead piping. Submersible brass well-water pumps can also leach lead.

Industrial/Political Solutions

What's Being Done

- There is a global move away from lead use in any products.
- Some European countries are limiting heavy metal use.
- Household hazardous waste is being collected in the United States and in many industrial countries.

- Leaded gas has been phased out in Japan, the United States, and much of Europe; Mexico City has introduced unleaded gasoline; all new cars made there must use it.
- Lead use in new drinking system construction has been phased out in the United States; permissible lead levels in public water have been lowered.
- Standards for airborne lead emitted by incinerators and secondary smelters have been tightened; pollutant emission standards for incinerators have been toughened in a number of countries. The EPA instituted tougher heavy metal emission regulations for hazardous waste incinerators in 1993.
- Lead in paint and food cans has been reduced and eliminated.
- The mining of virgin lead is being discouraged in the United States by the government.
- Recycling of lead, especially in batteries, is being encouraged by the U.S. government. A few major corporations are instituting recycling programs for lead batteries.
- The use of substitutes for lead is being encouraged by the U.S. government.
- Some states have set standards for lead-based paint removal.
- Some states require home sellers to disclose environmental toxics like lead and radon.
- The U.S. Department of Energy is studying reducing lead from coal plants.
- By July 1, 1994, Illinois, Maine, New Jersey, New York, and another dozen states will ban sales of products such as inks, dyes, pigments, adhesives, or stabilizers that intentionally have heavy metals (lead, cadmium, mercury, chromium) added; unintentional heavy metal content higher than 100 parts per million will be banned July 1, 1996.
- Under the 1992 U.S. Lead and Copper Rule, lead monitoring requirements were set for all public water utilities for large, medium, and small suppliers; if lead levels are over 0.015 milligrams per liter and/or copper levels are over 1.3 milligrams per liter in 10% of tap water samples, the public utility must install corrosion control within

24 months, educate the public, and replace service lines if corrosion control fails to work.

What Needs to Be Done

- Phase out industrial use of lead.
- Tax lead to encourage recycling.
- Prevent industry from disposing of lead in wastewater and air emissions.
- End fossil fuel subsidies; support renewable energy.
- Phase out incineration, and support recycling.
- Treat incinerator ash as hazardous waste; ban its use in any construction material; mandate battery removal from waste before incineration.
- Help developing countries gain access to unleaded gasoline.
- Worldwide, ban the use of lead in pipes and other areas where it is a direct source of environmental pollution. Require overhaul of water supply systems with lead piping, regardless of lead levels in the water.
- Keep lead out of household products, packaging, and other nonessential product uses.
- Set national standards for lead paint removal.

Also See

- Heavy metals

L

Livestock problems

Definition

Environmental damage caused by raising increasing numbers of cattle, sheep, pigs, and poultry worldwide.

Environmental Impact

Soil degradation: Overgrazing of cattle or livestock can remove so much plant material that the land becomes dried out. Their hooves trample and compact soils further, so water is not absorbed and instead runs off with eroded soil. This soil degradation is a major cause of desertification and erosion worldwide. Land ruined by overgrazing cattle in arid regions also contributes to loss of wildlife in those areas.

Deforestation: Cattle ranching is a major cause of deforestation, especially in the tropics, where forests are cut or burned to make room for increased cattle grazing. Soils of deforested land can quickly lose their nutrients and erode. Deforestation is a major contributor to extinction and loss of biodiversity.

Freshwater degradation: Manure runoff into rivers, lakes, streams, and oceans causes increased algae growth, leading to eutrophication. Oxygen is used up, and fish and aquatic plants die. Streams, rivers, and lakes can become clogged. Water usage for grains and feed for livestock further depletes water resources.

Greenhouse effect: Livestock contribute significant amounts of methane, a greenhouse gas, to the atmosphere through belching. Termite populations have exploded in deforested areas, and they are also a significant source of methane (which is produced during digestion in their digestive tract). Cut or burned trees release carbon dioxide, which is the major greenhouse gas. The greenhouse effect is part of global warming.

Wetlands: Wetlands have been drained for cattle grazing. This practice destroys wildlife habitat and groundwater filtration and recharge areas.

Major Sources

Cattle, pigs, sheep, goats, horses, and poultry (chicken and turkeys): Livestock are raised in nearly every country. Brazil, India, the former Soviet Union, and the United States account for roughly 1/2 of the world's cattle. China raises 40% of the world's pigs.

Manure problems: Manure problems occur when livestock is left to forage near streams and rivers; if the manure is not collected, the rain washes the manure into the water. If the soils become too saturated with manure, the excess phosphates and nitrogen percolate to groundwater. This undesirable percolation can also occur if farmers apply chemical fertilizers after spreading manure. Manure pollution is a problem in the United States and most countries with cattle production.

Deforestation: The loss of rain forests due to cattle ranching in equatorial countries has gained world attention. Temperate countries like the United States, however, have cleared just as much land for cattle. Rain forest deforestation for cattle production is still occurring at a rapid rate; in developing countries this process is occurring as living standards are raised and meat consumption increases—meat is often considered a status symbol. Termite growth as a result of deforestation is a problem mainly in tropical areas.

Soil degradation: Soil degradation due to overgrazing is a problem especially in arid and semiarid regions in Africa, Asia, Australia, and the western United States. Desertification can be a problem in any area if overgrazing occurs.

L

State of the Earth

More than 250 million acres of forest in the United States have been cleared to raise cattle; 4.2 million cattle currently graze on 280 million acres of U.S. public range. Cattle ranching is one of the major reasons for tropical deforestation and thus extinction; the United States imports about 300 million pounds of meat from Central and South America yearly. Cattle fencing in Africa threatens the last migration routes of animals such as wildebeest, buffalo, and zebra.

Overgrazing is a significant problem in Africa, Asia, Australia, Mexico, South America, and the western United States. More than 1/3 of the world's degraded soils are the result of overgrazing.

More than 1 1/4 billion cows worldwide are belching more than 70 million metric tons of methane into the air every year; termites may be releasing 150 million tons of methane each year. Methane contributes to nearly 1/5 of the greenhouse gases.

Burning methane derived from composted animal dung (in biogas digesters) is one of the principal forms of energy in some developing countries. Farm manure use in biogas digesters to supply biogas is being studied as a supplementary fuel for Illinois farms.

About 176 million tons of meat are produced for consumption yearly. Meat production requires 3 times as much fossil fuel as vegetable and grain production. One pound of wheat uses 25 gallons of water; one pound of meat uses 2,500 gallons. Livestock production consumes more than 1/2 of all water used in the United States; it uses 30% to 40% of world grain consumption and 70% of U.S. grain consumption; 80% of U.S. corn goes to livestock and poultry. The United States produces 25% of world beef, 11% of world pork, 40% of world corn, and 50% of world soybeans. The United States is ranked third in beef consumption at 97 pounds of beef per person per year; Uruguay consumes 123 pounds per person; and Argentina 154 pounds. Meanwhile, 20 million people die from malnutrition yearly.

India has 200 million cattle; Africa, nearly 200 million; the United States and the former Soviet Union, about 100 million each. China has nearly 350 million pigs. The world's farms must feed more than 5 1/2 billion people; more than 3 1/2 billion cattle, pigs, sheep, and goats; and more than 10 billion chickens. About 1/4 of the land, worldwide, is used for livestock pasture; 1/4 of all U.S. land (more than 1/2 of all farmland) is used for grazing livestock, mainly cattle.

Healthy animals produce 4 to 5 times their body weight in dry dung each year. It takes only 60 pigs and 40 cows together to produce 4,000 pounds of manure daily and nearly 1.5 million pounds yearly. Minnesota's cattle and pig feedlots annually produce manure equivalent to the human waste produced by 40 million people; the chicken and turkey feedlots produce manure equivalent to the human waste of 20 million people. U.S. livestock is estimated to produce up to 2 billion tons of manure yearly; India's cattle produce 700 million tons of manure annually—1/2 is used for fertilizer, the rest for heating. In India most cattle are fed foodstuffs that are inedible for humans and thus do not compete with people for grain.

In the United States cattle are given growth hormones, anabolic steroids, antibiotics, and corn and soybean feed saturated with pesticides. More than 90% of feedlot cattle are given growth hormones; more than 40,000 feedlots are operating in 13 cattle-feeding states. Beef has the highest herbicide contamination of any food, is third highest in insecticide contamination, and ranks second to tomatoes as a food cancer risk. Prostate cancer in men has now been linked to the fat in red (even lean) meat; the National Research Council has estimated that beef represents at least 10% of all food cancer risk to consumers.

Human Impact

Food: Land degraded by livestock results in less agricultural (or productive ranching) land and thus less food. The grain used to feed cattle and other livestock could feed 1 billion people. Freshwater pollution reduces fish resources.

Toxics: Current livestock practices often use antibodies and growth hormones, and the feed given to livestock often contains pesticides and other toxic chemicals. Many of these chemicals have not been tested for long-term health effects on humans and can increase cancer risks. The fat in red meat causes prostate cancer.

Drinking water: Manure can contaminate drinking water with toxics and pathogens. Wetlands loss also reduces drinking water and its purity, since wetlands act as filters when water drains through them to aquifers.

Medicine: Deforestation due to cattle ranching results in losses of plants that might be capable of healing current or future health problems.

Energy usage: Meat production requires many times more energy than would be used to create an equal caloric value of grain or other vegetarian crops.

Individual Solutions

- Eat less meat, especially red meat.
- Eat organically grown meats and poultry, which have not been given chemicals or growth hormones, and eat free-ranging meats and poultry, which are not kept in feedlots. These meats can be found at food cooperatives or through local, small farmers.
- Call or write your legislators to demand an end to the importing of rain forest beef and to require any imported rain forest beef used in any product to be labeled as such.
- Oppose the use of growth hormones and other chemicals in cattle, pigs, and livestock until they have been studied by independent researchers for long-term health effects and proven safe for the animals they are used on.

Industrial/Political Solutions

What's Being Done

- In a number of countries, including the United States, local programs help fund or educate farmers to better manage livestock problems.
- Sustainable farming practices are being taught and encouraged worldwide.
- Manure is being used, as a biomass fuel, to generate biogas.

- U.S. farmers can be fined or criminally prosecuted for allowing manure runoff to enter rivers, lakes, and streams.
- Proposed incentives of lower grazing fees may be offered to ranchers who take steps to maintain public grazing land.
- To limit fencing and overgrazing, research is being conducted on electronic eartags that give cattle a mild shock whenever they enter forbidden grazing areas.

What Needs to Be Done

- Increase the use of manure for fertilizer and fuel; improve enforcement of laws against manure runoff from farms.
- Ban beef grown on burned or cut rain forests.
- Mandate testing for long-term health effects for all chemicals given to animals whose meat is consumed; testing should be done by independent organizations.
- Minimize grazing in arid and semiarid regions; limit the number of cattle grazed per acre.
- Increase training in sustainable agriculture practices.

Also See

- Deforestation
- Soil degradation
- Sustainable agriculture

Mercury (Hg)

Definition

Silvery-white liquid that is a heavy metal.

Mercury is indestructible, does not biodegrade in the environment, and can evaporate at room temperature.

Environmental Impact

Environmental buildup: Mercury concentrations continue to build up over time in the environment, especially in salt and fresh water. Airborne mercury can travel thousands of miles.

Animals: Bacteria in the water convert mercury to methylmercury, which is taken up by fish and stored in fish muscle. Some fish can also metabolize inorganic mercury into methylmercury. Mercury is toxic to fish, especially freshwater species. Animals at the top of the food chain (big fish, birds of prey, and other predators) that eat fish tend to bioaccumulate and biomagnify mercury and thus have high concentrations of mercury in their fatty tissues. It causes thin eggs, reproductive failure, neurological problems, cancer, and death in young and adult animals. Mercury also can cause behavioral changes in animals and affect reproductive patterns. Mercury builds up in shellfish, which filter it out of the water and concentrate it in their tissues.

Acid rain: Acid rain can leach mercury out of the ground, allowing it to run into rivers and lakes. Acid rain also increases water acidity, which can help convert inorganic mercury into more dangerous forms.

Major Sources

Mining: Mining releases mercury into the environment during excavating, processing, and smelting. Pan mining for gold is also a major source of mercury emissions.

Fuels: Coal and crude oil both contain mercury, which is released when these fossil fuels are produced and burned. Mercury is also used to process nuclear fuels.

Incineration: Mercury is emitted during incineration of solid waste; this process is one of the largest sources of environmental mercury emissions.

Products: Mercury is used in plastics, batteries, electronic devices, paints, brushes, air conditioner filters, dental amalgams (50% of "silver" amalgrams are mercury), household appliances, children's toys, and thermometers. Industrial use of mercury for products adds to emissions during product production and later disposal.

Seafood: Fish and shellfish are the most common organisms that bioaccumulate mercury in the environment.

State of the Earth

In the Amazon region huge amounts of mercury have been released into the waterways during the past 2 decades by pan mining for gold. In a process called amalgamation, mercury is combined with gold in the silt to separate it out; excess mercury then escapes into the water. Gold rushes are also occurring in Bolivia, China, Colombia, Ecuador, Ghana, Guyana, Indonesia, the Philippines, Papua New Guinea, Venezuela, and Zimbabwe. It is estimated that more than 1/2 million miners in the Amazon have been poisoned with mercury; 650,000 miners release 90 to 120 tons of mercury a year in the Amazon, and mercury-contaminated fish have been found 360 miles downstream from miners. Twenty-five tons of mercury are released on one island alone by miners in the Philippines. Several million pan miners produce 1/4 of the world's gold.

M

The United States is the largest consumer of mercury, at more than 1,000 tons a year. The former Soviet Union is the largest producer; in some areas of Russia 70% of the freshwater fish have high levels of mercury.

One large incinerator can legally put out thousands of pounds of mercury each year; 13 million pounds of mercury from industrial emissions drop out of the atmosphere yearly in rain.

It takes only very small amounts of mercury to contaminate a whole lake. A concentration of even 1 part per million of mercury in fish is considered toxic; fish and fish products are responsible for nearly 95% of human mercury exposure. Since mercury can be airborne, mercury-poisoned lakes, and thus fish, are found everywhere, even in pristine wilderness areas such as the Boundary Waters Canoe Area in Superior National Forest of Minnesota and in remote lakes in Canada, Finland, and Sweden, as well as Florida and Wisconsin. Loons with mercury contamination in Minnesota appear to have lost courting instincts and to have developed reproductive problems.

In the early 1950s a chemical company, owned by Chisso Corporation, dumped methylmercury into Minamata Bay, Japan, near a small fishing village. This pollution left 800 people with a disfiguring paralysis and thousands more affected with other health problems. The villagers ate fish that had accumulated the mercury, and the result was blindness, brain damage, deformed babies, and death. It took nearly 30 years, from 1953 to 1983, to see the full effects of the poisoning.

The expression "mad as a hatter" originated from the fact that hat makers in the past used mercury in felt hats, which caused irreparable damage to their nervous systems.

Human Impact

Inorganic mercury: Inorganic mercury can be breathed in, ingested, or absorbed through the skin. It is not as toxic as other forms of mercury, but in large enough doses or in accumulations over time, it can cause neurological disorders, kidney problems, genetic changes, and birth defects. Airborne mercury can enter the bloodstream quickly.

Organic mercury: Compounds such as methylmercury are much more dangerous than inorganic mercury, for they are fat soluble and easily penetrate the nervous system and brain, disrupting cellular functions; affect coordination; and cause tremors, slurred speech, convulsions, coma, and death. The usual source is contaminated fish and seafood.

Children: Children are more susceptible than adults to mercury and its effects. Mercury can cause irreversible neurological damage to fetuses; cerebral palsy in infants has been linked to pregnant mothers eating mercury-contaminated seafood.

Individual Solutions

- Oppose incineration and recycle; practice energy conservation.
- Oppose fossil fuel subsidies.
- Don't put anything with mercury in it (such as batteries, fluorescent lights, and thermometers) into the garbage; if unsure, call the U.S. Environmental Protection Agency or a local environmental group.
- Don't use disposable batteries; avoid products that do; buy electric or hand-powered products.
- Be aware of fish advisories about high levels of mercury; consult your local Department of Natural Resources for information.
- Decide if you wish to see a dentist that does not use, or removes, mercury amalgams (see *It's All in Your Head: Diseases Caused by Silver-Mercury Fillings*, by Hal A. Huggins).
- If working with mercury, be aware that it evaporates easily, and use precautions.

Industrial/Political Solutions

What's Being Done

- Batteries are being recycled in the United States and many European and other industrial countries.
- Household hazardous waste is being collected in the United States and a number of European and other industrial countries.
- Europe is phasing out mercury in most consumer batteries.

- Incinerators in a number of countries have stricter pollutant emission standards than the United States; the EPA toughened heavy metal emission regulations for hazardous waste incinerators in 1993.
- The U.S. Department of Energy is studying reducing mercury emissions from coal plants.
- The U.S. Food and Drug Administration (FDA) does not allow fish with more than 1 part per million of mercury to be marketed commercially; a number of experts believe this level of exposure is still dangerously high.
- By July 1, 1994, Illinois, Maine, New Jersey, New York, and another dozen states will ban sales of products such as inks, dyes, pigments, adhesives, or stabilizers that have heavy metals (lead, cadmium, mercury, chromium) intentionally added; unintentional heavy metal content over 100 parts per million will be banned July 1, 1996.
- A number of U.S. states have set a deadline of 1996 for mercury-free batteries.
- Mercury has been eliminated from indoor paints in the United States; labeling is required for outdoor paints.
- Fluorescent bulbs and high-intensity lamps are considered hazardous waste in the United States for their mercury content and are being recycled.
- More dentists are replacing the use of silver-mercury amalgam fillings with composites, which last just as long and are less toxic.

What Needs to Be Done

- Ban mercury in batteries, and limit the content of mercury in other products. Make built-in batteries removable. Require uniform product labeling on batteries and products that contain mercury so they are not thrown away.
- Phase out incineration.
- Force existing incinerators to presort their garbage so that mercury is not burned and emitted. Treat all incinerator ash as toxic hazardous waste; ban its use in any construction material, and keep it out of solid waste landfills.

- Phase in zero-discharge mercury emission limits for industry, utilities, and incinerators; phase out industrial use of mercury.
- Keep sewage separate from industrial wastewater.
- Keep mercury out of household products; find alternatives that are safer.
- Pursue world agreements to limit mercury use and environmental exposure.
- Prevent industrial nations from selling outdated mercury-based technologies to developing countries.
- End fossil fuel subsidies; increase support for renewable energy, and energy conservation and efficiency.

Also See

- Heavy metals

M

Methane (CH₄)

Colorless, odorless hydrocarbon (compound of carbon and hydrogen) gas that burns easily.

In nature methane forms the bulk of natural gas, which is a fossil fuel. This methane formed from the bacterial decomposition of plant and animal matter over millions of years. Methane is still produced by natural sources every day, worldwide.

Environmental Impact

Greenhouse effect: Methane is a greenhouse gas and adds significantly to the greenhouse effect and possibly to global warming.

Atmosphere: Methane has a complicated role in the atmosphere. Naturally occurring hydroxyl radicals (OH) in the atmosphere combine with and eliminate methane, but excess methane and carbon monoxide (given off by cars) are lowering OH levels; thus, as OH decreases, methane increases. Methane also has a role in the stratosphere, acting as a sink for chlorine (thus lessening ozone depletion), but also breaking down to form hydrogen and water vapor (which can form ice crystals and speed up ozone depletion.)

Cleaner fuel: When methane is produced or captured from a landfill or biogas digester and used as a heating fuel, it burns cleanly in comparison with other fuels. When burned, natural gas, which is primarily methane, produces no soot and smaller amounts of hydrocarbons and carbon monoxide than gasoline. It yields, however, more nitrogen oxides, which are a factor in acid rain. Acid rain has acidified lakes and destroyed fish, animal, and plant populations.

Natural gas: Methane is the principal component of natural gas, which is found in reserves similar to oil reserves in the earth's crust. Methane is also a by-product of the production, transportation, and use of natural gas.

Wetlands: Underwater rotting plant material creates methane called marsh gas. Bacteria in rice paddies, a form of wetland, also create methane.

Livestock: Belching cows and horses give off large amounts of methane.

Biomass: Decomposition of organic materials creates methane; these materials are mainly plant matter, but they include solid waste in landfills, manure, and sewage. Biomass burning of savanna grasslands, agricultural wastes, forests, and fuelwood increases methane production in soils.

Termites: Termites feed on the remnants of slash-and-burn deforestation, producing large quantities of methane in their digestive tracts.

Industrial emissions: Industrial emissions can include methane.

Car emissions: Burning fuel emits methane.

Coal mines: Methane is often found in coal mines.

Environmental sinks: Huge quantities of methane are locked up in ice in hydrates (substances bound with water molecules) in the tundra and in the continental shelves of the oceans. Methane is also held in forest soils and released during deforestation.

Atmosphere: Methane is found naturally in the atmosphere and is removed by naturally occurring hydroxyl radicals (OH).

State of the Earth

The U.S. National Aeronautics and Space Administration (NASA) says methane is increasing in the atmosphere 3 times faster than carbon dioxide; atmospheric methane levels have tripled over the past 3 decades. About 500 million tons of methane are emitted each year, worldwide, from bacterial decomposition and fossil fuel burning. Natural yearly emissions of methane total 160 million tons: wetlands, 120 million tons; lakes and rivers, 20 million tons; oceans, 10 million tons;

and termites, 10 million tons. Human sources emit 340 million tons: fossil fuels, 100 million tons; cattle, 80 million tons; rice fields, 50 million tons; biomass burning, 30 million tons; landfills, 30 million tons; animal waste, 30 million tons; and domestic sewage, 20 million tons. Atmospheric hydroxyl radicals destroy 420 million tons, soils absorb 30 million tons, and the stratosphere removes 10 million tons.

More than 1 billion cattle are belching about 80 million metric tons of methane into the air every year, at a rate of 300 quarts each day per cow. New Zealand's 80 million sheep produce 2 to 3 million gallons of methane each week. It is estimated that U.S. farms could produce 1.3 trillion cubic feet of methane for fuel annually.

Termite populations are expanding rapidly due to slash-and-burn deforestation. Termites live on plant materials in earthen mounds; the termites give off methane as they digest their food. Researchers estimate that there are now several hundred to 1,000 pounds of termites for every person on the planet. One termite mound can give off more than 1 gallon of methane each minute.

Landfill methane production is being tested and is already used in Europe and the United States. Burning methane derived from composted animal dung in biogas converters is one of the principal forms of energy in some developing countries. Small biogas digesters are in wide use in India and China, and China is now developing larger systems. In one city, Nanyang, 20,000 households will replace their cooking coal with methane generated from 2 large biogas digesters. The University of Illinois Swine Research Center is studying the use of farm manure in biogas digesters to supply biogas as a supplementary fuel for Illinois farms.

Methane contributes to about 1/5 of the greenhouse gases. It holds in heat 20 to 30 times more efficiently than carbon dioxide. With continued global warming, the permafrost could eventually begin to thaw and enormous amounts of methane could be released from tundra and continental shelf ice, doubling current atmospheric levels.

Human Impact

Global warming: Rising methane concentrations in the atmosphere are probably contributing to increased global warming.

Acid rain: Acid rain from natural gas use can destroy fish resources and cause crop losses.

Worker safety: Methane combined with air can be very explosive and has caused numerous mining accidents in the past. Precautions are now taken to minimize such accidents.

Natural gas: Natural gas is often stored as a liquid, liquefied natural gas (LNG), for ease of storage and transport. LNG can be very volatile, vaporize, and explode into a ball of flame with a single spark. Accidents involving LNG have caused deaths in the United States and elsewhere.

Individual Solutions

- Eat less red meat; worldwide cattle production generates significant levels of methane and is also responsible for deforestation.
- Don't buy tropical hardwoods such as tropical teak, rosewood, mahogany, ramin, lauan, and maranti. Instead use oak, cherry, birch, pine, or maple.
- Investigate using solar or wind energy for your home energy needs.
- Support reforestation efforts and an end to deforestation worldwide.
- Practice energy conservation.
- Recycle, and reduce waste to cut landfill methane production.

Industrial/Political Solutions

What's Being Done

- Methane from biomass and landfills is being used by more countries as a fuel source.
- Help is being given to some countries to slow fuelwood use, and thus deforestation.
- Numerous strategies are being used to prevent deforestation, especially in rain forests.

- Natural gas is being used more often as an alternative to coal and oil; this may increase methane releases.
- The United States offers tax incentives for solar, wind, and geothermal energy.

What Needs to Be Done

- Initiate a worldwide ban on beef exports from rain forest–grown cattle; ban imports of rain forest hardwoods.
- End fossil fuel subsidies; phase in zero-emission vehicles.
- Increase investments in renewable energy sources and energy conservation instead of in fossil fuels.
- Make manufacturers responsible for the recovery of packaging material to reduce landfill waste; this step has already been taken in Europe.
- Help developing countries gain renewable energy technology.
- Initiate worldwide agreements and efforts to end deforestation and to encourage sustainable use of all forest resources.

Also See

- Biomass energy
- Greenhouse effect
- Livestock problems

Mining

Extraction of raw materials from the earth by drilling holes, digging open pits (open-cast mining), shafts, or strip-mining (removing layers of soil from the surface down until the ore is reached).

Some special cases of mining not considered in this section are gold pan mining, which occurs in rain forests and contributes large quantities of mercury pollution; "mining" of landfill solid waste and incinerator ash for the raw material they contain; and drilling for geothermal energy.

Some materials are mined pure, and some are mined as alloys (mixtures of two or more metals). Raw materials can be either organic, such as fossil fuels, or inorganic, such as metals, gems, and minerals. Minerals are used in almost every phase of our lives and in almost every product that is manufactured. Fossil fuels provide the majority of the world's energy.

Raw material consumption is increasing as populations grow, as developing countries increase their own industries, and as new technologies are developed. A number of minerals and raw materials are rapidly being used up worldwide.

Environmental Impact

Soil degradation: Land mining uses large amounts of earth and ore to obtain often very small quantities of mineral. The result is huge amounts of waste, such as tailings (waste left over when the ore is removed), which is often toxic; tailings are frequently buried. Toxics in soils inhibit plant growth and enter the food chain. Oil spills and wastes can render lands unusable. Mining large areas, such as with coal strip-mining, causes erosion and increases the chances of flooding.

Freshwater degradation: Acidic and toxic wastes or silt materials can leach into nearby rivers and groundwater, killing fish and aquatic plants. Water is also sprayed in mines to keep down dust and pumped out if it rains or seeps into a mine; either way it picks up heavy metals, silt, or salts. Discharged brine (water with large amounts of dissolved minerals or salts) increases the salinity of fresh water and can make it unfit for aquatic life. Toxics like heavy metals can enter the food chain; bioaccumulate in predators; and cause behavioral and reproductive problems, neurological problems, cancer, and death in young and adult animals. Mining can release radon, a radioactive gas that can also enter water sources. Using fresh water from aquifers for mine drainage can cause nearby salt water to seep into the aquifers as they are depleted.

Ocean degradation: Oil drilling brings up heavy metals and radioactive materials, and drilling rigs often dump their hazardous wastes directly into the ocean. This practice threatens coral reefs, estuaries and wetlands, fish, and other aquatic animals and plants.

Air pollution: Mining wastes (such as tailings) can be picked up by the wind and carried in the form of particulates, which can range in toxicity, depending on the material being mined. Particulates can add to smog, acid rain, and cloud formation and can attract toxics in the air. Radon released by mining increases environmental local background radiation, which can be taken up in the food chain. Radon causes cancer, reproductive problems, and death in animals. Mining equipment, such as big tractors and processing facilities, also causes air pollution and adds to smog, acid rain, and greenhouse gases.

Deforestation: Sometimes raw resources are located in heavily forested areas, which are destroyed to mine the ore. This deforestation results in extinctions and habitat loss.

Wildlife habitat loss: Ecologically sensitive areas, such as shorelines, tropical rain forests, plains, caves, and wetlands, can be ruined by large-scale mining operations. Mining can also threaten or kill local species by causing overcrowding or destroying their habitat altogether. As pressures to obtain more raw materials increase, the pressures on wildlife habitats also increase.

Major Sources

Minerals and other raw materials are mined in every country in the world in mountains, forests, plains, hills, valleys, lake beds, and oceanic continental shelves. Geologists usually conduct surveys and drilling to test for deposits, which might lie near the surface or be many miles deep. Major raw materials that are mined are:

Fossil fuels: Coal, oil, and natural gas are used for energy.

Heavy metals: Aluminum, arsenic, cadmium, chromium, cobalt, copper, iron, lead, mercury, and zinc are used in thousands of products.

Uranium: Uranium is used for nuclear energy.

Phosphate: Phosphate is used in fertilizers.

Alloys: Mixtures of two or more metals—such as bronze, an alloy of copper and tin—are used in cars, airplanes, and many other products.

Abrasives: Diamonds, sand, and pumice have construction and industrial uses.

Asbestos: Silicates are used in insulation.

Precious metals: Platinum, silver, and gold are stocked for currency support.

Gemstones: Jade, rubies, emeralds, and diamonds are used in jewelry.

Sulfur: Sulfuric acid has more industrial uses than almost any other chemical, including fertilizer and many manufactured products.

Limestone and shale: Limestone and shale are used for cement and lime.

Tungsten: Tungsten is used in light bulbs and drill bits.

Minerals: Fluorspar is used to make chlorofluorocarbons (CFCs). Feldspar is used to make glass and ceramics. Graphite is used in industrial molds and pencils. Iodine is used in animal feeds, inks, and dyes. Potassium is used in fertilizers. Nitrates are used in fertilizers. And quartz is used in watches.

Rock: Granite, gravel, gypsum, marble, and pumice are all used for construction. Salt and sandstone are used to make glass.

State of the Earth

The industrial countries consume more than 2/3 of the 3 most used metals: aluminum, iron, and copper. The 1/4 of the world's population in industrial countries consumes 3/4 of the world's resources. The United States alone holds 1/20 of the world's population and uses 1/3 of the world's raw materials. Known world reserves of copper, zinc, mercury, tin, and lead are all expected to be used up in about 50 years; high-grade phosphorus, in 50 years; and aluminum and iron, in 200 years. Estimated fossil fuel reserves vary widely but are probably at least 30 to 50 years for oil, 60 to 120 years for natural gas, and 130 to 200 years for coal.

Japan imports 100% of its key raw industrial materials; the United States imports more than 25 different types of minerals. Europe has used up many of its minerals, but Russia has large reserves of many minerals, as do Brazil and many developing countries. The U.S. Bureau of Mines has estimated that each American uses 40,000 pounds of new minerals yearly: 50 times what is used by the average Indian and 2 times what is used in other industrial countries. The mineral industry produces several billion tons of toxic waste yearly; 50 billion tons have already accumulated in mining areas in the United States. Cyanide heap leaching, in which huge mounds of ore are soaked with cyanide to leach out the gold, also leaches out heavy metals, nitrates, and other toxics; this method is used in Australia and North America.

The U.S. Bureau of Mines has estimated that 10,000 miles of rivers have been polluted from mining; 424,000 acres of federal lands have been degraded by hard-rock mining. Mining is responsible for 52 designated Superfund sites, including the largest such sites. One of the largest hazardous waste sites in the United States is a 125-mile stretch of the Clark Fork River, contaminated with metals from mining and smelting done at the now-closed Berkeley Pit copper mine in Montana. There are more than 1/2 million abandoned mine sites in the United

States. Abandoned Appalachian underground coal mines have leached acids into thousands of miles of streams, rivers, and aquifers. Many of the streams cannot support aquatic life. In Germany, topsoil and subsoil are saved in strip-mining so they can be replaced after mining, making reclamation easier and more complete.

One of the largest oil spills that ever occurred was a blowout from the Ixtoc 1 oil-drilling rig in the Gulf of Mexico; it dumped close to 4 million barrels of crude oil into the gulf. Several hundred million barrels of toxic oil-drilling wastes are dumped each year in unlined pits throughout U.S. oil-producing states. Often the wastes have high levels of heavy metals and organic chemicals. Millions of barrels of water contaminated by drilling processes are dumped yearly, untreated, into waterways and oceans.

The energy required for raw extraction is often very high—much higher than that required to recycle the same material from already produced minerals; recycling aluminum saves 95% of the energy required to produce new aluminum. It requires 2,500 tons of iron ore and 1,000 tons of coal to produce 1 ton of steel. In the United States 1 ton of gold ore contains 0.01 ounces of gold; 1 ton of copper ore contains less than 1% copper; 1 ton of uranium ore contains 0.2% uranium. In 1992, production of 2,170 tons of gold generated 650 million tons of waste.

Human Impact

Health: Mining often releases heavy metals, acids, and organic chemicals into the environment that can enter the food chain locally and be consumed with food or water. They can cause reproductive problems, neurological problems, birth defects, cancer, liver and kidney failure, and death. Mining-related particulates, smog, acid rain, and air pollution increase respiratory diseases. Miners can suffer from silicosis, a progressive lung disease caused by inhaling silicates, which are a large part of the earth's crust.

Radiation: Mining for uranium and other minerals releases radon, which causes lung cancer. Miners are especially susceptible, as are people living near mining operations. Uranium tailings have often been left above ground. All mining increases local background radiation levels, which can increase cancer risks for the local population.

Indigenous peoples displacement: Local inhabitants of rain forests and other areas have lost their homes to mining operations.

Food: Oil spills, silted water, or toxicity from mining can make cropland acidic and destroy inland and coastal fish and shellfish reserves.

Freshwater degradation: Mining can pollute water resources and make them dangerous to consume.

Recreational areas: Mining operations can acidify or destroy streams, rivers, and forests. Mined land may subside and be unsafe to live on.

Aesthetic values: Mining can destroy wildlife and wilderness areas, along with their biodiversity.

Individual Solutions

- Practice energy conservation; when less fuel is used, less mining is needed.
- Minimize use of plastic and foam materials, which are petroleum products.
- Buy recycled goods; avoid disposables; reuse containers; buy only used or durable goods.
- Use natural instead of artificial fertilizers, which are mining intensive.
- Oppose incineration, which destroys materials and increases raw material needs.
- Investigate solar and wind energy for home and business.
- Support an end to coastal oil drilling and a continued ban on oil exploration in Antarctica and in Alaska's Arctic National Wildlife Refuge.

Industrial/Political Solutions

What's Being Done

- The Antarctic Treaty of 1961 (updated in 1991) prevents any mineral extraction in Antarctica for at least 50 years, until 2041.

- The environmental impact of oil drilling is being taken into account during the permitting process for drilling in coastal areas, Alaska, the Antarctic, and the Arctic.
- Wetlands are sometimes constructed to filter metals from contaminated mining wastewater and to neutralize acidity. This is being researched in the United States and a few other countries.
- In the United States biohydrometallurgy research is being conducted in which biotechnology is used to create microorganisms to recover metals and purify contaminated water.
- Mining companies are slated to be charged royalties for mining gold and metals on U.S. public lands.
- The U.S. Surface Mining Control and Reclamation Act of 1977 requires coal-mining companies to restore strip-mined areas to their initial condition as nearly as possible; this objective is rarely fully achieved.
- The U.S. Coal Mine Health and Safety Act of 1969 reduced risks to miners by requiring better ventilation and safer tunnels; it was amended in 1977 by the Mine Safety and Health Act.
- More U.S. states are requiring recycled content in products.
- Used oil is being recycled in the United States.

What Needs to Be Done

- Rewrite the outdated 1872 U.S. General Mining Law with stringent up-to-date environmental provisions.
- End fossil fuel subsidies.
- Increase investments in renewable energy.
- Phase out incineration.
- Mandate that high-grading practices be ended; high-grading means that the richest areas of an ore bed are depleted, leaving lower-grade ore because the profits from them are not as high. Yet since the destruction of the environment is already done, mining companies should mine the low-grade ore rather than seeking other high-grade ore sites.
- Mandate that by-products of mining be mined as well as the principal

ore. For example, when copper is mined, rhenium, selenium, or tel-
lurium can be by-products. These substances could be stockpiled
while markets are developed, instead of left as waste.

- Limit mining to protect ecologically sensitive areas; end cyanide
  heap leaching.
- Maximize recycling, reuse of minerals, and use of products with recy-
  cled content.
- Back mass transit and light rail, which would lessen fossil fuel use.
- Phase in zero-emission automobiles, like electric cars, which use
  1/10 the energy of internal-combustion vehicles.
- End all offshore oil drilling.
- Phase out uranium mining, nuclear weapons, and nuclear power
  plants.
- Support alternatives to current uses of toxic raw materials such as
  lead and mercury.
- Tax raw materials to encourage efficiency in mining and raw material
  usage; factor the environmental costs into raw material costs to reflect
  true costs.

Also See

- Fossil fuels
- Heavy metals
- Oil

Nitrogen oxides (NOx)

Definition

Various combinations of nitrogen and oxygen molecules, including nitrogen dioxide (NO_2), nitric oxide (NO), and nitrous oxide (N_2O, also called laughing gas).

Nitrogen oxides are stable gases that do not break down quickly. Thus they can accumulate in larger and larger concentrations in the atmosphere over time. Nitrogen dioxide gives the sky a yellow-brown cast.

Environmental Impact

Nitrogen oxides contribute to:

Acid rain: Nitrogen combines with atmospheric water to form nitric acid, which can acidify lakes and soils, destroying fish and small animal populations, as well as forests. Nitrogen oxides can also precipitate as acidic particulates. Acid rain leaches heavy metals into water sources.

Smog: Nitrogen oxides are a major component of ground-level ozone and smog, which inhibit plant and tree growth.

Greenhouse effect: Nitrogen oxides are greenhouse gases and thus increase the greenhouse effect and possible global warming.

Ozone depletion: At high altitudes nitrogen oxides cause ozone depletion, which increases the ultraviolet radiation that reaches earth. This radiation inhibits plant growth and can harm animals. There is also evidence that nitrogen oxides can convert reactive chlorine and bromine to nonreactive forms, which would lessen ozone depletion.

Nitrogen levels in water and soil: There is evidence that increased atmospheric nitrogen, after falling out of the sky in rain, snow, dew, or fog, can cause increased algae growth in water, called eutrophication. Increased levels of nitrogen in soils also interfere with plant and tree growth.

Animals: Studies show nitrogen oxides cause lung damage in animals.

Major Sources

Fossil fuels: Nitrogen oxides are produced when oil or gas is burned at high temperatures, combining oxygen and nitrogen; the largest users of fossil fuels are utility plants, automobiles, and industry (such as nylon, fertilizer, and explosives factories).

Other sources: Nitrogen oxides are also produced by incinerators, landscaping equipment, ovens, kerosene heaters, gas ranges, deforestation and biomass burning (more than 20% of global emissions), leaf burning, aircraft engines, and cigarettes.

Natural sources: Lightning and biological sources in the soil also produce nitrogen oxides.

State of the Earth

Industrial nations create 65% to 80% of the greenhouse gases. The United States releases 1/4 of the world's nitrogen oxide (19 million tons per year in 1990); China, the former Soviet Union, and the United States produce 1/2 of all the nitrogen oxides in the world. Of industrial emissions, vehicles produce 40% of the nitrogen oxide pollution, electric utilities and factories over 50%; the remaining 10% comes from multiple smaller sources. One incinerator can put out more than 1,000 tons of nitrogen oxides each year.

In more than 70 major U.S. cities, residents are breathing air that exceeds health standards for smog for at least 50 days each year. Large industrial cities have smog levels 3 times higher than the level at which crop damage begins. In the United States 3 out of 5 Americans risk lung damage from smog; up to 76 million live in areas where the air quality is hazardous to their health. Smog due to biomass burning is a serious problem in some tropical and temperate areas. Ozone levels in tropical forests have been detected at near lethal levels to animals and plants.

Nitrogen oxides contribute to about 1/3 of acid rain problems in the northeastern United States. Nitrate levels have leveled off in some areas of the eastern United States in the past decade. Yet lifeless acidified lakes and rivers have been documented since the 1960s in nearly every country in the world. U.S. EPA studies concluded that 10% of Appalachian streams are acidic; out of several thousand lakes and streams tested across the nation, 75% of the lakes and 47% of the streams are acidic. Biomass burning releases nitrogen oxides and is responsible for acid rain in tropical and temperate forest areas.

Nitrogen oxides are contributing to thinning ozone; recent studies have linked thinning ozone with increased ultraviolet. Ozone depletion as high as 45% has been measured in some areas over North America, Europe, and Siberia. In 1989 chlorine levels in the upper atmosphere began leveling off, and with continued reductions, massive ozone depletion may not occur. The ozone layer will continue to thin for the next 50 years before it can heal.

Nitrogen oxides, which make up 5% to 10% of the greenhouse gases, trap heat much more efficiently than carbon dioxide. Nitrogen oxide levels in the atmosphere and soils may decrease methane uptake by soils and thus increase greenhouse gases; European forests are becoming nitrogen saturated. In 1991 the National Academy of Sciences concluded there was clear evidence of global warming and recommended immediate reductions of greenhouse gases. They and others have predicted temperature increases of 3° to 9°F in the next century at current rates of greenhouse gas emissions, with sea levels rising 3 to 25 feet.

Human Impact

Health: Nitrogen oxides contribute to bronchitis, pneumonia, emphysema, influenza, lung cancer, eye irritations, chest pains, coughing, lung tissue damage, and lowered resistance to other diseases.

Indoors: Nitrogen oxide levels are often greater indoors than outside, since fumes from gas ranges, kerosene heaters, and other indoor appliances are kept captive in small, confined areas.

N

Ozone depletion, acid rain, and smog: Nitrogen oxides contribute to ozone depletion, acid rain, and smog. All three harm crops and timber resources: acid rain destroys fish stocks; thinning ozone increases chances of skin cancer and cataracts.

Individual Solutions

- Make sure you have adequate ventilation in your house for gas ranges and kerosene heaters.
- Have water heaters and furnace flues serviced and inspected as recommended, and direct vents away from the house.
- Don't exercise in heavy traffic areas.
- Drive fuel-efficient cars; use mass transit; support higher fuel efficiency for all vehicles.
- Investigate solar or wind energy for your home.
- Practice energy conservation; oppose fossil fuel subsidies.
- Don't burn yard litter.

Industrial/Political Solutions

What's Being Done

- Research is being conducted on nitrogen's effects on forests and the environment in Europe and the United States.
- In 1988, 12 European nations agreed to cut nitrogen oxide emissions by 30% over the next decade; 30 other European countries have agreed to hold their emissions at 1990 levels.
- Canada, Europe, Japan, and the United States require all new cars to have catalytic converters; these devices reduce nitrogen oxide emissions by 60% to 75%.
- The U.S. Clean Air Act of 1990 required reductions in nitrogen oxides emitted by utilities beginning in 1994; the EPA will also be targeting heavy equipment such as farm tractors and bulldozers.
- In California pollution from nonvehicular engines must be cut 46% by 1995 and another 55% by 1999; federal regulations for 1996 aim

to reduce hydrocarbons by 32% and carbon monoxide by 14% by 2003.

- Technology now exists for reducing nitrogen oxide emissions at coal burners, such as utilities, by 70% to 97%; retrofitting would be needed at existing plants.
- A growing list of states is mandating use of cleaner automobile fuel such as ethanol or gasohol.
- California is mandating that 10% of cars sold there (200,000 cars) must be zero-emission vehicles by 2003; ten other states have indicated they will follow California's lead.
- California has begun to focus on mass transit planning, mainly light rail instead of highways.

What Needs to Be Done

- Raise fuel-efficiency standards; phase in zero-emission vehicles.
- Increase investment in solar and wind energy, especially in off-the-grid residential and commercial use.
- End fossil fuel subsidies; tax fossil fuels to reflect environmental and health costs.
- Fund and develop mass transit.
- Encourage worldwide use of catalytic converters and lead-free gasoline.
- Mandate wet scrubbers on smokestacks at coal- and oil-burning power plants, as well as at ore smelters and large industrial emitters; recent research will soon allow them to remove nitrogen oxides; phase in zero-discharge limits.
- End emission allowances, and instead focus on pollution prevention.
- Establish worldwide limits on nitrogen oxide emissions; help developing countries limit their emissions.

Also See

- Acid rain
- Greenhouse effect
- Smog

Nuclear energy

Definition

Energy produced either by splitting atomic nuclei with neutrons in nuclear fission or by joining two atomic nuclei in nuclear fusion.

Fission is already used in nuclear reactors and atom bombs, while fusion—the process that creates the sun's energy—is still in the early developmental stage for energy production. Both fission and fusion produce radioactivity and radioactive products.

Radioactive material is unstable and decays, or breaks down, into other material, called isotopes. The half-life of a radioactive substance is the time it takes for 1/2 of any amount of the substance to decay into another material. Some radioactive particles decay very slowly, and after changing into many different isotopes they eventually become nonradioactive. This process may take as long as billions of years—such material is said to have a long half-life. Some radioactive particles are more unstable and break down very quickly, with half-lives as short as milliseconds.

In nuclear reactors, fission of uranium fuel rods heats water under pressure to drive a generator and produce electricity; fission of uranium also produces plutonium. After the fuel has been used it remains highly radioactive and is stored in deep water pools on site; this fuel remains hazardous to humans, or any living creature, for thousands of years. Fuel reprocessing can be used to separate and recover weapons-grade plutonium and uranium from radioactive waste, such as spent fuel rods, so they can be used again. In the 1970s the United States considered reprocessing too expensive and risky and discontinued it; some countries have not.

The difference between a nuclear bomb and a nuclear reactor is that in a reactor the fission process is controlled; in a bomb it is not. Water is used in reactors to cool and control the fission process; when the amount or flow of water is reduced, sometimes because of poorly designed gauges, plumbing, valves, or backup safety systems, the temperature becomes too hot and the nuclear core can experience a meltdown. Release of radioactivity during a meltdown can be thousands of times worse than the amount released by the bomb dropped on Hiroshima.

When a nuclear power plant ages or needs to be retired because of inadequate design or high radioactivity, it is decommissioned (in other words, dismantled and decontaminated). This process can take many years, cost hundreds of millions of dollars, and result in large amounts of low-level radioactive waste.

Environmental Impact

Nuclear accidents: Nuclear accidents add to radiation in the air, which eventually precipitates onto the ground. There it is taken up by plants, which are eaten by animals, and it thus enters the food chain. Long-term effects of widespread radioactive emissions on ecosystems, such as what happened at Chernobyl, Hanford, and Hiroshima, are not clearly understood. Radiation can cause cancer, neurological and reproductive problems, mutations, birth defects, skin disorders, and other disorders in animals.

Nuclear waste: Nuclear waste is disposed of in pits, shafts, or pools; it can contaminate soils and groundwater and enter the food chain through plants and animals. Buried highly radioactive waste, which can remain dangerous for 10,000 years, is at risk for entering the environment through earthquakes, geologic disturbances, leakage, or container breakage. Thousands of canisters of radioactive waste have also been dumped into the oceans. It is now understood that there are violent storms and strong currents at the bottom of the sea; if the containers are broken, their radioactive contents could spread on currents for thousands of miles and enter the ocean food chains. Some waste has low levels of radioactivity and is dumped into municipal landfills, where it can leach out into nearby soils and water.

Mining uranium: Mining uranium releases radon, increases local radioactivity, and adds to local soil and water pollution. When mining wastes are left on top of the soil, they are picked up and transported by the wind.

Nuclear power plant operation: Many nuclear installations discharge some radioactivity into the environment; these emissions have thus far been rated at "acceptably safe levels."

Background radiation: All facets of nuclear power increase local environmental background radiation levels.

Major Sources

Nuclear power plants: Nuclear power plants are in use in many industrial countries, as well as in more than a dozen developing countries. Almost all countries have suspended plans for more reactors because of accident risk, low performance, and high costs.

Standard reactors: A standard reactor uses up its fuel, about 1/3 of which needs to be replaced about every 1 1/2 years; this is the most common type of nuclear reactor. In standard reactors, often called "light water" reactors, normal water slows down the fission process and either boils and turns into steam for use or transfers the heat without boiling to heat exchangers under pressurized conditions.

Breeder reactors: Breeder reactors actually generate more plutonium fuel than they use up; the plutonium is recovered in reprocessing and reused. Breeder reactors are in use in France, Japan, and the former Soviet Union; about a dozen are in operation.

Nuclear energy-related radioactive wastes include:

Tailings: Tailings are low-level radioactive sandy residues of uranium mining; in the past they were often left to lie in the open. Uranium is a naturally occurring radioactive material; the United States has at least 1 million tons, perhaps as much as 1/3 of the world's total. Uranium must be made more concentrated (enriched) to be used in nuclear reactors.

Low-level radioactive waste: Low-level radioactive waste includes contaminated clothing, hardware, dismantled reactors, research waste, and sludges from about 17,000 utilities, power plants (1/2 of low-level waste), industries (1/3 of low-level waste), hospitals, and labs. It can be radioactive for decades to hundreds of years; in the past it was stored on site or in three commercial sites; only two are now open, in South Carolina and Washington. As of 1993 states were to assume responsibility for storage; the U.S. Supreme Court declared it unconstitutional, however, to require states to assume full responsibility for the nuclear waste of private industry.

High-level waste: High-level radioactive waste includes spent fuel rods (used fuel from nuclear reactors) and liquid and solid waste from reprocessing spent nuclear fuel. Highly radioactive and hot for thousands of years, high-level waste contains uranium, plutonium, and some transuranic waste, and it must be shielded. It is currently stored on site at nuclear plants in deep water pools; the government is trying to find a "permanent" storage site. Several dozen nuclear plants in the United States are already close to running out of on-site storage space for high-level waste, and 3/4 of all nuclear plants in the United States will run out of on-site storage space by 2003.

Transuranic waste: Transuranic waste consists of materials from building nuclear weapons and nuclear power plants. Heavier than uranium, this waste is not as hot, but it is highly radioactive for thousands of years. It is stored on site. The government is building underground storage in the New Mexico badlands.

Mixed waste: Waste containing both hazardous and radioactive waste is produced by academic institutions, government, industry, nuclear power, and medical facilities; it is stored on site.

Below regulatory concern (BRC): Radioactive wastes considered safe enough not to need regulation come from labs, hospitals, and power plants. There are concerns that current standards of radioactivity are not low enough. BRC waste can be dumped into a municipal solid waste landfill.

State of the Earth

Nuclear energy supplies 15% of global commercially produced energy. The United States has more nuclear power plants than any other country (109), and 44 have permits expiring between 2002 and 2014; most are licensed for 40 years. France has the second highest number of nuclear plants, at 55. More than a dozen nations get at least 20% of their electricity from nuclear power: In France nuclear power plants provide 75% of the country's electricity; in Belgium, 60%; in Sweden, 50%; in Switzerland, 40%; in Japan, almost 30%; in the United States, 20% (5% of total energy); in England, 20%; in Germany, 30%; in the Ukraine, 30%; and in the former Soviet Union, 10%. France's heavily subsidized nuclear program produces only 1/3 of the electricity needed to pay for the costs of the plants. France is considering building 5 new plants; Japan, 40; and Indonesia, 12. There are about 423 nuclear power plants worldwide; many U.S. plants will soon need to be decommissioned because their permits will have expired; because of their aging structures they will not be able to get new permits. Decommissioning costs are running as high as $500 million.

Many nuclear plants are operating at only about 60% of their capacity, because they are often shut down for repairs and maintenance. Even though a pound of enriched uranium contains millions of times more energy than a pound of coal, nuclear power is more expensive than all forms of conventional electricity generation—it costs 2 times as much as electricity generated from coal, oil, or gas, and it is 10 to 20 times more expensive than energy efficiency programs.

Nuclear power plants generate only electricity and thus cannot supply the majority of energy demand, which is nonelectric. Thousands of nuclear power plants would be required to replace currently operating electricity-generating coal plants, costing trillions of dollars.

Every country with nuclear power plants has had serious accidents, near accidents, or shutdowns. Austria, Greece, and Sweden have already voted to discontinue nuclear power use; many countries have followed suit and have ended plans for more plants. In 1979 a partial core meltdown occurred at the U.S. Three Mile Island nuclear plant; a feedwater valve had been closed owing in part to human error. There is controversy about the amount of radiation that escaped, but some observers, citing an estimated increase in area infant deaths due to birth defects, believe it to be significant. There are deteriorating nuclear plants in the former Soviet Union and inadequately designed plants planned in Cuba and a number of other countries; many countries have poor nuclear waste disposal and safety systems. Less than 1% of developing countries' power comes from nuclear energy.

In 1986 in Chernobyl, Soviet Union, there occurred the largest release of radioactive emissions to date resulting from a nuclear reactor accident. The radiation and fallout, especially cesium-137, will contaminate food chains across Europe for many years, and an estimated 50,000 cancer deaths are expected over the next decades; an estimated 8,000 premature deaths have already been observed. Several million people have a high risk of obtaining cancer as a result of the accident. The Chernobyl accident was caused by human error; in 1992 the European Economic Commission estimated a 25% chance of another Chernobyl-type accident in the area in the next 5 years, because of other outdated and unsafe reactor sites still in operation.

The Nuclear Regulatory Commission (NRC), which oversees nuclear power plants in the United States, has concluded that several dozen plants have inadequate containment structures for major accidents, and many pose a 90% risk of leakage should such an accident occur. Many plants also have inadequate systems to deal with fires, earthquakes, or explosions. A former NRC commissioner has said there is a strong chance of a meltdown at a U.S. reactor by 2010; a recent NRC study found high chances of a major radiation leak because of design problems in every plant studied.

One nuclear power plant can generate more than 30 tons of highly radioactive waste yearly. A typical nuclear reactor will produce close to 500 pounds of plutonium each year; producing 1 pound of plutonium creates 150 gallons of high-level radioactive waste, 25,000 gallons of low-level waste, and more than 1 million gallons of contaminated cooling water. It takes 15 pounds of reactor-grade plutonium to make a 1-kiloton bomb; by 2003, nuclear power plants worldwide will have produced 330 tons of plutonium—enough for 44,000 nuclear bombs.

Currently there are 24,000 tons of high-level radioactive waste in the United States—with 40,000 expected by the year 2000—and 145 million cubic feet of liquid and solid low-level radioactive waste. One million cubic feet of low-level waste is produced yearly. Plans are to convert the high-level waste to glasslike solids through vitrification and then store them in deep vaults in Yucca Mountain, Nevada. Earthquakes have already cracked mine shafts there; no one knows if the vaults will be safe and secure for 10,000 years, which is the lifespan of the nuclear waste. There is no safe way to dispose of high-level nuclear wastes.

Human Impact

Exposure: Humans can experience direct exposure by handling radioactive materials or through windborne radiation due to leaks, accidents, mining, and transportation. In some cities (Salt Lake City, Utah, and Grand Junction, Colorado) radioactive tailings were left above ground and even used in road and pavement construction where they could blow away or erode into the environment. Indirect exposure can occur by drinking contaminated water or eating contaminated foods grown in contaminated soils. Radiation-contaminated groundwater and soils may be unfit to use for crops or drinking for decades or longer.

Health: Nuclear fission and its waste emits alpha, beta, or gamma radiation and other nuclear particles that can damage human tissues and lead to tumor growth, thyroid problems, immune system damage, birth defects, skin problems, digestive problems, neurological damage, genetic damage, leukemia, lung cancer, and death. A single radioactive particle could set into motion the possibility of cancer 10 to 20 years later. The very young and very old are especially susceptible. Plutonium, one of the products of fission, is one of the most toxic substances known.

Low-level radiation: Over a period of time low-level radiation may actually be more damaging to the body than high-level radiation. Low-level radiation weakly stimulates the body's repair mechanisms, which then operate imperfectly and allow damaged cells to survive, whereas high-level radiation triggers a stronger reaction, and the damaged cells are destroyed. Growing evidence suggests that current estimates of safe radiation doses may be too high.

Radon: Radon comes from the natural breakdown of uranium. Uranium mining releases radon, and uranium mill tailings contain radon. Radon causes bone, skin, lung, and blood cancer. People living in areas near uranium mines suffer higher incidences of birth defects. Whole body exposure to radiation can accelerate inherent weaknesses, and thus speed aging.

Worker safety: Since the early years of the nuclear industry, many employees of the industry have been exposed to low-level radiation doses.

Nuclear fallout: Past testing and accidents caused radioactive parti-cles, like strontium, to be distributed over large areas worldwide. Stron-tium-90 replaces calcium in the bones and causes leukemia. Iodine-131 concentrates in the thyroid and ovaries and is especially harmful to chil-dren. Cesium-137 is absorbed by the kidneys, liver, and reproductive organs.

Food irradiation: Cesium-137 is a by-product of nuclear power plants and is used with cobalt-60 to irradiate food to prevent spoilage; the food is not radioactive, but workers could be exposed. Long-term health effects of consuming irradiated food have not been studied. Transport-ing the cesium-137 and cobalt-60 also poses further risks of accidents.

Background radiation: Nuclear power plants, uranium mining, nuclear wastes, and nuclear accidents all increase the amount of local back-ground radiation people are exposed to and thus increase the risk of cancer and other related health problems.

Individual Solutions

- Oppose nuclear power.
- Investigate solar and wind energy for your home.
- Support labeling on all irradiated food; inform grocers you will not buy irradiated foods; buy organic at food cooperatives.
- Practice energy conservation.

- Demand that the U.S. Department of Defense, Department of Energy, and defense contractors be limited in the amount of nuclear waste they generate.
- Oppose the Yucca storage site and short-term storage like above-ground storage casks being proposed by federal regulators and nuclear power plant operators.

Industrial/Political Solutions

What's Being Done

- The International Atomic Energy Agency (IAEA) tracks uranium and plutonium worldwide and inspects nuclear power plants to ensure that spent fuel is not being used to make nuclear weapons.
- No new nuclear plants have been ordered in the United States since 1978; worldwide orders have declined significantly.
- The United States and other countries are helping retrofit Soviet nuclear power plants with up-to-date safety designs.
- The United States is considering burying high-level radioactive waste in Yucca Mountain, low-level waste in California, and transuranic waste in New Mexico. Earthquakes have caused heavy cracks in the mine shafts at Yucca Mountain. As on-site storage pools become full, some utilities use temporary monitored, retrievable, above-ground, dry-cask storage facilities to hold sealed cannisters of spent fuel rods. This relatively new technology may end up as long-term storage if a federal repository is not developed. The U.S. Department of Energy is test-burying 1,100 drums of transuranic waste in Carlsbad, New Mexico, in stable, bedded-salt deposits 2,150 feet deep.
- Outdated nuclear plants are being decommissioned.
- The National Research Council is studying transmutation (converting radioactive wastes to less dangerous materials); its report is due in 1994.
- The U.S. Environmental Protection Agency (EPA) has set radiation exposure limits of no more than 15 millirems per year (equivalent to 2 chest X rays); the Safe Drinking Water Act also has limits for radiation that cannot be exceeded.
- The U.S. Department of Agriculture has approved irradiation of pork

and poultry and is considering approval of beef; meat must be labeled "Treated with irradiation."

- New York has banned irradiated foods until 1995.
- The Massachusetts Institute of Technology (MIT) has developed a plasma arc furnace to destroy transuranic wastes.
- A microbe called *Desulfovibrio sulfuricans* can remove dissolved uranium from wastewater; it may be used in the uranium mining and milling operations, groundwater, and solvents used in the nuclear fuel processing cycle.
- Fusion research is expected to produce commercial fusion power reactors by as early as 2030.

What Needs to Be Done

- End nuclear energy subsidies; factor environmental costs into nuclear power costs.
- End the U.S. Rice-Anderson Indemnity Act, which limits the liability of the nuclear industry to the public in the event of a major nuclear accident.
- Phase out nuclear power plants, and shift to renewable energy and energy conservation and efficiency.
- Ban irradiated foods; the poultry and beef industries need to prevent food contamination by means other than irradiation. Label any irradiated food.
- End proposed plans for Yucca Mountain and "short-term" storage facilities of nuclear waste; store the waste on site, and close down aging reactors.
- End dumping of BRC nuclear waste; reevaluate levels of acceptable doses.
- Sign international agreements to limit nuclear waste generation.
- Sign international agreements to end fuel reprocessing to produce weapons-grade plutonium.

Also See

- Warfare effects on the environment

Ocean degradation

Definition

Pollution, erosion, or destruction of coastal shorelines, coral reefs, or deep ocean water, all of which affects ocean biology and harms aquatic plants and animals.

Oceans contain more than 97% of the planet's water supply and have heavy concentrations of dissolved salts and minerals. Most ocean life, including most of the important fisheries and shellfish, originates either in coastal areas and estuaries or on coral reefs.

Many things about the oceans are poorly understood, although it is known that oceans affect the planet's climate through such disturbances as El Niño (see glossary), through their temperatures and currents, and through their capacity to act as sinks for carbon dioxide and frozen methane (as hydrates in the continental shelves); carbon dioxide and methane are major contributors to the greenhouse effect. Phytoplankton and coral reefs are also large carbon sinks.

Environmental Impact

Environmental buildup: Parts of the oceans that are enclosed, such as the Baltic, Caribbean, Mediterranean, and North seas, all concentrate more pollutants and clean up much more slowly than open waters. River basins, estuaries, and bays also tend to accumulate more nutrients and toxics than they may give off, acting like catch basins. These sensitive areas are therefore prone to pollution problems. Sediments, sewage, and sludge can fill in estuaries and bays, harming minnows, larvae, small organisms, and fish populations, as well as destroying coastal shellfish beds.

Animals, plants, and toxics: Industrial wastes, sewage containing heavy metals, agricultural runoff containing pesticides, air pollution, and municipal waste can enter the oceans from rivers and the coastline. Toxics enter the food chain in small organisms or in plant uptake of water. Birds, fish, and aquatic mammals that inhabit shorelines can then ingest these same plants, small organisms, or the water itself and bioaccumulate the toxics. Often these toxics can cause neurological problems, kidney and liver failure, reproductive or behavior problems, cancer, and death in animals. Toxics can also lower animals' resistance to diseases, like viruses. Coastal development not only contributes to runoff pollution, but also destroys habitat and breeding grounds for fish, shellfish, and other aquatic organisms.

Coral reefs: Pollution is suspected of destroying coral reefs. Chief contributors are believed to be dredge sediments; heated water (thermal pollution); pesticide and chemical runoff from rivers to ocean currents; silt runoff from erosion owing to deforestation; and nutrients such as fertilizer, sewage phosphates, and nitrogen, which increase algae growth and block sunlight in a process called eutrophication. Often these algae blooms, called red, brown, or green tides because of their color, can be toxic and kill fish, small organisms, and reefs. Coral reef bleaching occurs with heated water, pollution, and sedimentation. The algae that live on the coral and give it color either lose their pigment, leave the coral, or are expelled by it. The coral then cannot grow or reproduce; some may eventually die. Fishing with cyanide or explosives has also damaged reefs. Some countries are mining reefs for construction material. Excessive tourism has damaged a number of reefs worldwide. Reef damage or death leads to lowered fish and aquatic animal populations and eroded shorelines.

Oil spills: Oil spills destroy coastal regions, killing fish, birds, mammals, small organisms, and aquatic plants. It can take years for the effects of an oil spill to be cleansed from the environment. Oil drilling brings up heavy metals, radioactive wastes, and other toxic organic chemicals that can enter the food chain and harm fish and plant life.

Surface waters: Surface waters of the open ocean concentrate pollutants through wave action and buoyancy, killing fish embryos and microorganisms. Many species of fish and ocean life at the top of the food chain also tend to live in surface waters, and thus are threatened by surface pollution. Ozone depletion also threatens the Antarctic food chain as increased ultraviolet kills plankton in the surface waters there.

Garbage: Discarded plastic, beer can rings, nylon fishing nets, and driftnets entangle sharks, dolphins, seals, turtles, otters, birds, and other animals until they suffocate or starve. Often the plastic or nylon is not seen or is mistaken for food.

Hazardous wastes: At one time deep ocean sites were considered safe and tranquil places to dump canisters of highly dangerous hazardous and radioactive waste. Oceans are now known to be very volatile, with powerful underwater storms sweeping huge amounts of sediment over long distances in some deep regions, violent volcanic activity occurring in others, and powerful surface and deep currents present in all oceans. Just like air currents, ocean currents can carry ocean pollution world-wide. Thus no one knows where hundreds of dumped toxic canisters may end up or how long they will stay intact. These chemicals are extremely dangerous to any form of life.

Major Sources

Rivers: Rivers contribute sediments, agricultural runoff (pesticides, fertilizers, manure, heavy metals), logging and deforestation runoff, industrial and municipal wastewater, urban runoff, used automotive fluids and grease, sewage, and road salts.

Coastal development: Mining, construction, cities, and other coastal development contributes sediments, nutrients, wastewater, sewage, garbage, municipal and industrial waste, and pollutants. Coral reefs have been mined for construction material.

Oil: Inland and ocean accidental spills, discarded oil from cars, equipment, and power tools, and leaking oil from boat use all damage oceans. Offshore drilling also results in accidental spills, leaks, and the dumping of drilling wastes into the water.

Air pollution: Acid rain, particulates, toxic chemicals, mercury, lead, heavy metals from industrial and power plant emissions, incinerators, motor vehicles, and airplanes pollute ocean waters.

Ships: Cargo, tourist, navy, and other ships add to discarded garbage, wastewater, used oils and fuels, and toxic wastes; heavily trafficked areas contribute to surface pollution in the open ocean and along coastlines. Several nuclear submarines have had accidents in which nuclear waste was leaked.

Dredge spoils: Dredge spoils are silt from harbors, rivers, and waterways that is dug up and carried by barges out to sea. Up to 1/10 of it is toxic with oil, organic chemicals, heavy metals, nutrients, and industrial and municipal wastes.

Dams and canals on rivers upstream: Dams and canals prevent silt, fresh water, and nutrients from reaching coastal estuaries, raising salinity and depleting coastal marshland nutrients.

Plastic: One of the most prevalent waste items is plastic; it enters the oceans from inland solid waste, disposables, tourists, cargo ships, yachts, fishing vessels, and coastal cities.

Hazardous and radioactive wastes: Discarded canisters and nuclear warheads and reactors from military accidents are left on the ocean floor.

Aquaculture: Aquaculture may destroy coastlines, mangroves, and estuaries to grow shrimp, shellfish, or fish.

Ozone depletion: Ozone depletion is destroying plankton in the Antarctic Ocean and may affect the whole ocean food chain.

Tourism: Tourists can decimate reefs by gathering coral relics.

Fishing: Cyanide and explosives have been used on coral reefs for fishing.

State of the Earth

Each year millions of tons of plastic and old fishing nets and gear are dumped into the ocean—up to 30 million tons of litter. The U.S. Office of Technology Assessment says that discarded plastic kills more than 1 million birds and more than 100,000 seals, sea lions, sea otters, whales, dolphins, sharks, porpoises, and turtles annually. The United Nations Environment Programme estimated in 1990 that the major contributors to ocean pollution were coastal and river runoff (44% of all ocean pollution), air polllution (33%), shipping (12%), garbage dumping (10%), and offshore oil production (1%).

The lead content of the ocean has increased severalfold since the introduction of leaded gasoline, which is still in use in much of the world. About 5 to 6 trillion gallons of industrial waste are dumped into U.S. rivers every year. Soils carried by the wind and containing heavy metals were tracked from Asia to the central northern Pacific Ocean.

Worldwide, 2.5 billion people live within 60 miles of coastlines; 2/3 of all humans live in areas near coastlines and rivers draining into coastal waters; 75% of the U.S. population lives within 50 miles of the coast; 70% of Israel's population lives close to coastlines. The United States has developed more than 50% of its coastal wetlands; Italy, nearly 100%; tropical areas, 50%. Depending on the area of the world, 75% to 90% of all commercial marine fish caught are species that depend on coastal estuaries for reproduction, food, migration, or nurseries; most of the rest depend on coral reef systems.

Reefs, which are very sensitive to changes, support 1/3 of all fish species. Coral bleaching is being observed with the greatest frequency in the Atlantic and Pacific oceans. In some tropical countries, such as India and Sri Lanka, coral has been mined for construction; in other countries dynamite and cyanide poison have been used to "fish" near reefs. On many reefs exotic fish are collected for sale. Recreational and commercial boaters damage reefs with anchors and collisions, and tourists are disturbing reefs with careless snorkeling and diving. Coral reefs are one of the most endangered ecosystems in the world.

Oil drilling wastes have high levels of heavy metals and organic chemicals. Millions of gallons of water contaminated by the drilling processes are dumped yearly, untreated, into waterways and oceans. About 1 billion gallons of oil pollute the oceans each year: 1/2 from the continents (rivers and coastal runoff), 1/2 from deliberate dumping of fuels at sea, accidents, and production platforms. Shipping is responsible for about 12% of marine pollution. In the 1989 oil spill by the *Exxon Valdez* in Alaska's Prince William Sound, it is estimated that 350,000 birds died, as well as killer whales, dolphins, seals, otters, fish, and sea grasses along a 1,000-mile coastline; as of 1993 some bird, seal, and fish populations were still in decline.

Every year human activities release into rivers and oceans millions of tons of nitrogen and phosphorus, which feed algae blooms such as red tides. Red tides are increasing worldwide and have been reported in coastal areas along China, Japan, and North and South America, as well as in the Baltic Sea, the Black Sea, and the North Sea. It is estimated that up to 2,500 dolphins died from 1987 to 1989 from eating fish contaminated by toxic algae.

Ocean sediments in the North Sea were found to contain toxic chemicals whose source was a sea-based incineration program. Polychlorinated biphenyls (PCBs), DDT, and other toxic chemicals have been found in ocean sediments and animal populations worldwide. In 1987–1988, 700 dolphins were found dead on beaches, and all had significant amounts of toxics, such as PCBs and pesticides, in their bodies. Dolphins that died from serious viral infections are suspected to have had their resistance lowered by such toxics.

Globally, 95% of all sewage in developing countries is untreated. Even in industrial countries sewage treatment generally removes only 1/3 to 2/3 of the toxics, 1/3 of the phosphorus, and 1/2 of the nitrogen. In the United States, 1/3 of coastal shellfish beds are closed because of toxic and bacterial pollution; in 1992 beaches were closed or had swimming advisories 2,600 times as a result of bacterial contamination from human and animal waste.

More than 70 countries have set aside about 1,000 marine reserves for protection. The Baltic Sea (the world's largest body of brackish water), the Black Sea, the Mediterranean Sea, and the North Sea all show symptoms of dying from pollution; they may not be able to recover. Nearly 2/3 of the Mediterranean pollution is from France, Italy, and Spain; 70% of urban wastewater goes into the Mediterranean untreated. Oil pollution in the Mediterranean from heavy shipping, accidents, and polluted rivers and coastal runoff is equivalent to 17 *Exxon Valdez* tankers emptying their tanks every year. The former Soviet Union dumped radioactive wastes into the Arctic Ocean.

In the Antarctic, scientists have reported up to 12% reductions in phytoplankton growth in areas where ultraviolet radiation has doubled. A 1994 study found that ultraviolet B radiation kills the eggs of frogs and toads in Oregon; this raises concerns for fish larvae and other small creatures living in ocean surface waters.

Human Impact

Health: There is risk of toxic and heavy metal buildup in human tissues from the presence of toxic pollutants in ocean seafood. Toxic chemicals such as PCBs, lead, mercury, heavy metals, and organic chemicals can cause cancer, reproductive problems, birth defects, neurological problems, skin disorders, and death. Bacteria and viruses in contaminated shellfish can cause respiratory paralysis, diarrhea, gastrointestinal disorders, and death. In the United States, 30,000 to 60,000 illnesses are caused by contaminated seafood each year. Swimming in sewage-polluted waters can cause respiratory infections and gastroenteritis. England, Mediterranean countries, and developing countries have many beaches with high bacteriological counts due to poor sewage disposal methods. Untreated sewage can spread numerous diseases and parasites.

Food: Food sources are threatened as ocean fish populations become toxic, estuaries are polluted and damaged, and food chains are interrupted. As fish and shellfish populations continue to decline, serious food shortages could emerge.

Greenhouse effect: No one understands what will happen if the oceans warm further, but some experts believe that changes in ocean temperatures might accelerate the greenhouse effect, especially if the oceans' capacity to hold methane or carbon dioxide is changed by human activity. This change in turn could affect fish, mammals, and other aquatic wildlife.

Aesthetic: Aesthetic concerns increase as swimming and walking beaches are fouled and species like the gray whale become threatened by coastal pollution. Also, tropical beaches may be lost to erosion if their protecting reefs are damaged or destroyed.

Individual Solutions

- Never discard anything on a beach or from a boat.
- Don't throw any toxic materials down the drain, on the soil, or into the sewers; it all ends up in the water supply, rivers, and eventually the oceans. Contact the U.S. Environmental Protection Agency (EPA) or local officials for information on hazardous waste collections.
- Practice energy conservation; conservation lessens fossil fuel use, air pollution, and acid rain.
- Recycle; buy recycled products; buy in bulk; avoid packaging; reuse containers.
- Buy organic foods, whose production uses no pesticides or chemical fertilizers and creates little runoff.
- Join a coastal group dedicated to keeping the ocean and beaches clean.
- Buy as little plastic as possible; plastic is difficult to recycle, is toxic to produce, and uses fossil fuels, which are nonrenewable.
- Conserve water, which minimizes sewage output.
- Don't buy rare shells or coral products, especially black coral.

Industrial/Political Solutions

What's Being Done

- There are world agreements to limit and end ocean dumping; 71 countries have agreed to phase out industrial waste dumping by 1995 in the London Dumping Convention Treaty, which also bans dumping nuclear waste at sea. Several dozen countries have agreed to end plastic dumping at sea. There are concerns, though, that the London agreement will simply increase land, river, and coastal dumping.
- Ocean incineration is being abandoned; the United States has already suspended incinerating industrial wastes at sea.
- The 1990 Mediterranean Environmental Technical Assistance Program (METAP) will seek to improve wastewater treatment and solid waste management; 85% of METAP is aimed at developing countries bordering the southern and eastern Mediterranean.

- Countries have claimed 200-mile territorial coastal zones to protect their coastal waters.
- The U.S. Act to Prevent Pollution from Ships (APPS) prevents ships from dumping oil, toxic liquids, plastics, or garbage; this is the U.S. version of an international treaty to prevent pollution by ships.
- The U.S. Ocean Dumping Ban Act, passed in 1988, set 1991 as the last year ocean dumping of sludge, sewage, municipal waste, and industrial waste would be allowed.
- The U.S. Oil Pollution Act of 1990 governs oil spills; violators must clean up spills and pay any costs of cleanup and other damages to affected states and local businesses.
- The U.S. Clean Water Act comes up for reauthorization in 1993–1994.
- The EPA has announced support for integrated pest management, instead of chemical-intensive farming.
- The Food and Drug Administration will implement new seafood safety controls in 1995 that will make processors responsible for ensuring seafood is from clean water and is properly maintained, processed, and chilled until it reaches retailers.
- The environmental impact of oil drilling is being taken into account in approval of drilling in coastal areas in Alaska, the Antarctic, and the Arctic.
- Coastal states in the United States receive funding and assistance to develop required nonpoint source pollution programs and coastal protection plans.
- Watershed protection programs are being instituted in a number of countries.
- Oil tankers and barges larger than 5,000 tons and using U.S. ports will be required to have double hulls by 2010; smaller barges and tankers have until 2015.
- Used oil is being recycled in the United States.
- The Florida Keys Marine Sanctuary Protection Act designates 2,600 square miles along the reefs as a marine sanctuary to prevent reef damage, pollution, and commercial development. The National Marine

Sanctuary Program protects 13 designated marine sanctuaries; the National Estuary Research Reserve System protects 21 estuaries.

What Needs to Be Done

- End offshore coastal oil drilling.
- Set higher penalties for oil spills.
- End the dumping of coastal effluent into the oceans.
- End all industrial and commercial dumping of chemicals, toxics, wastes, or hazardous materials into sewer systems, rivers, and oceans.
- End all discharges of raw sewage, including overflow; upgrade sewage systems to better purify waste; detoxify and recycle more sewage as fertilizer; help developing countries develop sewage systems.
- Establish a worldwide agreement to mandate double hulls on oil barges and tankers.
- Mandate double hulls on barges and vessels transporting hazardous materials or oil products on inland rivers.
- Phase out all incineration.
- Initiate programs to end the dumping of fishing equipment, lines, and nets into oceans.
- End all ocean ship dumping of any wastes into the ocean.
- Fund sustainable agriculture to end high-intensity chemical farming; mandate that all farm chemicals be tested by independent researchers for long-term health effects before use; remove chemicals from markets until such testing has occurred. Agribusiness should pay for all testing.
- Reduce the production of plastic packaging and plastic products, especially disposable goods.
- Establish worldwide agreements for estuary, coral reef, and coastal protection.

Also See

- Freshwater degradation
- Overfishing
- Wetlands degradation

O

Oil

Definition

A fossil fuel formed by millions of years of compression of organic matter (plant and animal remains) at the bottom of ancient lakes or seabeds.

Oil, or petroleum, is a nonrenewable resource; once used up it cannot be replaced. It is made up of hydrocarbons, which are organic chemicals, and traces of sulfur, nitrogen, and oxygen.

Oil drilled from the ground is called crude oil and can vary from low (sweet crude) to high (sour crude) sulfur content. Crude can be brown, black, or even green. It is transported overland by pipelines and shipped by barge and tanker over water. There are 42 gallons of oil in a barrel and about 7 1/2 barrels, or 315 gallons, in a ton. Oil is used to make a number of products, primarily gasoline and plastic.

Environmental Impact

Burned oil: Burned oil releases greenhouse gases, which contribute to the greenhouse effect, and hydrocarbons, which contribute to smog. Burned gas in motor vehicles releases carbon monoxide, which diminishes concentrations of atmospheric hydroxyl radicals; these radicals naturally decrease methane (a greenhouse gas) and ground-level ozone in smog. Other toxic pollutants and particulates add to air pollution, acid rain, and soil and water pollution. These problems inhibit plant growth, can damage or destroy forests, and, in the case of acid rain, can kill aquatic life in lakes and rivers. Particulates from burned oil can increase the chemical reactions between chlorine and ozone, further depleting the ozone layer. Recycled oil often contains toxic chemicals, which also escape into the air when it is burned.

Oil production: Oil spills from refineries, pipes, platforms, and tankers threaten and ruin ecosystems. Beaches; coral reefs; and bird, mammal, shellfish, plankton, and fish populations are all threatened in areas where spills have occurred, and the cleanups are long, difficult, and sometimes impossible. Oil can kill marine life by its toxicity and by smothering; it also taints the taste of the flesh of shellfish and finfish. Oil exploration and drilling can threaten ecologically sensitive areas. Accidents can happen during drilling if cracks in the drilling shafts occur, allowing the oil to rise to the surface in a nearby area. Dumped oil-drilling wastes leach into the soils and waters. Drilling for oil also brings to the surface heavy metals, organic chemicals, toxic chemicals, and radioactive material. These chemicals can be incorporated into the food chain, first in plants or small organisms, and then in fish and higher predators. Heavy metals can cause reproductive problems, cancer, and death in animals. Sulfuric acid is used to wash gasoline and petroleum products during refining and adds to the acid rain problem.

Petrochemical products: Plastic and other oil-based products yield toxic production waste, and plastics do not biodegrade in the environment. When these products are burned in solid waste incinerators, they yield dioxins and other toxic chemicals; dioxin threatens wildlife and accumulates in the food chain. Often plastics are made with heavy metals, which are released when they are burned. Petrochemicals are often toxic and are responsible for a wide range of problems, including ozone depletion and air and water pollution.

Commercial use: Gasoline can leak or drip from gasoline station nozzles, as well as from gas and fuel storage tanks. Underground storage tanks can leak and contaminate groundwater. Used oil poured down sewers or onto land is a major cause of soil and water pollution.

Major Sources

Oil reserves: Oil is usually found in reservoirs or pools, often 1 to 5 miles below the surface. Half of all reserves are found offshore, on continental shelves beneath the ocean. Both land and sea reserves are brought to the surface by drilling. Often natural gas is found with oil reserves.

Shale: Oil can also be found in oil shale, a porous rock that can be strip- or shaft-mined and then processed to remove the oil. These deposits are huge, but environmental concerns, costs, and technological limitations are preventing their recovery. Tar sand is sandstone impregnated with thick oil that can be refined into liquid oil, an expensive and difficult process.

Products: Crude oil is used to make fuel and lubricants for gasoline and diesel engines, and it is the main ingredient for plastic and other oil-based products and chemicals, including solvents, pesticides, fertilizers, synthetic organic chemicals, and chlorofluorocarbons (CFCs). Synthetic clothing fibers are made from oil.

State of the Earth

Oil reserves are found in a number of countries, including the European countries, Indonesia, Japan, Libya, Mexico, the Middle Eastern countries (with more than 2/3 of world reserves), Nigeria, the former Soviet Union, the United States, and Venezuela. Oil reserves are estimated at 30 to 50 years. New technology could increase reserve estimates, but environmental problems and cost might make other energy sources more appealing.

The more than 500 oil wells set ablaze in Kuwait in 1991 by Iraq generated enormous amounts of air and soil pollution (precipitated from the air), and the long-term effects still remain to be seen. Diesel engines emit 50 to 100 times more particulate matter than gasoline-powered engines. In the United States transportation (mainly cars) is responsible for 2/3 of all oil used (3 billion barrels used in 1993). Worldwide, transportation accounts for 1/3 of all oil usage; the rest is used mainly for heating, power plants, and manufacturing. Twenty percent of all oil and natural gas consumed in the United States is used to produce petrochemical feedstocks for the plastics industry. Oil provides 40% of world commercial fuel use; the United States imports 45% to 50% of its oil.

According to the U.S. Environmental Protection Agency (EPA), 179 million tons of oily wastes are buried at refinery sites yearly, 18,000 tons of waste go to community landfills, and 57 million tons of oil field waste go to landfills. Millions of gallons of water contaminated by the drilling processes are dumped yearly, untreated, into waterways and oceans. About 1 billion gallons of oil pollute the oceans each year: half from the continents (from rivers and coastal runoff) and 1/2 from deliberate dumping of fuels at sea, accidents, and production platforms. The EPA estimates that of more than 2 million underground storage tanks, as many as 20% could be leaking.

Since the early 1980s there have been several hundred oil spills, releasing an average of 1 million gallons of oil each month. In Minnesota, since the early 1970s, oil pipelines have ruptured and leaked 9 million gallons of petroleum—nearly the amount the *Exxon Valdez* spilled in Prince William Sound. In the 1989 Prince William Sound spill it is estimated that 350,000 birds died, as well as killer whales, more than 50 gray whales, dolphins, seals, otters, fish, eagles, and sea grasses along a 1,000-mile coastline; 4 years later, in 1993, fish were not reproducing normally, and seals and some birds were still in decline. Some oil spills have not been as detrimental, but since tankers are often near coastlines, spills are always a potential risk. Some of the more recent oil spills were the Persian Gulf War tankers, 1991, 420 million gallons; the *Aegean Sea*, 1992, Spain, 22 million gallons; and the *Braer*, 1993, North Atlantic, 26 million gallons.

Human Impact

Health: Burning oil wells, oil refineries, and oil-based fuels give off particulates, heavy metals, and toxics, such as benzene, fluorides, and toluene, which cause cancer, respiratory diseases, birth defects, and other health problems. Secondary pollution from manufactured, and later incinerated, petrochemical products can also cause cancer, lung problems, immune problems, and neurological diseases.

Food: Burned oil reduces food crop yields when it contributes to smog, acid rain, and air pollution. Ocean spills harm finfish and shellfish populations.

Fresh water: Dumped used oil and spills can contaminate drinking water supplies.

Aesthetic damage: Spilled oil ruins beaches and other scenic areas.

Individual Solutions

- Recycle all used oil; don't dump any oil-based products down the sink, in the yard, or in the street.
- Practice energy conservation, which reduces fuel needs.
- Drive fuel-efficient cars; keep them tuned; use mass transit.
- Minimize the use of plastic and foam materials.
- Consider buying an electric or reel (push) lawn mower; use electric or hand-powered motors.
- Investigate solar and wind energy for your home needs.
- Support higher fuel efficiency for automobiles and an end to fossil fuel subsidies.
- Purchase cotton, silk, or wool products instead of synthetics.
- Support an end to coastal oil drilling and a continued ban on oil exploration in Antarctica and in Alaska's Arctic National Wildlife Refuge.

Industrial/Political Solutions

What's Being Done

- The Antarctic Treaty of 1961 (updated in 1991) prevents any mineral extraction in Antarctica for at least 50 years, until 2041.
- Electric vehicles are being researched and marketed in Europe, Japan, and the United States. California will have 200,000 by 2003; 10 other states say they will follow California's lead.
- The environmental impact of oil drilling is being taken into account in approval of drilling in Alaska, the Antarctic, the Arctic, and coastal areas.
- The 1990 U.S. Clean Air Act is mandating use of cleaner fuel and fewer air emissions from automobiles, as well as fuel-recovery nozzles to recapture fuel vapors during refueling.
- The U.S. Oil Pollution Act of 1990 governs oil spills; violators must clean up spills and pay any costs of cleanup and other damages to affected states and local businesses.
- Oil tankers and barges larger than 5,000 tons and using U.S. ports

will be required to have double hulls by 2010; smaller barges and tankers have until 2015.

- Fifteen states require used oil to be managed as special or hazardous waste.
- Used oil is being recycled in the United States.
- In California, pollution from nonvehicular engines must be cut 46% by 1995, another 55% by 1999. Federal regulations for 1996 aim to reduce hydrocarbons by 32% and carbon monoxide by 14% by 2003.

What Needs to Be Done
- Tax oil to pay for cleanup costs of oil pollution; end oil subsidies.
- Increase investment in solar, wind, and other renewable energy.
- Increase the efficiency of the oil industry, which wastes several hundred million barrels of oil yearly.
- Define oil wastes as hazardous, instead of using the current nonhazardous label.
- End the use of civil damages and cleanup costs as tax deductions by the oil industry.
- Require all oil-related businesses to report a Toxic Release Inventory (TRI).
- Mandate higher fuel efficiency and lower tailpipe emissions for automobiles.
- Support mass transit and light rail; phase in zero-emission vehicles.
- Enforce safe maintenance of underground and above-ground storage tanks.
- Tighten regulations and penalties for inland oil pipeline ruptures and spills.
- Tighten regulation of oil transport on inland rivers and waterways, including quicker mandating of double hulls on all barges that transport oil and petroleum products.
- End offshore coastal oil drilling.

Also See

- Fossil fuels

O

Organic chemicals, synthetic

Human-created chemicals containing the element carbon.

All living creatures are made of organic chemicals; the vast majority of these chemicals are not toxic and are broken down into other substances in the environment by natural forces (bacteria, fungi, and the elements). Many human-created, or synthetic, organic chemicals, however, are very toxic and often do not break down in the environment very quickly.

Synthetic organic chemicals include a number of large groups of chemicals, including some well-known toxic chemicals. Examples are volatile organic compounds (VOCs); hydrocarbons; organophosphates (pesticides); organochlorines, also called chlorinated hydrocarbons (chlorofluorocarbons, or CFCs; DDT; polychlorinated biphenyls, or PCBs; polyvinyl chloride, or PVC); herbicides; benzene; dioxins; furans; and other petrochemicals (chemicals derived from petroleum, or oil, including most chemical solvents).

Environmental Impact

Plants and animals: For every natural organic compound there is an enzyme in nature that will break it down. This is not true for a number of synthetic organic chemicals. Thus, since they do not break down in the environment quickly, many organic chemicals build up in the food chain and are toxic to wildlife and to plants. Pesticides and many other organic chemicals can wash off of soils or precipitate out of the air and be taken up by living organisms. They can be responsible for wildlife reproductive problems, birth defects, cancer, and the death of young animals. Chemicals such as PCBs, dioxin, and furans have been found in the environment worldwide in animals, soils, and aquatic sediments. They are all harmful to wildlife. High levels of PCBs have been implicated in Atlantic dolphin die-offs; dissected dolphins have had very high levels of PCBs in their tissues.

Air pollution: VOCs and hydrocarbons are components of air pollution and trigger the occurrence of smog; other organic chemicals make smog more toxic. Smog destroys plants and inhibits their growth.

Ozone depletion: CFCs are one of the major gases destroying the ozone layer as well as adding to the greenhouse effect. A thinning ozone layer increases ultraviolet radiation reaching the earth. Ultraviolet inhibits plant growth and destroys phytoplankton, which may affect the ocean food chain.

Major Sources

Fossil fuels: One of the largest sources of organic chemicals is fossil fuels, which are used in the production of most synthetic organic chemicals. Aromatic hydrocarbons (complex organic chemicals such as benzene, toluene, and napthalene) are found in most petroleum products and are also given off in car exhaust. Hydrocarbons are produced by automotive and transportation emissions and landscaping equipment.

Products: Organic chemicals are used in manufacturing a wide variety of products such as pesticides, herbicides, nylon, dry-cleaning chemicals, household construction products, embalming products, refrigerants, solvents, plastics, pharmaceuticals, and cleaning agents. They are also used in a wide variety of industrial processes to manufacture other products.

By-products: Organic chemicals are often produced as by-products of manufacturing, incineration (of municipal solid waste, medical waste, or hazardous waste), burning of fossil fuels, or the mixing of several chemicals, as in a chemical dump, landfill, or other toxic site. Dioxin is an example of a chemical that is never made intentionally, but rather is a by-product of incineration and a number of other processes. When any number or type of organic chemicals are mixed together in waste, industrial processes, or incineration, they may form new organic chemicals, the effects of which are often unknown.

Transported chemicals: Some of these chemicals can be carried thousands of miles by wind or water. Thus many chemicals like PCBs, DDT, dioxins, and furans have been found in wildlife populations, soils, and aquatic sediments worldwide.

State of the Earth

Ending hydrocarbon emissions would eliminate most smog. Use of the catalytic converter, increased fuel efficiency, and improved engine design can yield 90% reductions in the emissions of hydrocarbons. Most countries, however, do not require catalytic converters on their cars. Thus hydrocarbon emissions and smog are increasing worldwide because of increases in the number of cars. Canada, Japan, the United States, and much of Europe require catalytic converters.

Chlorofluorocarbons (CFCs), which are ozone depleters, are set to be phased out worldwide in the mid-1990s. Yet a 10% to 12% thinning of the ozone layer has already been documented in many areas. Thinning will worsen as the concentrations of CFCs build in the atmosphere; planned alternatives to CFCs such as hydrochlorofluorocarbons (HCFCs) will also continue to thin the ozone layer.

Some organic chemicals break down to form other toxic chemicals. For example, at high temperatures carbon tetrachloride converts to phosgene, as does chloroform in sunlight. Chlorinated phenols can break down to form dioxins.

Estimates are that several hundred million tons of organic chemicals are produced worldwide each year; the United States released 18 million tons of VOCs alone in 1990. Most chemicals, even if toxic, are not regulated. More than 90% of all pesticides have not been tested for long-term health effects, even though many are believed to be carcinogenic.

Studies have shown that human breast milk contains dozens, in some cases more than 100, synthetic organic chemicals. Traces of PCBs and a few other chemicals are believed to be in nearly every person in the United States. Long-term health effects are known for very few chemicals; no one knows the long-term effects of exposure to dozens, or hundreds, of chemicals at the same time, over several years or a lifetime. The effects of some toxic chemicals are cumulative, and effects show up in succeeding generations through genetic damage. Steadily increasing cancer rates are believed to be directly linked to the amount of toxic organic chemicals in the environment.

Human Impact

Health: Many synthetic organic chemicals can be inhaled, taken in with food or water, or absorbed through the skin. These chemicals can be stored in the fatty tissues, where they can bioaccumulate. Since synthetic organic chemicals can mimic the organic chemistry of living organisms, they can often fit the enzyme niches that normal molecules take. Thus some chemicals may gain entry into the cellular DNA structure; disrupt normal biochemistry; and then cause mutations, cancer, and death. They can disrupt the reproductive, endocrine, and glandular systems and can cause respiratory, skin, and blood problems. Minute amounts of some chemicals are very dangerous. For furans and dioxins there is no known safe level of exposure.

Air pollution: Organic chemicals that are part of smog are responsible for respiratory, skin, and cardiovascular diseases. They can irritate the eyes and nose and increase allergy and asthma problems. Smog also eats away at building surfaces.

Ozone depletion: As the ozone layer thins, more ultraviolet radiation reaches earth, causing skin cancer and cataracts and inhibiting plant growth.

Food: Organic chemicals in smog are responsible for food crop losses; toxic organics in fish reduce available food and food quality.

Individual Solutions

- Contact the U.S. Environmental Protection Agency (EPA) or local environmental groups for lists of toxic chemicals and their sources in household products; obtain a list of safe alternatives.

- Never dispose of toxics down the drain, on the soil, or in the garbage; contact local officials for hazardous waste disposal sites.
- Buy organic foods; they are not grown with chemicals such as toxic pesticides. Don't use pesticides; contact an organic lawn care company.
- Minimize your use of plastics.
- Oppose incineration; support recycling.
- Recycle Freon in car and home air conditioners and old appliances such as refrigerators.

Industrial/Political Solutions

What's Being Done

- Numerous technologies are rapidly being developed in Europe, Japan, and the United States to destroy organic compounds, including ultraviolet light, bioremediation, and supercritical fluids that can destroy toxics and give off safer by-products than incineration.
- Household hazardous waste is being collected in Europe, Japan, and the United States.
- There is an emphasis in many industrial countries on recycling solid and hazardous waste instead of landfilling it. The waste is reused in industry.
- CFCs are being phased out worldwide.
- Some known organic toxics, like PCBs and DDT, have been phased out in most of Europe and the United States.
- The U.S. Clean Air Act Amendments of 1990 mandate regulation of emissions on about 200 toxic chemicals. As of February 1994, the EPA required companies to reduce toxic chemical emissions by 506,000 tons yearly, within 3 years; this will reduce at least 1 million tons of by-product volatile organic compounds (VOCs) yearly.
- The EPA is now supporting integrated pest management (IPM) instead of high-input pesticide farming.
- California is mandating that by 2003, 10% of all new cars (roughly 200,000) be zero-emission cars (electric). Ten other states have said they will follow California's lead.

- Toxic use reduction programs are attempting to replace toxics in U.S. industries with safe alternatives; for example, VOCs in solvent-based contact cements are being replaced by water-based cements.
- Emission standards for some chemicals have been tightened in the United States.
- There is a moratorium until 1995 on building any new hazardous waste incinerators in the United States.
- Toxic "trading" is taking place between companies that produce chemicals as waste and companies that can use such chemicals in manufacturing.

What Needs to Be Done

- Mandate further reductions in the use of organic toxics; phase organic toxics out of all nonessential manufacturing; aim for zero discharge of organic toxics.
- Separate sewage waste from industrial wastewater.
- Tax toxics to minimize their use.
- Require that all chemicals (especially pesticides) be tested for long-term health effects before they are used in household products or agriculture or emitted into the environment.
- Mandate that industry have plans for the safe disposal and destruction of their waste before they create it.
- Phase out incineration.
- Prevent U.S. companies from selling banned pesticides to developing countries.
- Support sustainable organic agriculture.
- Phase in zero-emission vehicles.
- Phase out CFCs and their replacements, hydrochlorofluorocarbons (HCFCs—an ozone depleter and greenhouse gas) and hydrofluoro-carbons (HFCs—a greenhouse gas), immediately.

Also See

- Dioxins
- Toxic chemicals

Overfishing

Definition

Depleting a population of fish, shellfish, squid, or other aquatic species to the point where the population is threatened with extinction.

Beginning signs of a fish population in trouble are: accelerated growth rates (owing to lack of competition within the species); earlier sexual maturity; and gaps in generations of fish. As these symptoms increase, total harvest may remain the same until the population is completely overfished and its numbers plummet. A population being overfished, with the catch tonnage remaining stable, is said to be in decline.

Environmental Impact

Extinction: Driftnets, fishing factory boats, and trawlers all bring in catches that lead to overfishing. The numbers of fish they take in daily will eventually exhaust current ocean stocks. A depleted stock may not recover if its population is too low; disease or current and temperature changes may wipe it out, or another fish population that competes for the same food may fill the depleted fish's ecological niche and prevent it from ever recovering. This possibility is especially great if a larger fish is overfished; a smaller prey or competitor species may reproduce faster and thus become the dominant species.

When one fish species becomes less plentiful or is exhausted the ships and fishermen turn to more plentiful fish; this pattern may repeat itself until the populations of all major stocks are lowered.

Food chain: The depletion of even one fish species affects the whole food chain. Other animals, such as sharks, dolphins, whales, and bigger fish that feed on the depleted fish species may die off or may turn to other species of fish, depleting them faster. This process can create havoc in the food chain, and the mechanisms are not well enough understood to predict what effect this process has on ocean aquatic populations as a whole.

Nontarget species: Some of the tools of overfishing, such as driftnets and trawlers, also indiscriminately kill many other species of fish, dolphins, shellfish, sharks, sea birds, and other aquatic animals. The numbers of incidental kills are very large and will deplete these species populations so that they, too, may be overfished. Seals, sea lions, and other predators that rely on shellfish and other depleted animals may decline as their food sources are eliminated.

Major Sources

Nonsustainable yields: It is very hard to obtain exact estimates of fish populations, but estimates of sustainable yields (fish harvests the oceans can support without depleting populations) are given at roughly 100 million metric tons per year (includes mollusks and crustaceans). That figure has already been reached for a number of years; overfishing has occurred, and is occurring, in most common fish stocks in the Atlantic and Pacific oceans. Overfishing is occurring along coastal areas and in island reef systems, inland lakes and rivers, and the open ocean.

Driftnets: Driftnets are long, monofilament, fine-mesh nets that can be dragged behind a single fishing boat. They can be 3 to 30 miles long and are usually used to catch a single type of fish, such as salmon, tuna, or squid. They are floated at the surface and weighted at the bottom, and thus hang vertically in the water 10 to 40 feet deep. Their nearly invisible mesh is designed to collapse around anything that touches it, and they ensnare fish, dolphins, seals, sharks, turtles, manatees, whales, and squid, as well as the wings and legs of sea birds. Driftnets cover large areas—a single fleet of boats may cover thousands of miles of ocean in a single sweep.

Trawlers: Fleets of more than 70 multimillion-dollar processing trawlers are scouring the North Pacific and taking billions of pounds of fish out of the ocean each year. Trawlers use long nets that have much the same effect as driftnets on fish and other ocean life.

Increased fishing pressure: In many areas of the world, high-tech boats with radar, fathometers, and other expensive equipment are overfishing grounds continually. More boats, and bigger boats, are taking more fish for more days of the year than ever before.

Target species: Of some 20,000 species of fish, only about 20 make up 2/3 of the annual world catch. With such pressure on just a few species, it is easy to deplete open ocean stocks. Heavy pressure on a few species can also deplete them in localized areas such as lakes, rivers, or coral reefs, where spearfishing or exotic species capture occurs.

Illegal catches: A number of countries have fishing fleets that roam open ocean waters in constant search of fish stocks. Some of them are pirate fleets, operating illegally in other countries' 200-mile economic fishing zones, illegally using driftnets, or exceeding recommended catches. Some fish species—salmon, for example—breed upstream in rivers, say in the United States, and then roam in open ocean waters as adults. Thus whatever measures are taken in coastal or river areas for salmon protection may be nullified by fishing fleets that illegally overfish the salmon in open waters.

Pollution and degradation: Pollution and degradation of estuaries and reefs place more stress on existing ocean fish populations, which breed, reproduce, or feed there. As fish populations dwindle, it becomes easier to overfish them. Also, a number of ocean fish and inland lake fish cannot be eaten because of high mercury or organic chemical content; this situation again increases the pressure to catch more fish, since contaminated fish may be discarded.

Dams: Dams have also contributed to the declines of a number of major fish species, like salmon, that have trouble traveling upstream, past the dams, for breeding purposes; dams also affect downstream fish by reducing water flow. It takes much less fishing pressure on these species to overfish them.

The North Atlantic, South Pacific, and North Pacific supply about 2/3 of the world's fish catch. The Central Pacific, Indian Ocean, and South and Central Atlantic supply most of the rest. Some of the major species that are sought are squid, king crab, Peruvian anchovy, herring, cod, haddock, pollack, capelin, tuna, mackerel, and South African pilchard.

In 1993 the UN FAO estimated that fish provides 16% of world animal protein consumed by humans. Thus in the 1980s the United Nations encouraged driftnet use to help world hunger problems. In 1988 more than 1,000 driftnet boats in the Atlantic were using enough nets to encircle the world. In 1989 a U.N. resolution banned driftnet use in all open waters when it became clear how dangerous they were. It is estimated that as many as 1 million sea birds were killed yearly by driftnets. Millions of target fish, such as tuna, were wasted after being caught in driftnets, which often allowed injured fish to escape; millions of nontarget fish were also killed. It is estimated that several million dolphins have been killed by driftnets.

One-thousand-foot trawler nets for shrimp and other catches are having much the same effect on food chains in the North Pacific. There, Steller's sea lion populations are threatened as their fish food is exterminated. Each year trawlers take in millions of tons of shrimp or other fish, and they catch and release, dead or injured, several times that amount of nontarget fish and shellfish species. It is estimated that as much as 10% of all fish caught are nontarget fish, called bycatch, and thrown away—10 to 20 billion pounds yearly. In Alaska alone, 500 million pounds of fish are wasted every year.

About 1/4 to 1/3 of the world's ocean fish catch is used to feed animals (mainly poultry and other livestock) and fertilize crops. Ocean fish provide 80% of the world fish catch; aquaculture, 15%; and inland fish, 5%. Europe, Japan, and the former Soviet Union each have yearly marine fish catches of 10% to 13% of the world total; Peru takes 8%; the United States, 7%; Chile, 6%; and Canada, Denmark, Iceland, Norway, and Spain, each 2%. Ninety-five percent of the world fish catch occurs in coastal waters.

Aquaculture is expanding rapidly throughout the world and accounts for 10% to 15% of the global fish catch—15.3 million metric tons in 1990. Production should double by the year 2000. It is currently used in close to 100 countries all over the world; Asia produces 84% of the total aquaculture catch, and China alone accounts for 1/2 of that. Currently finfish provide 60% of aquaculture output; crustaceans, 18%; and mollusks, aquatic plants, and others, the rest. In 1992 the United States supplied 25% of the world's salmon (71 million pounds) and shrimp (390 million pounds). In many countries aquaculture is a rural activity, and production takes place in inland ponds. In Asia and Latin America, however, coastal production is increasing, often in mangrove areas. This practice reduces the native breeding grounds for wild fish species and accelerates overfishing.

Depending on the area of the world, 75% to 90% of all commercial marine fish caught are species that depend on coastal estuaries for reproduction, food, migration, or nurseries. The United States has developed 50% of its coastal wetlands; Italy, nearly 100%; and tropical areas, 50%. Reefs are very sensitive to changes and support 1/3 of all fish species. Coral bleaching is occurring with more frequency in the Atlantic and Pacific oceans. In some tropical countries coral is being mined for construction, and some reefs are being "fished" with the use of cyanide poison and explosives.

More than 90% of all fisheries are overfished. A number of ocean fish stocks have collapsed in the past decades, including Peruvian anchovy, major Atlantic fish stocks such as cod and salmon, Alaskan pollack, North Sea cod, South African anchovy, Alaska king crab, and California sardine. Dogfish and skate populations have exploded off the Grand Bank shoals in the Atlantic since the decline of cod from overfishing. Canada has closed cod fishing off Newfoundland because of overfishing; this area used to be one of the world's richest cod fisheries.

The United Nations reports that 4 of 17 major fishing areas are already overfished, 10 areas show declining harvests, and 40 to 60 fish species are in decline or overfished. The U.S. National Marine fisheries

Service estimates 1/3 of important U.S. fish stocks are overfished, and 30% of U.S. stocks have been in decline since the late 1970s. If all fishing stopped, Atlantic salmon would take 20 years to recover. Taiwan's waters are so overfished that they are nearly dead. It is estimated that in the waters around Europe more than 100 species are overfished.

Shark fishing worldwide is threatening a number of species with extinction; more than 200 million sharks may be killed each year. The practice of "finning," cutting off the fin which is a delicacy in some Asian cultures, and discarding the rest of the shark, is used in many parts of the world. Some countries, however, are beginning to protect sharks. Gray nurse sharks are protected off southern Australia; South Africa has banned killing great white sharks. The United States is now considering regulating shark fishing; only about 100 shark attacks on humans are reported yearly worldwide, with about 25 resulting in death.

Thousands of inland lakes in the United States now carry fish advisories because of mercury and other contamination. More lakes are added every year, and more fish species, both ocean and inland, are coming under inspection for this hazard. About 1/3 of all U.S. shellfish beds are closed because of pollution and contamination problems. In the past decade more than 3,500 fish kills were reported along the coasts of more than 20 states, totaling more than 400 million fish; the main causes were wastewater, chemical pollutants, and dissolved oxygen depletion due to nutrient discharges. The Columbia River system used to support 15 to 16 million adult salmon and trout every year; as a result of damming, the river now supports about 2.5 million salmon and trout yearly.

A driving force behind overfishing, besides profit and the need for more food for more people, is the financial investment in fishing boats, whether they are high-tech boats, trawlers, or driftnet boats. This investment creates a vicious cycle: the more money that is invested, the greater the catch must be to pay it off, and thus the more depleted the stocks of fish become, requiring more high-tech boats or driftnets, etc. In short, fishermen are ruining their own livelihoods, along with the world's fishing grounds.

Human Impact

Food: Overfishing reduces food supplies. Reduced supplies put pressure on other food sources, which could lead to famine in some areas.

Ecosystem imbalances: If a species of fish is overfished and cannot make a comeback, that species is for the most part lost to human consumption. No one knows what overfishing is doing to the ocean food chains or to the balance of species in the oceans. Imbalances could affect other fish food sources.

Nontarget species: Driftnets, trawlers, and fish factories can seriously harm other species, including seals, dolphins, sea lions, and whales, which are valuable aesthetically and as an important part of the ecological web. Each time a species is lost, the whole food web is changed; the impact this may have on ocean ecology is not known.

Economics: Lowered fishing yields has resulted in large economic losses for individual fishermen and their countries.

Individual Solutions

- Ask your congressional representatives to support the Magnuson fisheries Conservation and Management Act and to extend it to include reducing bycatch (nontarget fish that are discarded) to insignificant levels.

- Write your congressperson to protest a provision of the General Agreement on Tariffs and Trade (GATT) that could have forced the United States to buy tuna from Mexico caught in driftnets in 1991 (Mexico backed down, for the moment, to help get the North American Free Trade Agreement passed). Be aware that a can that says "dolphin safe" may not in fact be so; current laws still allow 20,000 dolphin deaths through fishing practices.

- Don't throw anything toxic down the drains, on soils, or in sewers; contact local officials or the U.S. Environmental Protection Agency (EPA) for help in identifying and disposing of toxics and hazardous waste.

- Purchase nontoxic products; get a list of nontoxic household products from the EPA or a local environmental group.

- Don't use chemical fertilizer or pesticides; organic products are available.
- Eat less meat; poultry and livestock are major sources of fish consumption.
- Oppose incineration, which adds mercury to lakes and oceans and makes fish inedible.

Industrial/Political Solutions

What's Being Done

- Driftnets have been banned by most countries as of June 1992; Japan, Korea, and Taiwan—which have the largest driftnet fleets in the world—promised to end driftnetting by the end of 1993. Smaller driftnets will still be used by coastal fishermen. France also promised to end driftnetting by the end of 1993. All boats using driftnets will be considered pirate vessels.
- The United Nations recommended an immediate end to all largescale fishing in the South Pacific to protect depleted stocks.
- Countries have 200-mile coastal territorial waters to protect fishing stocks. In the United States this zone was established by the Magnuson Fishery Conservation and Management Act; the Fisheries Conservation Amendments to the act strengthen fisheries management. The act, which sets U.S. fishing quotas, is due for reauthorization during 1994–1995.
- There are world discussions and some agreements on limiting catches to protect major fish stocks; salmon are protected in international waters and can be caught only in coastal waters.
- Europe and the United States have quotas for most fish catches; both regions are still showing significant declines in catches. It is hoped that strict quotas will return fish stocks to healthy limits in 5 to 10 years.
- In November 1991 an international organization that manages the marine area of Antarctica, which is the largest fishery in the Atlantic, set limits on krill fishing.
- Since its fisheries plummeted owing to coral reef degredation, Thai-

land began a National Coral Reef Management Program; other countries are pursuing similar programs.

- The United States has banned imports of fish or marine life caught in driftnets.
- The U.S. Clean Water Act comes up for reauthorization in 1994.
- States are required to have nonpoint pollution programs (see glossary).
- U.S. shrimp trawlers must have turtle excluder devices in nets by December 1994.
- Because of heavy regulation, New England lobster fishing has remained steady, and East Coast striped bass and mackerel have made a comeback; salmon fishing in Alaska is also booming because of good management practices.

What Needs to Be Done

- Pressure countries that support pirate fishing fleets using driftnets or illegal catches.
- Determine sustainable fishing yields for the oceans, and get world agreement to achieve these levels.
- Aim for insignificant bycatch, especially from trawlers.
- Support the U.S. National Marine Fisheries Service so it can regulate sustainable fishing yields.
- Reduce the use of fish for animal feed.
- Bring all fishing fleets, trawlers, and high-tech boats under outside, independent regulation.
- Reduce the number of fishing boats and the number of days they can fish.
- Protect coastal waters from development and pollution; get world agreements on protection of estuaries, mangroves, coral reefs, and coasts.
- End the allowed yearly quota of 20,000 incidental dolphin deaths.

Also See

- Freshwater degradation
- Ocean degradation

Ozone (O₃) depletion, stratospheric

Definition

Thinning of the protective layer of the upper atmosphere, known as the ozone layer, by chlorofluorocarbons (CFCs) and other human-created ozone-depleting chemicals (ODCs).

The ozone layer is a strong-smelling, slightly bluish gas layer in the stratosphere, 10 to 25 miles above the earth, that shields us from the burning ultraviolet rays of the sun. It also helps to trap reflected solar heat around the earth, keeping the earth warm and adding to the natural greenhouse effect.

Ozone is formed as ultraviolet radiation splits apart oxygen (O_2) molecules, which recombine as ozone (O_3)—three joined oxygen atoms. Ozone is eventually broken down into oxygen again by ultraviolet radiation and by naturally occurring nitrogen, hydrogen, and chlorine. The amount of ozone in the stratosphere undergoes seasonal fluctuations (varying as much as 40%), which are influenced by the sun and natural occurrences such as volcanic eruptions. Overall there is a balance between the ozone being created and destroyed. This balance is changed, however, by ODCs. When ODCs rise from the earth's surface, the ultraviolet radiation breaks them down and releases their bromine and chlorine molecules, which destroy ozone molecules.

There are numerous complex chemical reactions and processes that occur in the atmosphere, and many of them are still being researched. Large amounts of chlorine are released by volcanoes and by the evaporation of seawater, but this supply of chlorine is thought to be washed out of the atmosphere before most of it reaches the upper atmosphere; the amount reaching the upper atmosphere is not known. It may be decades before scientists really understand the complexity of the ozone layer and related atmospheric chemical reactions and conditions.

Ozone created at ground level is a major component of smog, which is considered as a separate topic.

Environmental Impact

Plants: As ozone thins, more ultraviolet radiation from the sun strikes the earth. The increased ultraviolet interferes with plant photosynthesis, and even small increases stunt plant growth. It can damage such crops as peas, beans, melons, cabbage, and soybeans. Ultraviolet might also destroy microorganisms in the soil that help plants use atmospheric nitrogen—this would also reduce plant growth.

Oceans: Increases in ultraviolet could affect the growth of coral reefs and ocean life near the surface of the water, including zooplankton, fish larvae and juveniles, shrimp and crab larvae, and other organisms essential to the food chain. The photosynthesis process of phytoplankton in the Antarctic oceans has been affected. Plankton are at the bottom of the food chain in the ocean, and their depletion could affect many species of fish and mammals that depend on it for food. Phytoplankton also absorb large amounts of carbon dioxide, and their depletion may produce changes in the greenhouse effect. Phytoplankton produce dimethyl sulphate (DMS), which helps to form clouds; reductions in this chemical could affect global weather patterns.

Animals: Ultraviolet may cause animals to go blind and cause immune system breakdowns. It can kill the eggs of frogs, toads, salamanders, and other amphibians as well as freshwater fish eggs, minnows, and invertebrates such as insect larvae.

Greenhouse effect: Thinning ozone will trap less heat, which may influence the greenhouse effect. Some evidence shows thinned ozone may cool the stratosphere and affect wind patterns. Ozone-depleting chemicals are also greenhouse gases and thus major contributors to global warming.

Smog: Smog and airborne particulates may help block ultraviolet in industrial areas of the world and temporarily decrease the effects of increased ultraviolet in localized areas.

CFCs: All CFCs are human-created organic chemicals. They are used as cooling agents (often called Freon) in car and home air conditioners, refrigerators, medical solvents, styrofoam, house insulation, foam, microchip solvents, and aerosol cans (banned in the United States but still used in many areas around the world). Use of CFCs in air conditioners is not harmful in itself; the problems result when the conditioners leak or are vented during servicing. CFCs release chlorine.

Halons: Halons are used in fire extinguishers, moth crystals, and solvents. They release bromine.

Carbon tetrachloride: Carbon tetrachloride is used in dry cleaning, metal cleaning, and industrial processes. It releases chlorine.

Methyl bromide: The second most widely used pesticide and soil fumigant in the world, methyl bromide is lethal to all living organisms. It releases bromine.

Methyl chloroform: Methyl chloroform is used as a solvent. It releases chlorine.

Particulates: Sulfates from volcanic eruptions and fossil fuel burning can speed up ozone thinning, as can the ice crystals in the air in polar regions. These particles increase the reactions between chlorine and bromine and ozone molecules. Particulates can also prevent methane from reacting with reactive chlorine, which would render it nonreactive.

Nitrogen oxides: There is evidence that nitrogen oxides cause ozone depletion, especially when they are emitted high in the atmosphere by supersonic jets such as the Concorde. Other evidence shows that stratospheric nitrogen oxides help convert reactive chlorine and bromine to nonreactive forms, preventing them from damaging ozone.

Natural chlorine: Volcanic eruptions, evaporation of seawater, and biomass burning emit natural chlorine. These sources are large, and their contribution to natural ozone thinning is not known.

The ozone hole above Antarctica has been as large as 1.2 million square miles in size. It is created by global winds, which concentrate CFCs in this area. When Antarctic polar stratospheric clouds (PSCs) of micro-scopic ice crystals form, they allow more chlorine and bromine to react with the ozone, which is rapidly thinned by as much as 50% in spring and early summer; PSCs rarely occur in the Arctic. The Antarctic hole has been observed to be increasing by an average of 1.7% to 5% each year over the past decade. Summer 1992 was too warm for PSCs to form, but ozone depletion still occurred, probably because of sulfuric acid droplets owing to the June 1991 Mount Pinatubo eruption in the Phillip-ines. In 1993 a very cold winter and high chlorine levels in the atmos-phere destroyed and thinned parts of the ozone layer in Antarctica in an area 3.4 miles thick (between 8.4 and 11.8 miles high).

Antarctic thinning is thought to have occurred to some degree natu-rally before the widespread use of CFCs caused by PSCs and natural sources of chlorine. We do not know, however, what percentage of the current thinning is due to natural causes and what percentage is due solely to ODC use. In 1989 chlorine levels started leveling off in the ozone layer, at 3.4 parts per billion, thanks to decreased production, and scientists are now hopeful that as more CFCs are phased out massive ozone depletion may not occur.

For every 1% decrease in ozone, experts believe there may be a 2% increase in ultraviolet radiation (UV-B) and, according to the U.S. Envi-ronmental Protection Agency (EPA), an estimated 20,000 additional deaths yearly from related skin cancer. A 12% to 15% thinning of the ozone layer has been detected over North America—in temperate areas this thinning has been thought to be due to the Mount Pinatubo erup-tion. Thus far, large increases in UV levels on the ground have not been detected everywhere, though 45% increases in UV have been detected over the southern part of Argentina in the summer, significant decreases have been detected in the Antarctic, and smaller increases have been detected elsewhere. A Canadian study measuring UV levels over Toronto from 1989 to 1993 confirmed a strong, consistant correlation of thinning ozone to increases in UV. In some areas, atmospheric pollu-tants and smog absorb increased UV, preventing it from reaching the ground and making it difficult to measure. Up to 20% ozone thinning has been reported over Alaska, northern Canada, Greenland, Norway, and Siberia; the EPA estimated this thinning could lead to 200,000 deaths and 12 million cases of skin cancer over the next 50 years, as well as more than 1 million new cataract cases each year. Currently there are 600,000 cases of skin cancer each year in the United States, and 9,000 deaths (these cases are believed to be mostly due to increased outdoor leisure activities).

The United States is the largest producer of CFCs and uses 30% of the world CFC stockpile and 43% of the world's methyl bromide. Car air conditioners account for 20% of all CFC use in the United States; 60 to 90 million cars made before 1994 may need air conditioner retrofitting or a replacement chemical for the CFCs.

One atom of chlorine, from one molecule of CFC, can destroy 100,000 molecules of ozone. Halon and methyl bromide contain bromine instead of chlorine; bromine is 30 to 120 times more efficient than chlorine at destroying ozone. Halons alone contribute to 25% of ozone depletion. Methyl bromide is the second most used pesticide in the world; 4.3 million pounds were used in 1990 in California. A 10% reduction in methyl bromide would be equivalent to a CFC phaseout 3 years earlier. One proposed set of CFC alternatives, hydrochlorofluorocarbons (HCFCs), is also an ozone depleter.

In January, February, and March 1993, ozone depletion over parts of North America was measured to be as high as 45% on some days. Ozone depletion of the same magnitude was measured in 1992 and 1993 over parts of Siberia and Europe. In some small regions of Antarctica, ozone depletion has been measured as high as 90%; in some small narrow regions, it has vanished completely.

In the Antarctic, scientists have reported up to 12% reductions in phytoplankton growth in areas where ultraviolet radiation has doubled. A 1994 study found that ultraviolet B radiation kills the eggs of frogs and toads; it linked declining numbers of eggs in the Cascade Mountains of Oregon to increased UV as a result of ozone thinning.

Scientists estimate that chlorine levels will peak in the atmosphere by 2000, at about 4 parts per billion, even with immediate ending of production of all ODCs. The lags occur because it can take from 2 to 20 years for ODCs to reach the ozone layer. They can last up to 100 years in the lower atmosphere, and the high volume of ODCs produced in the past will not yet have reached the upper atmosphere. Depletion may continue until 2050, and then the atmosphere will begin to heal. Once ODC use and production ends, it will take anywhere from 70 to 100 years for chlorine levels in the atmosphere to return to normal—0.5 parts per billion; the Antarctic hole may last through most of the next century. All these predictions worsen if ozone-depleting alternatives to ODCs, such as HCFCs, are allowed to be used.

Industry and scientists suspected as early as the 1970s that CFCs could destroy ozone; they have known for many years that HCFCs also destroy ozone.

Human Impact

Health: If ultraviolet levels increase, they will increase skin cancer (and related deaths), cataracts, blindness, and suppression of immune systems, causing a lowered resistance to diseases and infectious illness. All of these problems may be more serious in developing countries where people lack adequate health care. Ultraviolet also increases the number of chemical reactions in the air and will increase formation of acids and other toxic chemicals, leading to respiratory problems and other diseases.

Food: fish populations dependent on phytoplankton, which is destroyed by ultraviolet exposure, may decline. Increased ultraviolet will cause more crop losses.

Material damage: Ultraviolet weakens wood, rubber, and plastic materials.

Individual Solutions

- Limit sun exposure, especially between 10 A.M. and 3 P.M.; wear sunscreen with a sun protection factor of 15 or more for UV-A and UV-B, hats, long-sleeved shirts, and sunglasses, or use an umbrella.

- Make sure refrigerants in your air conditioners (house and car) and refrigerators are recycled by service professionals.
- Have home and car air conditioners checked for leaks regularly by professionals; don't continue to use leaking systems.
- Don't buy halon fire extinguishers.
- Don't buy products that use CFCs or HCFCs; safe alternatives, such as CFC-free insulation and halon-free fire extinguishers, are available.
- Don't buy methyl chloroform products. Read labels; methyl chloroform is found in pest sprays, spot removers, and other products.
- Call or write your legislators to demand an immediate phaseout of all ozone-depleting chemicals; insist that hydrochlorofluorocarbons (HCFCs) and hydrofluorocarbons (HFCs) not be used as substitutes. There are already safe, reliable alternatives available for all of the uses of all of the ODCs already in use. Some work even better than currently used ODCs, and some are less expensive. One example is a water-based cooling system for vehicles, already used in buses.

Industrial/Political Solutions

What's Being Done

- Some of the safe alternatives to CFCs and ODCs are: water, hydrogen, and ammonia-based cooling systems to replace CFC air conditioning and cooling use; water-based and solvent-free cleaning solutions, and electronic circuit boards that do not require CFC cleaning solvents; pump bottles instead of aerosols to replace CFC propellants; different insulation material, or water and carbon dioxide as a blowing agent to replace CFC use in making foams; carbon dioxide, water, and foam to replace halon in fire extinguishers, and organic farming techniques to replace use of methyl bromide.
- The Montreal Protocol, an international agreement, was amended and signed by major CFC producers and consumers and calls for phasing out CFCs by 2000. A $240 million fund was established to assist developing nations meet this goal. Most countries are phasing out CFCs earlier. CFCs can be destroyed by incineration, ultraviolet, and oxidation.

- In 1989 the European Community agreed to cut CFC consumption by 85% as soon as possible. Several major European CFC producers will end production in 1995, and Germany will cease production by 1994. The EC has decided to phase out HCFCs by 2014 and is cutting methyl bromide use by 25% by January 1, 1996.
- Israel, which produces about 1/3 of the methyl bromide used world-wide, has pledged to phase out its use and find alternatives.
- Nissan and other car companies pledged to replace CFCs in auto air conditioners by 1993.
- Upper atmosphere research satellites (UARS) will measure and track ozone, clouds, pollutants, rainfall, gases, glaciers, ocean circulation, and sea ice.
- Research on alternatives to replace CFCs is being conducted; in all cases safe alternatives already exist.
- Dupont, the world's largest producer of CFCs, is making alternatives called HFCs and HCFCs; HCFCs deplete ozone, and both HFCs and HCFCs are greenhouse gases. Dupont currently has permission to make these chemicals until 2030.
- Imported products with CFCs must be labeled.
- The 1990 U.S. Clean Air Act Amendments ban CFCs by the year 2000, methyl chloroform by 2002, and HCFCs by 2030. The United States is phasing out production of all CFCs (not ODCs) by 1995, halons by 1994, carbon tetrachloride by 1994, and methyl chloroform by 1996; methyl bromide will be phased out by 2000, with production and consumption frozen at 1991 levels by 1994.
- AT&T, one of the world's largest CFC users, pledged to end all CFC use by 1994. The company helped found the Industry Cooperative for Ozone Layer Protection to help other industries end use of CFCs.
- A U.S. ban on venting CFCs during appliance repair began in 1992–1993.
- The EPA set requirements in 1992 for the recycling, sale, and disposal of CFCs; it has already assessed $500 million in fines for illegal venting and sale of CFCs.

- Some states are banning registration of cars, beginning with 1993 models, that use CFCs.
- Refrigerators and air conditioners using propane, butane, and other alternatives as refrigerant chemicals are being manufactured; water-based cooling systems are already in use in buses in the United States.
- Research is being conducted on the effects of increased UV on plants and animals.
- The U.S. excise tax on ozone-depleting chemicals has been increased to encourage phaseout.

What Needs to Be Done

- Mandate a quicker worldwide ban on CFCs and all ozone-depleting chemicals, including HCFCs and HFCs.
- End exemptions on all products that use CFCs and other ozone-depleting chemicals.
- Ensure that developing countries have access to safe alternatives to CFCs—not HCFCs or HFCs.
- Mandate recycling, storing, and regulated destruction of CFCs currently in use and production.
- Mandate that alternatives to CFCs be safe and not pose other toxic threats.

Also See

- Chlorine (Cl)
- Greenhouse effect

Particulates

P

Definition

Liquid or solid particles in air or gas, also sometimes called aerosols.

Primary particulates are those emitted directly into the atmosphere. Secondary particulates are created when primary particulates react with other materials in the atmosphere. Moisture and gravity take particles out of the air.

Environmental Impact

Air pollution: Any particulates in the air contribute to air pollution, which often inhibits plant growth and can destroy forests.

Acid rain: Sulfur dioxide and nitrogen oxides change to acidic particulates. These particulates join with moisture to create acid rain, which destroys lakes, fish, and aquatic animal populations and also inhibits tree and plant growth.

Greenhouse effect: Particulates in the atmosphere block sunlight and can thereby temporarily cool the planet's surface temperature. Particulates also serve as seeds, called condensation nuclei, for cloud formation. Water vapor collects around these particles and forms fog or clouds. Low clouds hold heat in the air and warm the earth. Higher clouds can reflect heat and cause a cooling effect. Both processes can alter the greenhouse effect slightly.

Ozone depletion: Fine sulfate particles in the upper atmosphere can increase the chemical reactions between chlorine and bromine molecules and ozone, speeding up ozone depletion. Ozone depletion allows more ultraviolet radiation to reach the earth. Ultraviolet inhibits plant growth and destroys phy-

toplankton in the Antarctic; the destruction of phytoplankton could affect the ocean food chain. Sulfates may also temporarily block incoming ultraviolet; increased pollution control aimed at reducing particulates may therefore increase ultraviolet exposure in some local areas.
Smog: Fine particulates attract and hold more toxic chemicals, including heavy metals, radon, and organic chemicals. Thus the result is smog that is even more deadly to plants and animals.

Major Sources

Fine particulates: Particulates smaller than 10 microns in diameter can remain in the air for weeks, months, or years; can travel great distances in the wind; and have more relative surface area so more chemical reactions can occur on their surfaces and more toxics can stick to them. Fine particulates are created by minute dust or liquid droplets from fossil fuel combustion, or by changes of gas pollutants into secondary particulates, such as when sulfur dioxide changes into sulfate particles. Examples of fine particulate sources are electric utility plants, motor vehicles, smelters, and oil refineries.

Coarse particulates: Particulates larger than 10 microns fall to the ground faster than fine particulates, don't travel far from an emission source, and have less surface area for chemical reactions and toxics. Coarse particulates are soon precipitated out of the air in rain, snow, or other moisture or as dry soot. Coarse particulates are created by stone-crushing operations, soil erosion, mining, and volcanic eruptions. Biomass burning, such as forest fires and the burning of savannas, agricultural wastes, and fuelwood, is a large global source of both coarse and fine particulates; it contributes up to 1/3 of global organic carbon aerosols.

State of the Earth

Dangerous air particulates are used in many industries. For instance, silica, used in the silicon chip computer industry, can be as dangerous as asbestos if breathed or ingested. In the United States, 7 million tons of particulates were released in 1990. Many U.S. cities have particulate levels 2 to 5 times what it takes to cause childhood respiratory problems and increase death rates for people over 50; particulates threaten the health of 23 million Americans. Less than 1/3 of air pollution control efforts are aimed at removing fine particulates from air emissions.

In June 1991, Mount Pinatubo in the Philippines erupted and spewed 15 to 30 million tons of sulfur dioxide into the upper atmosphere. The coarse particulates, silicate dust, have already fallen out of the atmosphere. But the fine particulates, sulfuric acid droplets, will remain in the air about 3 years; they absorb the sunlight in the upper atmosphere, warming it but scattering light into space, thus cooling the lower atmosphere and earth's surface. About 1/3 of the sulfuric acid is washed out of the atmosphere each year as acid rain.

Sulfate levels have decreased in some areas in the eastern United States, mainly through power plant pollution control. Still, lifeless lakes and rivers have been documented in past decades in nearly every country in the world. Acid rain has damaged trees along nearly the whole length of the Appalachian Mountains. In Europe, 15% of the forests are damaged, and in the United Kingdom, nearly 65%. Acid rain threatens tropical areas as a result of biomass burning; many species could be threatened by acidification of tropical water.

Smog causes $3 billion to $5 billion in U.S. crop losses yearly; Europe suffered crop and timber losses above $20 billion to pollution in 1987. Worldwide, cars are the biggest cause of smog; though China is mainly a country of bicycles, its big city smog problems rival those of industrial nations because of its heavy industrial air emissions. Smog is increasing in many countries because the number of cars and trucks, and miles traveled, is rising. Smog is also prevalent in many tropical and temperate areas owing to global biomass burning of savannas, agricultural wastes, fuelwood, and forests.

Particulates caused by dust (from construction sites), forest fires, and especially burning fuels cause premature deaths of some 50,000 Americans yearly; in cities in developing countries, dust and soot cause about 500,000 premature deaths yearly. Worldwide, it is estimated that industry, the burning of fossil fuels, deforestation, biomass burning, and other emissions create several hundred million tons of particulates yearly. Airplanes release millions of pounds of particulates to the air yearly. About 1/2 of all human-created particulates are sulfates, created when sulfur dioxide is emitted in the atmosphere.

Human Impact

Health: Coarse particulates are easier to see and avoid, but they can cause coughing, lung damage, and nasal and eye irritation. Fine particulates are a more serious threat, because they can penetrate deep into the lungs and bloodstream, carry more toxics, and can cause respiratory illness and blood vessel diseases. A 1994 report by the Centers for Disease Control and Prevention found 1.9 million asthmatic children, 3.5 million asthmatic adults, and 6.2 million people with chronic pulmonary disease living in counties with the worst particulate air pollution. Other associated effects of particulates are aggravation of bronchitis, chronic respiratory disease, significant lung impairment of children, deaths in the elderly or those with preexisting heart or lung disease, asthmatic and allergic reactions, cancer, and premature death. Increased levels of particulates increase the severity and frequency of symptoms.

Children: Children often inhale particulates more deeply into their lungs, since they tend to breathe more deeply than adults and are more active. They also spend more time outside, where particulates tend to be more concentrated in polluted areas. Particulates will be concentrated indoors, however, if someone smokes inside the home or if particulate toxics like radon, lead, or asbestos are present.

Smoking: Cigarette smoke is the largest source of exposure to particulates for smokers and for nonsmokers exposed to repeated second-hand cigarette smoke.

Miners: Respiratory diseases from mining-related particulates have increased; miners can suffer from silicosis, a progressive lung disease caused by inhaling silicates, which are a large part of the earth's crust.

Food: Particulate problems such as air pollution, acid rain, and smog all limit food crop growth.

P

Individual Solutions

- Use exhaust fans when cooking over a gas stove.
- Limit heavy exercise when pollutant levels are high outside; don't exercise near roadways where traffic is high.
- Drive less; walk or bicycle, use mass transit, and combine errands.
- Practice energy conservation.
- Investigate solar or wind energy for home use.
- Support zero emissions for electric utilities and automobiles.
- If you work in industry, find out what particulates may be emitted inside your working area, and take precautions.
- If you live in a polluted area, wear a breathing mask when you go outside.

Industrial/Political Solutions

What's Being Done

- Programs worldwide are helping limit the burning of fuelwood and forests.
- Catalytic converters are required for new cars in Canada, Europe, Japan, and the United States.
- The 1990 U.S. Clean Air Act is setting tougher limits on sulfur dioxide and nitrogen oxide emissions; it is also requiring cleaner car fuels in some major cities.
- The U.S. Environmental Protection Agency (EPA) particulate standards emphasize monitoring and reducing emissions of fine particulates.
- By October 1993, the EPA required highway buses and trucks to burn low-sulfur diesel fuel (with 80% less sulfur) in order to reduce toxic particulate emissions by 90%.
- Many states are setting their own air emission and clean fuel requirements.
- California is requiring 10% of all new cars sold by 2003 to be zero-emission (electric) cars—about 200,000 by 2003. Ten other states have said they will follow California's lead. Also in California,

pollution from nonvehicular engines must be cut 46% by 1995, another 55% by 1999. Federal regulations for 1996 aim to reduce hydrocarbons by 32% and carbon monoxide by 14% by 2003.

What Needs to Be Done

- Mandate the use of scrubbers by electric utilities and industry; focus on pollution prevention instead of emissions trading.
- Encourage and set incentives for the use of catalytic converters worldwide.
- Mandate cleaner fuel use.
- Set stricter emission regulations on diesel engines; increase support for trains.
- Mandate higher fuel-efficiency standards for cars; phase in zero-emission vehicles.
- Tax fossil fuels to reflect health costs and environmental damage caused by their pollution.
- Increase investments in solar, wind, and other off-the-grid renewable energy.
- Make world agreements to reduce and set limits for particulate emissions.

Also See

- Acid rain
- Air pollution
- Smog

Pesticides

Definition

Chemicals or biocides used to kill and control unwanted insects (insecticides); spiders (acaricides); worms (nematicides); bacteria (bactericides); plants (herbicides); fungi, molds, and mildew (fungicides); fish (pisicides); birds (avicides); snails and slugs (molluscicides); and rodents (rodenticides).

Pesticides are usually synthetic organic chemicals, often derived from petroleum. They include organochlorines (chlorinated hydrocarbons), organophosphates, carbamates, and botanicals. Pesticides may be powders, granules, or liquids.

Pesticides can act on a broad spectrum, killing any unwanted weeds or pests, or they can act selectively and target one particular pest. Sometimes combinations of pesticides are used for several different effects. Pesticides can act on contact, killing the unwanted weed or insect immediately, or systemically, by entering the plant through the roots and then killing the plant or the insects that eat the plant. They can act as defoliants, repellents, disinfectants, sterilants, baits, or growth regulators.

Examples of pesticides, some of which have been phased out of use in the United States, are Agent Orange, lindane/HCH, captan, 2,4,5-T, dieldrin, 2,4-D, carbaryl, methyl bromide, aldrin, endrin, aldicarb, DDT, chlordane, heptachlor, pentachlorophenol, camphechlor, malathion, parathion, ethylene dibromide, methyl bromide, and chlordimeform. Major classifications are organophosphates, chlorinated hydrocarbons (most of which have been banned), botanicals, arsenicals, pyrethroids, phenol derivatives, carbamates, and phenoxyaliphatic acids. The most commonly used herbicides are alachlor (Lasso), atrazine, 2,4-D, paraquat (Gramoxone), glyphosate (Roundup), trifluralin (Treflan), and simazine. Methyl bromide, the most common fumigant, adds to ozone depletion.

Environmental Impact

Nontarget species: Pesticides are washed from the ground into lakes, rivers, and oceans; often they do not biodegrade and remain in the environment for long periods of time. They may be taken up by plants and small organisms and enter the food chain. Pesticides bioaccumulate and can become more toxic as they increase in concentration. They are lethal to a wide variety of plants, birds, insects, fish, and mammals. Through biomagnification, pesticides often affect top predators the most; bird and mammal populations have been decimated with their use. Many pesticides disrupt reproductive behavior, cause cancer and cell mutations, lead to abnormal eggs or young, and harm the nervous system.

Soil: Pesticides kill small insects and organisms in the soil that help the decay process of plant material. When these organisms are killed, soil nutrients are not replaced by the decay process, and farmers therefore need to increase their use of fertilizers. Soils eventually become sterile and lifeless.

Pesticide resistance: The target insects of pesticides often develop resistance to a chemical, and thus stronger doses, or different chemicals, must be used to secure the same results, yielding more environmental damage. Eventually the insect population can become immune to different pesticides.

Population explosions: Pesticides can kill insect predators. When that happens a population of prey insects may explode. Insect populations may also explode as a result of pesticide resistance. Killing one pest may open up a new ecological niche to a different insect, one that was not a pest before but that is able to thrive with reduced competition.

Transported pesticides: Pesticides can dry or vaporize and be carried around the world on air currents until they are eventually precipitated in rain or other moisture.

P

Agricultural: Three-fourths of all pesticide use is designed to limit pest destruction of crops and control weeds that compete with food crop plants. Pesticide use is usually a result of monoculture farming, where one crop is grown repeatedly on large acreage; this type of farming allows pest populations to explode among a large, readily available food source. Pesticides may be more necessary for hybrid crops, such as seedless grapes, which do not have as much resistance to fungi or other pests as natural varieties do. Pesticides are also often used to enhance and maintain cosmetic features of fruit and produce and to defoliate cotton before harvesting.

Other uses: Pesticides are used for roadside ditch spraying, mosquito spraying, lawns, golf courses, swimming pools, and lakes and in a number of products, including paints, carpets, wood products (to prevent rot and termites), household pests, and dog and cat flea collars.

State of the Earth

Researchers have shown that DDT used in Asia and Europe has been carried by wind to the United States. DDT has even been found in the body fat and eggs of Antarctic penguins. In 1993 DDT was still in use in southern Mexico; 20 million acres of land in Russia still have DDT on them. DDT was used on Allied troops in World War II to prevent a typhus epidemic in Italy. Other pesticide residues have also been found all over the world in animal populations. A number of major cities and several dozen states in the United States have had pesticides from rural areas precipitated on them in rain or snow. In California pesticide residues have been found in winter fog.

Less than 1% of pesticides reach their target pest; two-thirds are applied aerially. There are millions of different insects, and though pesticides are aimed at less than 1/10 of 1% of them, these chemicals are lethal to most of them. For example, mosquito and other government pest control programs use chemicals like malathion. Malathion and carbaryl also kill bees, fish, birds, frogs, salamanders, and other beneficial insects.

Multiple food studies have shown that pesticide use creates pest problems instead of eliminating them. Crop losses due to pests have increased from 7% to 13% despite increases in pesticide use; a growing number of insects (more than 500) and weeds are showing resistance to pesticides and herbicides. Sweden cut its pesticide use by 50%, and its crop yields have remained the same. A number of other European and Asian countries are also cutting pesticide use. Integrated pest management (IPM) is being used in Indonesia and other Asian countries. Canadian farmers in Quebec use 2/3 less pesticides than U.S. farmers.

The National Research Council of the National Academy of Sciences says 20,000 cancer cases a year in the United States are caused by pesticide residues, which are present in nearly every food. Israeli women showed sharp declines in breast cancer when use of dangerous pesticides was discontinued. Dozens of pesticides now in use in the United States are carcinogenic in animals. Nineteen pesticides that disrupt the human hormone system are in wide use in the United States today (220 million pounds are used on 68 different crops yearly). Nearly 90% of pesticides have not been tested for long-term health effects; 2/3 of the 600 in use now have never been subjected to any health regulations.

The United States is the largest user of pesticides worldwide; it now uses about 1 billion pounds of pesticides each year for agriculture and conventional uses and another billion pounds for wood preservation (mainly creosote on railroad ties). Brazil is the second largest user; together Argentina, Brazil, Colombia, and Mexico use 90% of Latin America's usage. China is a major user, followed by India, Indonesia, and South Korea; Egypt and Sudan also use significant amounts of pesticides. Worldwide more than 4 billion pounds are used. It is estimated that 40 million pounds of pesticides are used for lawn care in the United States. Three-fourths of all pesticide use is for agriculture; two-thirds of all agricultural pesticides are herbicides used mainly on corn, soybeans, cotton, and wheat. Thirty percent of all fruits, vegetables, and meat in Russia are contaminated with high levels of pesticides and nitrates.

Many foreign countries are using pesticides banned in the United States, such as DDT, aldrin, lindane, BHC, and dieldrin; U.S. companies still are exporting some of these chemicals to them. DDT and DBCP

(1,2-dibromochloropropane) was sold to Central American countries; DBCP was banned in the United States in 1979 because it causes cancer and sterility. Chlordane and heptachlor, also carcinogenic, are sold abroad as well. Foods grown in developing countries with these pesticides are imported by the United States. Over 70% of the world's coffee is sprayed with malathion, DDT, and other chemicals; the chemicals are destroyed when coffee is roasted, but they endanger workers and contaminate soils. The U.S. Food and Drug Administration (FDA) samples less than 1% of imported produce; DDT is the third most detected residue. The U.S. Environmental Protection Agency (EPA) approves dozens of pesticide residues on foods; many are suspected carcinogens.

There are about 7,000 agrichemical companies in the United States. The contaminated soil around them is usually treated as hazardous waste and landfilled; the companies, however, want to sell the soil as a pesticide treatment. Biodegradable herbicides were supposed to be safe, but current studies show they can break down into even more toxic components that live longer in soil and water than the original herbicide.

In Iowa, Minnesota, and Wisconsin (all heavily farmed states), nearly 1/3 of all wells tested are toxic with pesticides; in 1992 71 southern Wisconsin wells were too contaminated with atrazine to drink from. Pesticides have fouled drinking wells in more than 40 states in the United States; an EPA study found that 10% of all community drinking wells and 4% of domestic wells have pesticide residues. In a survey of Midwestern streams, 90% tested had herbicide residues. The whole navigable length of the Mississippi River is contaminated with herbicides; in 1991 the U.S. Geological Survey found the river dumping 160 tons of atrazine, 56 tons of metachlor, and 18 tons of alachlor into the Gulf of Mexico.

Human Impact

Exposure: Pesticides can be absorbed through the skin (often by workers) or taken in with food (of plant or animal origin), air, or water. These chemicals leave residues on the surface of the foods we eat, and often they leave systemic residues inside the food. Pesticides are also washed into the ground and thus enter drinking water. When dried they can become windborne and later precipitate in snow or rain.

Health: The effects of pesticides vary widely, from mild to severe, though long-term health effects have been studied for few of them. Pesticides can cause hair loss, stomach cancer, kidney and liver problems, neurological problems, heart and lung damage, immune problems, birth defects, sterility, miscarriages, reproductive organ problems, hormone imbalances, genetic mutations, and skin disorders. Women with high pesticide levels in their blood and fatty tissues have cancer risks 10 times the normal level. No one understands how several of these chemicals might affect the body if ingested at the same time. And no one understands how these chemicals will affect long-term health and succeeding generations. The EPA does not even test for a number of health effects, such as hormone disruption, which many pesticides cause.

Children: Children are much more sensitive than adults to the effects of these chemicals, and they eat more foods that may be heavy in pesticides; children also obtain pesticides from contaminated water. It is estimated that children receive about 1/3 of their entire lifetime dose of carcinogenic pesticides by age 5; the average 1-year-old has already received a lifetime dose of 8 pesticides from just 20 commonly eaten foods.

Circle of poison: U.S. companies are allowed to sell a number of banned U.S. pesticides to developing countries, which use them on produce that is sold to the United States; thus the pesticides complete a full circle.

Worker safety: Fruit and produce field workers have heavy exposure to these chemicals and suffer high incidences of poisoning. Rates of poisoning are especially high in undereducated workers in developing countries. Each year there are an estimated 10,000 pesticide-related deaths and 1 million poisonings among farm workers. Farm workers, often migrant laborers, suffer the highest rates of occupational illness of any group because of their exposure to pesticides. Many children work in these fields.

Food: Because soils are sterilized, the nutritional quality of food is reduced. Food crop wipeouts due to insect population explosions are more likely with monoculture farming and pesticide use.

Individual Solutions

- Buy organically grown food (food grown without any pesticide use).
- Wash your produce and peel your fruits and vegetables if they are not organic; skins tend to absorb pesticides.
- If you drink well water, have it tested.
- Don't use pesticides on your lawns; use native grasses and organic (chemical-free) lawn care products.
- Use natural insect repellents.
- Keep any pesticides locked up and away from children.
- Put up a bat house; bats, which eat up to 3,000 bugs each night, are harmless.
- Oppose government spraying programs.
- Take old pesticide containers to hazardous waste collection sites.
- Call or write your legislators to demand an end to U.S. companies' practice of selling banned pesticides to developing countries. Support strengthening the Delaney Clause to cover the use of pesticides on raw foods; currently the clause prohibits FDA approval of any food additive in processed foods that causes cancer in humans or animals. Demand long-term health testing before any chemical is allowed to be used in agriculture.

Industrial/Political Solutions

What's Being Done

- Worldwide, countries are using pesticides less as they realize the dangers they pose to the soil and humans.
- The U.S. Federal Insecticide, Fungicide, and Rodenticide Act (FIFRA) of 1947 regulates pesticide use; in 1972 it was amended to protect human health and the environment.
- The EPA is reregistering some 20,000 pesticides for new stricter standards; California farmers are losing 150 pesticides (and 50% of

active pesticide ingredients) because manufacturers are not pursuing reregistration.

- Research is being conducted on nontoxic, biologically safe pest controls and farming methods.
- Small farming associations in California and other states are using pesticide-free farming practices.
- A larger U.S. organic industry is growing around pesticide-free cotton and other products.
- Organic food grocery chains are spreading around the country. In 1989 organic food markets could not keep up with the demand for organic foods.
- Texas has one of the most comprehensive organic certification programs in the United States; sales of Texas organic products are nearly $250 million annually.
- Biotechnology (genetic engineering) is being used to create herbicide-resistant food crops, which will lead to continued use of herbicides.
- Farmers from Central and South America are suing U.S. agrocompanies in U.S. courts for selling banned U.S. pesticides that cause serious health problems.
- A hormone is being studied that may interrupt mosquito reproduction; it may be used instead of pesticides.
- The EPA is setting lower limits for exposure of food pesticide residues for children and is also considering total exposure, not just food exposure; The EPA has stated that children's health will be placed ahead of agricultural concerns. The FDA, Department of Agriculture, and National Academy of Sciences have also supported the use of fewer chemicals on food crops.
- The EPA is encouraging integrated pest management and biologically based nontoxic pesticides; it no longer allows emergency use of pesticides on crops if the pesticide poses any risks to animals or humans.
- The EPA set new regulations in 1992 to protect migrant and crop workers: workers cannot be present during pesticide applications or for an interval afterward; workers must be provided with protective equipment; workers must be provided with necessary emergency care

and decontamination; and workers must receive pesticide safety training and information.

- The National Academy of Sciences issued a 1993 report, "Pesticides in the Diets of Infants and Children," stating that current regulatory procedures are inadequate to guarantee the safety of the food supply.

What Needs to Be Done

- Research and fund natural methods of pest control.
- Mandate testing of all pesticides for long-term health effects before they are used. Do not allow chemicals to be used while they are being tested. Companies should pay for all testing, which should be done by independent organizations.
- Discourage all use of pesticides for cosmetic purposes.
- Ban U.S. companies from exporting pesticides, or their constituent chemicals, that are already banned in the United States.
- Ban government pest control programs that use chemicals that have not been tested for long-term health or environmental effects.
- Support sustainable agriculture.
- End subsidies of pesticide-based agriculture.
- Discontinue use of university and federal research funds to create herbicide-resistant crops using biotechnology.
- Exempt organic farmers from mandated pesticide applications. In Minnesota in 1989 and 1990, organic farmers had to suffer pesticide spraying, without choice, to control grasshopper populations, which were threatening nonorganic farmers.
- Phase out lawncare pesticide use.

Also See

- Biotechnology
- Sustainable agriculture

Population pressures

Stress on the environment, people, wildlife, or resources from the demands of increasing human populations.

Any population in nature—of insects, animals, plants, birds, or other living organisms—is affected by the environment's carrying capacity. This term refers to the optimal population that a given environment can sustain without suffering damage to its basic structure. For example, if there are too many rabbits a meadow's grass becomes overgrazed, and the rabbit population plummets, or migrates, until the meadow can again grow back and support more rabbits. A population that uses local resources at a pace that allows the resources to be replenished, without depletion or destruction, is a sustainable population.

The human environment includes both local areas and the world as a whole, and both have finite, limited resources—they will provide only so much food, so much water, so much energy, and so much space, or absorb so much pollution, before they are exhausted or destroyed. Therefore if the human population continues to increase at its present pace, it will eventually reach the carrying capacity of the planet, and the human population will also plummet. In a number of areas worldwide, in both industrial and developing countries, increasing human populations have already used up local resources.

Environmental Impact

Population pressures contribute to and accelerate all major environmental problems, including:

Air pollution: Air pollutants, smog, and acid rain increase with rising industrial pollution, automobile use, and fossil fuel use, especially coal use.

Freshwater degradation: Freshwater pollution results from increased sewage, agricultural and urban runoff, and industrial waste and pollutants; shortages of fresh water come with increased use by cities, industries, and agriculture, and with degradation of wetlands.

Soil degradation: Erosion and pollution occur with increased deforestation, overgrazing, industry, and urban waste.

Deforestation: Increased fuelwood needs, agriculture, ranching, mining, and logging lead to deforestation.

Extinctions: Extinctions are caused by increases in habitat loss, toxic chemicals, overhunting, and overfishing. Increased human population leads to decreased animal habitats and animal populations. Birds, mammals, and other wildlife also become more stressed, neurotic, and diseased when they lose habitat, become crowded, lose food sources, or are contaminated with pollutants. These stresses can result in breeding and reproductive problems and a lower overall survival rate of young and old.

Ecosystem breakdowns: As more species are lost, the food chain is interrupted and other species may become stressed, causing a breakdown that spreads along the food chain.

Ozone depletion: Chlorofluorocarbons (CFCs) cause ozone depletion, and the seriousness of the problem is partly related to the amount of CFCs used in air conditioners and industry, which is related to the size of the population.

Climate changes: Global weather changes are caused by increased fossil fuel use and resultant greenhouse gas emissions. Local changes stem from increased deforestation and soil degradation, which result in more water evaporation and exacerbate drought conditions.

Major Sources

Factors that lead to population growth, and thus population pressures, are:

Medical care: Medical care has improved worldwide, lowering infant mortality and extending lifespans; this improvement is the single biggest cause of population increases.

Poverty: Parents often raise many children as a form of "social security" in poor countries or developing regions.

Family planning availability: Countries that lack family planning resources have the highest birth rates.

Education of women: In general, the lower the education of women, the more children they have. Countries that have increased education of women have lowered birth rates.

Social patterns: In some countries women are pressured to have many children as a status symbol.

Religion: In some countries, regions, or social units, religious beliefs prevent the use of birth control or strongly support large families; this pattern is changing in many areas, particularly in Latin America.

Population pressures are felt most strongly by:

Developing countries: In developing countries resources are not as developed, agricultural abilities are not maximized, and pollution is not regulated or prevented. Poor developing countries also have few natural resources to use or export and must import much of their food, energy, and water. As populations expand, these countries often go into debt to meet their needs.

Indigenous peoples: Indigenous peoples often live in areas desired by expanding populations to satisfy their increasing energy, food, and land needs.

Wilderness areas: Wilderness areas are viewed as sources of timber, land, and food.

Some population pressures are made more severe by:

Nonsustainable lifestyles: Throwaway societies and energy-intensive lifestyles use large amounts of raw resources and generate a great deal of pollution—especially in consumer-oriented industrial countries.

Developing countries: Developing countries are beginning to develop the same technologies (often without pollution control) and lifestyles as industrial countries, thereby intensifying resource use and pollution.

Poor planning: Governments and local communities have inadequate powers to deal with increasing populations.

P

Unfairness: Some governments are unwilling to allocate resources fairly.

Limited resources: Limited water, agricultural land, and energy resources exacerbate population problems.

Pollution: Some countries practice no pollution prevention or focus on regulation instead of prevention.

State of the Earth

Three people are born every second, 1/4 million every day, and about 95 million every year. The annual rate of population increase is just under 2%; from 1950 to 1990 world population growth rates were estimated by the United Nations Population Division to be 1.8% and 1.6%, respectively. The fertility rate averages 2% worldwide: for Rwanda it is 8.5%; Saudi Arabia, 6.4%; Bolivia, 4.6%; Mongolia, 4.6%; Argentina, 2.8%; Sweden, 2.1%; the United States, 2.1%; Germany, 1.5%; Hong Kong, 1.4%; and Italy, 1.3%. Thirty percent to 50% of U.S. population growth is a result of immigration. Most of the world increase and 85% of the population in the next few decades will be in developing countries; currently 73% (4 billion people) live in developing countries and 37% in cities. By 2000 there will be several dozen cities with more than 10 million people and nearly 50 cities with more than 5 million people. Almost 1/2 of these 50 cities will be located in Asia. By 2000, 17 of the 20 largest cities will be in developing countries. More than 1/2 the inhabitants of the largest cities in developing countries live in slums.

In 1650 the world population was 1/2 billion; in 1850, 1 billion; 1925, 2 billion; 1960, 3 billion; 1974, 4 billion; 1987, 5 billion; 1990, 5.3 billion; and 1993, 5.5 billion. By 2000 it will be 6.3 billion and by 2035, 11 billion. In 1993, 60% of all people lived in Asia, 12% in Africa, 9% in Europe, 8% in Latin America, 5% in North America, 5% in the former Soviet Union, and 0.5% in Oceania. By 2000 megacity projections are Mexico City, 25 million; São Paulo, 22 million; Tokyo, 20 million; Shanghai and New York, 17 million each; Calcutta and Bombay, 16 million each; and Beijing and Los Angeles, 14 million each. More than 100 million people emigrate yearly.

Brazil, China, India, Indonesia, Sri Lanka, and Thailand have strong family planning campaigns. Contraception is used by less than 50% of people in developing countries overall: 60% in Brazil, 70% in China and Thailand, and 80% in Korea and Taiwan. Contraception is much less common in developing African countries—sub-Sahara Africa has the fastest growing population of developing countries. China's strong family planning program could lessen its population expansion by as much as 1/2 billion by 2000. China offers public health, education to women, and job programs; some of its measures, however, have been viewed as strongly coercive. In contrast, to keep its economy growing, Japan is encouraging its women to have more children; the educated women are resisting this campaign. More than 1/2 billion women around the globe who want fewer children do not have access to family planning services. Worldwide, women perform 2/3 of the work, earn 1/10 of the income, and own less than 1% of the property.

Out of about 150 nations, Singapore has the highest population density per square mile and Mongolia the lowest; the United States falls in between. Japan has the highest average life expectancy, at 78. Half the world population has an average expected life span greater than 65, and 1/2 has an average expected life span between 40 and 65. The upper life span limit for men is 82.5, and for women 87.5; both have been steadily increasing over the past decades. About 1/2 billion to 1 billion people suffer from malnutrition. More than 10 million children under 5 die yearly as a result of pollution, disease, and malnutrition; 2 million die from diarrhea. Famine is decreasing worldwide, except in Africa where more than 10 million children may be orphaned owing to maternal deaths related to AIDS in the 1990s.

Worldwide, 60% of all forested lands have been cleared. Half of all tropical rain forests have already been cut down; 40 million acres of tropical forests are destroyed annually. Tropical forest soil supports crops for only 2 to 5 years. Five thousand species are going extinct each year. More than 90% of all commercial fisheries are overfished. Currently 1/2 the population in the United States lives in areas close to coastlines; worldwide, 2/3 of all humans live in areas near coastlines and rivers draining into coastal waters. The United States has developed 50% of its coastal wetlands; Italy, nearly 100%; and tropical areas, 50%.

P

The Green Revolution, designed to increase food production in the 1960s and 1970s, relied on hybrid crops and heavy fertilizer and pesticide use to increase crop yields. Hybrid crops are not as hardy as native varieties and require more chemical input; pesticides deplete and destroy soils and, with fertilizers, pollute rivers, lakes, and drinking water. The initial burst of high crop yields has ended, and depleted soils from this type of farming now require ever more fertilizer and pesticides to maintain their yields. Crop loss to insects has gone from 7% to 14% in the past 3 or 4 decades, even with the emphasis on pesticides and monoculture farming.

Earth's farmers must feed 90 million more mouths annually, with 25 billion tons less topsoil; by 2035, to feed everyone, food production will have to double. By 2000 about 95% of the arable land in developing countries will have been plowed. Production of food crops, fish, and meat has fallen, or peaked, over the past decade. Grain production has decreased almost 10% since the mid-1980s, fish harvests for the past 3 years have been 97 million tons (with maximum sustainable ocean fishing limits estimated at 100 million tons), and meat production per person is at its lowest level per capita in 3 decades. It is estimated that by 2100, with the current population growth trends and resource use, the earth's depleted resources will be able to comfortably support a population of no more than 2 billion people.

Population increases in industrial countries, though much smaller than in developing countries, pose much greater threats to the environment; the average citizen in an industrial country uses up to 100 times the resources and contributes up to 2,000 times the toxic waste generated by a farmer in a developing country. Japan went past its carrying capacity long ago and now imports about 1/3 of its food and nearly all of its key industrial raw materials. Japan is also one of the largest consumers of tropical forest hardwood, along with the United States. As developing countries such as China and India begin to compete for the earth's resources, they will each contribute as much to the global warming problem, and other environmental problems, as the United States does now.

Population increases create pressures associated with:

Famine: Lack of agricultural land, soil degradation, crop wipeouts, overfishing, overhunting, poverty, and drought exacerbated by poor land and water use can lead to famine.

Freshwater shortages: Water shortages can be due to overuse and waste, exacerbated drought conditions brought on by poor land management, pollution, and the destruction of wetlands and forests, which aid the soils absorption of water and groundwater replenishment.

Health problems: Crowding, poor waste disposal, raw sewage dumping, and industrial wastes and toxics from increased emissions can result in health problems.

Poverty: Increased populations often result in more limited jobs, resources, and education. Countries may devalue their agricultural resources to feed urban populations, thus increasing rural poverty. Rural populations may move to urban areas, which are not equipped to handle increased numbers.

Violence: Increased warfare, murders, or other types of conflict can be generated over land ownership issues and rights, political demonstrations, and conflicts between urban and rural sectors and within crowded cities.

Noise pollution: Noise pollution increases directly with the number of people, automobiles, and airplanes and with the amount of construction and industrial production.

Aesthetics: Aesthetic values are reduced especially in poor, crowded urban areas.

Individual Solutions

- Support family planning education.
- Support state maternal and child health services and funding.
- Write to legislators to support U.S. family planning support for developing countries.

P

Industrial/Political Solutions

What's Being Done

- A number of countries are trying to limit their populations through education and family planning.
- The U.S. National Academy of Sciences and the Royal Society of London have stated that current population growth and unchanged human activity may result in irreversible environmental degradation.
- In September 1994 the United Nations is sponsoring a Conference on Population and Development in Cairo.
- The United States is resuming funding for the U.N. Population Fund and is supporting international family planning efforts, as well as worldwide basic education, child and maternal health, empowerment of women, sustainable development, and preservation of the natural environment. One specific goal is to provide birth control to every woman in the developing world who wants it by the year 2000.
- A number of organizations are offering family planning assistance, as well as organization on this issue, to developing countries.

What Needs to Be Done

- Encourage recognition by governments worldwide that increasing populations are a major problem that needs to be addressed.
- Increase funding for contraceptive research.
- Provide state funding for reproductive health care services.
- Increase funding for maternal and child health programs.
- Make family planning services available to all families worldwide.
- Reduce poverty; one reason for overpopulation in developing countries is that parents see large families as a social security measure.
- Support education, especially of women, in developing countries. In general, the higher the status of women in a country, the fewer children they have; this pattern is especially true when women can realize personal gains, economic and otherwise, from their education.
- Support land reform in developing countries to equitably distribute land to more citizens instead of just a wealthy few.

- Urge that national population policies reflect a sustainable relationship with a country's assets, resources, and environment.
- Share technology and information with developing countries to help them stabilize their populations, reduce waste, and prevent pollution.
- Institute fair world market prices and trading practices that help poor nations reduce debt and dependence on foreign markets.
- Reduce consumption of red meat; increase grain production.
- Ensure that the North American Free Trade Agreement (NAFTA) does not supersede government, state, local, and tribal laws.
- End military and other aid to governments that neglect their poor.
- Pressure governments and lending institutions like the World Bank to focus on sustainable development.

Also See

- Sustainable development

Radon (Rn)

Definition

Odorless, tasteless, almost inert, invisible radioactive gas.

Radon is a natural product of the radioactive breakdown of radium, thorium, and uranium in soil, rock, and water; once emitted, it enters the air or water or is trapped in pockets in the soil or rocks. Radon decays, or breaks down, into a series of dangerous, short-lived radioactive particles, called radon daughters (polonium, bismuth, and lead), which emit alpha particles. Therefore, when radon is found in an area, it is generally assumed that radon daughters are also present.

Environmental Impact

Animals: If radon is released into the air, it can be precipitated onto soil, water, and plants in rain or snow or as dust or smoke. It can then enter the food chain after being taken up by plant roots or being eaten or consumed in drinking water by fish and animals. Radon causes cancer, breathing problems, and a shortened life span in animals.

Background radiation: Radon released during mining increases the overall level of local background radiation in the environment.

Major Sources

Environment: Since uranium and radium are present in most mineral deposits and soils, some level of radon can be found in nearly all soils and water. Globally, soils emit small quantities of radon daughters into the air. Radon is most concentrated in rock formations like granite and rock that have radium and uranium ores; the highest levels are associated with concentrated uranium ores. Radon is also found in phosphate rock, whose phosphorus is used in chemical fertilizers. Radon can also be transported by cold water and methane gas and can be found in many caves. Volcanic eruptions can emit large quantities of radon daughters into the air, and biomass burning has also been shown to release radon daughters. Unlike volcanic eruptions, the majority of global biomass burning (savanna burning, agricultural waste burning, forest fires, and fuelwood use) is human in origin.

Mining: Mining of any type can release radon, and thus any miner can be exposed to it. Radon can be attached to dust and carried on the wind. Construction materials derived from the earth's crust, such as concrete, can have low levels of radon. The fly ash of coal-generated plants can have very high levels of radon; fossil fuels like natural gas and oil also contain radon.

Indoor: Radon can seep through air pockets in the ground and then through cracks and openings into homes, where it becomes one of the primary indoor pollutants. When a house is sealed during cold weather, the heating system draws radon indoors more rapidly; radon can be found in any house, old or new, with or without a basement.

State of the Earth

Radon was used to treat numerous diseases during the 1930s and 1940s. Now, however, radon causes 7,000 to 30,000 cancer deaths each year; 75% are combinations of smoking and radon exposure. The U.S. surgeon general has warned that radon is the second leading cause of lung cancer in the United States; smoking ranks as number one, and second-hand cigarette smoke ranks as number one among nonsmokers.

Indoor home radon levels in some areas may exceed the sum of all natural background radiation. About 7% of the homes in the United States may have dangerous radon levels. Radon, first discovered in tap water in 1902, is very soluble in cold water and is released into the air when water is heated, as in a shower or bath.

It is estimated that 80% of radon and associated daughters come from natural sources and 20% from human-made sources (X rays and nuclear-related activities supply 10% each). The U.S. Environmental Protection Agency (EPA) states that average outdoor radon levels are 0.4 pCi (0.4 picocuries in 1 liter of air) and suggests that homes have maximum radon levels between 2 and 4 pCi per liter. The EPA estimates that for 1,000 people exposed, radon levels of 20 pCi could lead to lung cancer in 8 nonsmokers and 135 smokers; 10 pCi, lung cancer in 4 non-smokers and 71 smokers; 8 pCi, 3 nonsmokers and 57 smokers; 4 pCi, 2 nonsmokers and 29 smokers; 2pCi, 1 nonsmoker and 15 smokers; 1.3 pCi, fewer than 1 nonsmoker and 9 smokers; 0.4 pCi, fewer than 1 non-smoker and 3 smokers.

The Reading Prong, a rock formation under the junction of New York, New Jersey, and Pennsylvania has high concentrations of radium, and 40% of the homes in this area have high radon levels. A number of other sites with high concentrations of radon have been found in the United States and worldwide.

Early uranium miners rested and ate their lunches in mine shafts filled with smoke and radon gas; high levels of lung cancer were first noted among miners in 1597. The first link between radon and lung cancer was discovered in the 1940s. The average age of miners who die of lung cancer is less than 50. Uranium miners are several times more likely to die of lung cancer than the general population.

There are more than 100 million tons of uranium mill tailings being stored at various mining sites. These tailings emit the carcinogenic radon-222 gas, which decays to the radon daughters in a half-life of just under 4 days; about 80% of U.S. uranium production occurs in New Mexico and Wyoming. Phosphogypsum waste piles, from phosphate mining, also can have high levels of radon; more than 90% of U.S. phosphate mine rock production occurs in Florida.

Human Impact

Health: Radon daughter particles often attach themselves to dust or particulates that are inhaled. Radon daughters give off alpha radiation; even a single alpha particle hitting a single cell may set into motion the beginnings of a lung cancer that might not be detectable for 10 to 20 years. These particles, if inhaled or ingested, will settle into the lungs or stomach tissues and irradiate them; this process can result in tumors, lung diseases, and death. Radon risks increase according to how much radon one is exposed to, the amount of time one is exposed, and whether one is or was ever a smoker. Low exposures to radon over long periods pose more risk of cancer than high exposures over short periods. Children are more susceptible than adults to the effects of radon.

Mining: Uranium miners have a high incidence of lung cancer, owing in part to inhalation of radon gas. Over a long period of time radon can also cause cancer in bones, skin, and blood cells. People living in uranium and radium mining areas are at high risk for these diseases. Eating vegetation and meat from these areas also poses a high risk, since radon can be present on the crops and plants and thus in the tissues of animals that eat the plants.

Background radiation: Radon released from mining increases the overall level of background radiation in the local environment, and thus increases the chances of cancer and related diseases.

Individual Solutions

- Have your house tested for radon, or purchase a home test kit; check it over a period of 60 to 90 days in summer and winter. Radon tests are reliable and easy to perform. Radon varies from home to home; your neighbor's test results will not be accurate for your house. Sometimes it is easy to reduce radon entering a building; contact the EPA for recommendations. The EPA suggests taking action to reduce radon when levels are between 2 and 4 pCi per liter; however, it is difficult to reduce radon levels to lower than 2 pCi.
- Test your well water for radon if your home has high radon levels; water is usually a less likely source of concern. Radon can be removed from water before it enters the house.
- Don't smoke in your house; radon attaches to smoke and dust particles.
- Don't use artificial fertilizers; mining phosphate can release radon.

- Keep your house well ventilated; indoor air pollution is often greater than outdoor air pollution.
- If you live near a mining area, there may be a risk that your house has high radon levels. Check with your state health board to see what areas are known to be a high radon risk; all states have radon information centers.
- When buying or renting, check with state officials or the EPA to see if the area has elevated radon levels. Ask the seller or landlord if the building has been checked for radon levels; if so, ask to see the results; ask if corrections were made to reduce radon levels. If the house has not been checked, insist on a radon test. If it is a new home, ask if it was built with radon-reducing specifications.
- Call or write your legislators to demand an end to nuclear energy; mining uranium for any type of nuclear facility releases radon.

Industrial/Political Solutions

What's Being Done

- The EPA has published voluntary guidelines for home builders on reducing radon risks; it has established the National Radon Contractor Proficiency Program (RCP) and Radon Measurement Proficiency (RMP) program to certify contractors who fix and measure radon problems in residences.
- The EPA has proposed permanent radon barriers for uranium mill tailings.
- Studies are being performed to increase understanding of radon exposure risks.
- The United States maps high-risk geological areas.
- Some U.S. states have mandated environmental disclosure requirements for home sellers.
- Building construction techniques that can reduce radon have been devised.
- Kits are available to test for radon.
- Uranium miners wear respiration masks, and mines are better ventilated and monitored than they once were.

What Needs to Be Done

- Require environmental disclosure by home sellers in all states.
- Set more stringent standards for radon exposure for mine workers and people who live near mines.
- Mandate that mining companies contain radon gases that they are releasing.
- Phase out nuclear energy.
- End nuclear energy and fossil fuel subsidies.
- Increase investments in energy conservation and solar and wind energy.

Also See

- Mining
- Nuclear energy

Renewable energy

R

Definition

Energy sources that can be repeatedly tapped without using them up; either the source is inexhaustible or it can be renewed or regrown over and over again.

Renewable energy sources are usually separately from nonrenewable fossil fuels, such as coal, natural gas, or oil.

Environmental Impact

Solar and wind: Solar energy causes no major environmental damage and emits no pollution. Wind energy poses some risk to birds, which can fly into the turbines and be killed. Energy from the sun or wind is very clean in comparison with fossil fuels.

Biomass energy: Depending on the source, biomass energy can be very clean or polluting; still, it is a much cleaner energy source than fossil fuels. Burning methane produced from crop-grown biomass emits less acidity than even the cleanest coal plant; burning fuelwood, however, gives off greenhouse gases and particulates, adding to acid rain and air pollution. Regrowing plant source material takes the emitted carbon dioxide back out of the atmosphere.

Hydroelectric energy: Large dams flood local land, starve downstream ecologies of water, cause erosion, and pollute nearby water. Smaller, run-of-river dams are less disruptive to local ecology.

Geothermal energy: Some geothermal energy designs produce hazardous waste and air emissions; designs that reinject the wastes below ground are much cleaner. Geothermal heat energy is difficult to transport; electricity can be transported more easily. Overall, geothermal energy is a much cleaner energy source than fossil fuels.

Tidal dams: Tidal dams force incoming and outgoing tides to move over turbines that generate electricity; such dams can disrupt coastal ecologies. Tidal power is being developed for seabed use, and the environmental impact is being studied.

Wave stations: Wave stations rely on incoming ocean waves to fill an air chamber and push air over turbines; they suck air back over those same turbines when the waves recede. Wave stations disrupt local ecologies but are less disruptive than tidal dams.

Ocean current power stations: Ocean current power stations are stationary and simply allow ocean currents to pass over turbines. The stations would emit carbon dioxide into the water, and the effects of such emissions on ocean ecology need to be researched further.

Ocean thermal energy stations: Ocean thermal energy stations use the temperature differences between surface and deep ocean water to create electricity. Research on them is ongoing.

Hydrogen: Hydrogen, which can be used as a liquid or gas, burns cleanly, yielding mainly water. It gives off no carbon dioxide, sulfur, hydrocarbons, carbon monoxide, or particulates. The little nitrogen oxide it gives off could be minimized with catalytic heaters or eliminated in fuel cells.

Major Sources

Solar energy: Solar energy can be generated almost anywhere on the planet but still works best in warm, sunny, and often arid climates. Solar energy is being used in every country; it can be used with wind energy.

Wind energy: Wind energy works best in areas that have constant windy local conditions, such as coastal areas or high hills. Some older wind turbines are active only 30% of the time; newer turbines will be active 95% of the time. The West, Northeast, and Great Plains are the most optimal locations in the United States. Wind energy is used and available on every continent and is being used to some degree in more than 100 countries. Residential areas are also using wind energy generated by small turbines or large turbines shared by a neighborhood; wind energy can be used with solar energy.

Geothermal energy: Geothermal energy is derived from hot water, steam, or hot rocks deep within the earth. Often these sources are located in volcanic and earthquake-prone areas such as California, Hawaii, Japan, and New Zealand. Geothermal energy is difficult to recover; new technologies making it more accessible are being developed. Methane is sometimes a by-product of geothermal drilling.

Ocean power: Wave, tidal, and current energy rely on local conditions favoring strong water movements. Wave energy is best where strong winds blow between latitudes 40° and 60° north and south and at latitudes of 30° where the trade winds blow. An example of a location for ocean current power generation would be the Florida Current, the most powerful current in the world. Ocean thermal energy works best in tropical and subtropical oceans, where the temperature gradient is largest; tropical oceans can have water as warm as 25°C on the surface and 5°C at a depth of 1,000 meters.

Biomass energy: Biomass energy can be derived from manure (which produces methane gas), burning wood, plants grown for fuel, or synthetic fuels made from plants. Experiments are under way in a number of countries to find fast-growing plants that yield high-energy outputs when burned or synthesized into fuels.

Hydrogen: Hydrogen is thought to be the renewable energy of the future since it can be created from water (virtually limitless) using renewable energy. Hydrogen is created by splitting apart water molecules (which have two atoms of hydrogen and one atom of oxygen: H_2O) through electrolysis, a process that can be powered by wind, solar, or hydroelectric energy or by the energy derived from the gasification of renewably grown biomass. Hydrogen can also be made from fossil fuels, but then it is no longer a renewable energy, since fossil fuels are not renewable.

State of the Earth

Geothermal energy is currently used in about 30 countries around the world and will be used in 40 countries by 2000; in 10 to 20 years geothermal could be accessible across the United States and provide much of U.S. energy needs. Hydroelectric is used in nearly 100 countries; in Brazil 90% of all electricity comes from hydroelectric.

Wind energy use is significant in a number of countries, including Canada, Denmark, Germany, India, the Netherlands, Sweden, the United Kingdom, and the United States; more than 100 countries are expanding their investments in wind energy sources. Smaller turbines are in use in rural areas in many developing countries. Residential areas in some countries use either small turbines or larger turbines that supply neighborhood electricity, with excess electricity sold to local utilities.

Solar energy is expected to undergo a rapid expansion in the next 10 years, as its technology becomes more advanced and its price becomes competitive with the cost of fossil fuels; the price of solar energy is already competitive if hidden fossil fuel costs, such as pollution and related health problems, are factored in. Large-scale research is currently going on in Europe, Japan, and the United States to make solar energy available for large-scale utility projects. More than 1 million Americans already use solar water-heating systems.

Biogas digesters convert manure or other organic materials into methane fuel for cooking and lighting. They are in wide use in China (more than 6 million biogas digesters) and India. They require little work and yield rich fertilizer from the converted manure. Biomass—dung and crop waste—serves as the main cooking fuel for more than 50% of the world's population. In Brazil, more than 1/2 of fuel has come from sugarcane-produced ethanol since 1986; nine out of 10 cars in Brazil run on ethanol and have since 1983.

Currently about 25% of California's electricity is from renewable sources: nearly 20% is hydroelectric; 6.5%, geothermal; 3%, biomass; 1%, wind; and 0.3%, solar. Renewables currently supply 8% to 10% of U.S. energy needs and are rapidly increasing output; it is estimated that wind energy alone could meet all U.S. electricity needs. Many people worldwide find it makes sense to combine renewable energy sources, like wind and solar or wind and wave energy. Europe has stated its objective of advancing nonnuclear renewable energy sources in the next decades. Germany and Japan are giving large subsidies to research on renewable energy and are not far away from debuting hydrogen-powered vehicles. Germany built a solar-hydrogen prototype power plant in 1991.

In France a 240-megawatt tidal facility has been operating since 1966; Nova Scotia, Canada, has had one since 1984. China, Russia, and the United Kingdom are exploring tidal power; China has built a number of small tidal facilities, as well as small wave stations. India, Norway, Portugal, Sweden, and the United Kingdom have also built wave-power stations. Norway and the United Kingdom have tested wave generation extensively in the North Sea, and the first U.K. wave-power station opened on the island of Islay in 1991. It is estimated that offshore wave-power stations in the Atlantic could meet 50% of the electricity demands of England and Wales; seabed-mounted tidal stations could meet 1/5 of the United Kingdom's electricity needs. Japan is doing extensive research on ocean wave, tidal, thermal, and current energy; it has been using small wave-power units for navigation and weather monitoring buoys for years. Ocean thermal power plants are being developed and built by France, Japan, the Netherlands, Taiwan, the United Kingdom, and the United States.

Human Impact

Health: Two of the major sources of pollution are the use of fossil fuels in cars and power stations. The resultant smog, air pollution, global warming, acid rain, and water pollution contribute to lung diseases, cancer, eye and skin problems, asthma, and other health hazards. Many of these risks would be greatly reduced if we invested in clean, renewable energy sources like hydrogen, solar, and wind. Geothermal energy has had negligible effects on human populations, and ocean power is still being explored; its major impact could be on coastal fish and marine populations.

Hydroelectric energy: Among renewable energy sources, hydroelectric energy has the most severe impact on human populations. Dams create huge reservoirs that often flood large areas and displace indigenous populations. These reservoirs are also breeding grounds for parasites and malarial mosquitoes and thus increase the incidence of both in tropical latitudes. Fish supplies downstream are often decimated.

Biomass energy: Burned biomass can emit air particulates that contribute to respiratory diseases such as asthma and bronchitis.

Economics: Wind and solar energy are already as cost efficient as fossil fuels (cheaper when considering pollution and health-related problems from fossil fuels), and hydrogen in a fuel-cell vehicle is 3 times more efficient than a gasoline internal-combustion engine.

Individual Solutions

- Investigate solar and wind energy for your home electricity needs.
- Build with solar energy for cooling and heating needs.
- Purchase solar-powered products.
- Practice energy conservation; get a home energy audit.
- Ask your congressional representatives to strengthen support for renewable energy and to end fossil fuel subsidies.

Industrial/Political Solutions

What's Being Done

- Renewable energy research is gaining funding worldwide, because of environmental pollution concerns and lowering costs. Solar, wind, and biomass are undergoing rapid expansion; hydrogen and ocean energy are being researched; hydroelectric is expanding in developing countries, but not in the United States, where it faces opposition.
- Electric utility plants must purchase energy developed by small renewable producers (qualifying facilities certified by the Energy Department), and must do so at a rate equivalent to the avoided cost of new or replaced generation. This requirement began with passage of the Public Utilities Regulatory Policies Act (PURPA) in 1978.
- Electric utilities in the Midwest, Northwest, and Northeast are now investing in wind energy; utilities on the West Coast have already invested in wind energy.
- In a number of states, utility regulators must include environmental costs when deciding whether to approve new power plants, and they must approve the least expensive option presented.
- The National Energy Policy Act of 1992 offers variable incentives to renewables in the form of investment and production tax credits.

What Needs to Be Done

- Raise taxes on fossil fuels, to encourage renewable energy sources.
- End fossil fuel and nuclear subsidies; phase out nuclear energy.
- Increase investments in solar and wind, especially in off-the-grid residential, commercial, and industrial applications.
- Increase the U.S. Department of Energy's research and development budget for energy conservation and renewable energy.
- Mandate that environmental costs be figured into coal and nuclear energies.
- Establish joint technology development programs between developing and industrial countries to facilitate renewable energy research and development.
- Research ways to combine systems of wind and solar energy or other renewable energy sources.
- Strengthen local and state energy policies to reflect environmental impact and sustainability of energy generation.
- Reform the electric utility industry; demand that it move in the direction of decentralized renewable energies that are environmentally sound and sustainable.

Also See

- Biomass energy
- Geothermal energy
- Hydroelectric energy
- Solar energy
- Wind energy

Smog

Definition

A gas formed when nitrogen oxides and organic hydrocarbons (chains of hydrogen and carbon molecules) combine in sunlight. This gas is called photochemical smog. It is easily recognized as a yellow, brown, or gray haze, usually hanging over urban areas.

One of the main components of smog is ozone, the same gas that protects the earth from the sun in the upper atmosphere. Smog ozone forms in the lower atmosphere, called the troposphere; it is a combination of sunlight, nitrogen dioxide, and volatile organic compounds (VOCs—a type of hydrocarbon). Tropospheric ozone also forms naturally in nature; ozone is extremely toxic to living organisms.

More than 100 different chemical compounds have been found in smog, including organic chemicals such as VOCs, carbon monoxide, formaldehyde, heavy metals, hydrogen sulfide, lead, particulates, benzene, aldehydes, sulfur dioxide, peroxyacetyl nitrate (PAN), and other toxic chemicals. As more toxic chemicals are used in industry and emitted, smog becomes even more toxic.

Environmental Impact

Plants: Smog damages vegetation and kills pine forests. Ozone, nitrogen oxides, and PAN damage plant tissues, prevent photosynthesis, and render plants more susceptible to disease, drought, and other pollutants. As a gas, smog can enter leaf pores and burn out the cell membranes. Because of wind-transported smog, environmental damage is becoming increasingly serious. Most vegetation is sensitive to smog levels well below the human health–based standard.

Greenhouse effect: Smog holds heat near the earth's surface and thus increases the greenhouse effect.

Acid rain: Smog acidity adds to acid rain, which inhibits plant growth and destroys lakes, fish, small aquatic animal populations, and forests.

Toxics: Tiny solid particulates or liquid droplets in smog can attract toxic chemicals and hold them. This quality makes smog even more toxic and more harmful to plants and wildlife.

Major Sources

Motor vehicles: Cars, buses, and trucks are responsible for more than 50% of the smog-producing chemicals, including ozone, nitrogen dioxide, hydrocarbons, PAN, benzene, lead, and VOCs. Motor vehicle emissions also give off carbon monoxide (CO); CO combines with naturally occurring atmospheric hydroxyl radicals (OH), which destroy ground-level ozone. Thus car emissions not only add to smog, but also reduce quantities of a naturally occurring smog inhibitor.

Industrial and household sources: Industrial sources consist mostly of industries and electric utilities, but other sources include aerosol sprays, pesticides, wood-burning stoves, waste treatment, oil and gas production, gas stations, industrial solvents, paints, coatings, and aircraft engines. These sources give off VOCs, toxics, sulfur dioxide, nitrogen oxides, particulates, and heavy metals.

Biomass burning: Biomass burning occurs to clear forest for agriculture, to control brush and weeds, to burn off dead brush and wood in a forest, to add nutrients to land, to cook and heat, and to produce charcoal. It includes the burning of savannas (43%), agricultural wastes (23%), and forests (18%), and the use of fuelwood (16%). Biomass burning is occurring on a large scale in temperate areas as well as tropical; it gives off hydrocarbons, nitrogen oxides, and carbon monoxide. Some estimate that biomass burning accounts for up to 40% of world tropospheric ozone.

Urban areas: Most smog is centered in urban areas, where there are high concentrations of automobiles and industry. Smog is especially serious in developing countries that still burn leaded gas or that have few pollution control devices for automobile pollutants or industrial emissions. But even pollution control devices on cars will not prevent smog if the concentration of traffic is too high. Thus most large cities have problems with smog.

Weather: Smog worsens with low cloud cover, with little or no wind, and in valleys surrounded by hills or mountains. In a process called a temperature inversion, smog is sometimes trapped near the ground by a layer of warmer air resting above a layer of cooler air. Smog can be transported with wind and can accumulate over areas ranging from several hundred to several thousand square miles. As a result rural areas (especially those downwind of urban areas) also have high ozone levels.

Indoor air: Indoor air is often much more toxic than outdoor air, especially in high-smog areas. In buildings and vehicles, where people spend about 90% of their time, the air circulates less and gases and toxics accumulate.

Natural ozone: At ground level, ozone is formed when lightning strikes the earth; it separates the oxygen molecules (O_2), which combine with a third to form ozone (O_3). Sunlight also combines with methane to form ozone. During photosynthesis trees and vegetation release hydrocarbons, which can react with sunlight and pollutants to form ozone. In a natural forest the pollutant levels would be very low, and the hydrocarbons would be harmless and eventually changed into other harmless substances.

State of the Earth

Large industrial cities have smog levels 3 times higher than the level at which crop damage begins. In the United States 3 out of 5 Americans risk lung damage from smog; up to 76 million live in areas where the air quality is hazardous to their health. In the past, Los Angeles has had frequent smog alerts, in which people are advised to wear breathing masks or stay inside; in 1992 the federal ozone standard was exceeded 140 times in Los Angeles, where children suffer lung function impairment that is 25% worse than normal. Many large U.S. cities have never met the Clean Air Act's initial air quality standards.

Smog causes $3 billion to $5 billion in crop losses yearly, mainly of soybeans, wheat, peanuts, and corn; yields are reduced as much as 12% to 30%. It is estimated that California loses 20% of its cotton and grape crops to ozone damage; Europe suffered crop and timber losses greater than $20 billion to pollution in 1987.

Smog and ozone pollution are very high in tropical areas and in temperate areas where biomass burning occurs, since burning gives off hydrocarbons, nitrogen oxides, and carbon monoxide. Ozone has been detected at near lethal levels in tropical areas; the acidity of the smog is high and may threaten numerous tropical life forms, since most tropical species spend some stage in their life cycle in water, which can be acidified.

Mexico has some of the worst smog in the world, exceeding World Health Organization limits more than 300 days each year. Mexico City extended the school Christmas vacation in 1988–1989 so children would have less exposure to winter smog. The city's smog is created by 5.5 million tons of contaminants from its 2.5 million cars, 200,000 buses, 35,000 taxis, and nearly 40,000 factories; surrounding mountains hold the pollutants captive. Currently Mexico requires all new cars made there to use unleaded gas, and it is considering converting all public transportation vehicles to natural gas use.

Worldwide, cars are the biggest source of smog; 80% of driving is for transportation to work and running errands; 15% of the cars on the road are responsible for more than 50% of car pollutants. Catalytic converters can reduce hydrocarbons by 90% and nitrogen oxides by 75%; most developing countries, however, still do not use them. Though China is mainly a country of bicycles, its big-city smog problems rival those of industrial nations because of its heavy industrial air emissions.

Smog is increasing in many places because the number of cars and trucks, and miles traveled, is increasing. Europe, North America, and many other areas frequently exceed the World Health Organization's ceilings for ozone by wide margins at numerous locations; in October 1993, Athens banned all driving and ordered industry to reduce output by 30% after hundreds of people were hospitalized with smog-related illness. Smog and ozone can be 2 to 3 times more concentrated at increased elevations, and as a result both have damaged trees along much of the Appalachian Trail.

The U.S. Environmental Protection Agency (EPA) has done a study that found that using a riding mower for 1 hour creates as many smog pollutants as driving a car for 20 miles. The 80 to 100 million gas-powered lawn mowers in the United States produce pollution equivalent to 3.5 million newer cars. Electric mowers are more than 70 times less polluting.

In December 1952, London experienced "killer fogs"—black, thick smog created by heavy industrialization with no pollution controls and a lack of wind that allowed the pollution to settle into London's streets. Four thousand people died before the wind cleared the air again, and 8,000 people died in the following weeks from aftereffects.

Human Impact

Health: Smog can lower the body's resistance to colds and pneumonia. It can also aggravate or cause asthma, cancer, chest pains, death for people with cardiovascular problems, drowsiness, eye and skin irritations, heart diseases, lead poisoning, respiratory problems, and scarring of the lungs.

Children: Smog health threats are greater to children and fetuses than to adults. Smog with lead, arsenic, mercury, or other toxics can cause irreversible neurological, brain, kidney, and liver damage, as well as high blood pressure.

Agriculture: Smog causes significant damage to crops such as soybeans, wheat, cotton, lettuce, tomatoes, peanuts, and timber resources.

Individual Solutions

- Keep the air in your house circulating as much as possible.
- Walk or bicycle; combine errands; use mass transit.
- Use ethanol, gasohol, or other cleaner-burning fuels in your car.
- Keep toxics out of your house; call the EPA or a local environmental

group for a list of safe alternatives. Never dispose of toxics in the garbage; call local officials for information on hazardous waste disposal.
- Write your legislators to support zero-emission vehicles and higher fuel efficiency for all vehicles.
- Practice energy conservation; investigate wind and solar energy for your home.
- Use electric or hand-powered motors; buy an electric or reel (push) lawn mower.

Industrial/Political Solutions

What's Being Done
- Canada, Japan, the United States, and most of Europe require all new cars to have catalytic converters.
- Leaded gas is no longer used in most industial countries.
- The U.S. Clean Air Act of 1990 is requiring reductions in car and truck tailpipe emissions of hydrocarbons, carbon monoxide, and various nitrogen oxides beginning in 1994. Auto manufacturers must also reduce vehicle emissions from gasoline during refueling. Nine cities with the worst ozone problems must use cleaner gasoline by 1995; the 22 worst areas must monitor smog levels. The Clean Air Act is also reducing nitrous oxide emissions from utilities.
- In February 1994, the EPA began enforcing the Chemical Manufacturing Rule of the 1990 Clean Air Act, which requires the EPA to regulate 189 toxic air pollutants by the year 2000. The first step by the EPA was to require chemical plants to cut toxic air emissions on 112 pollutants by 88% within 3 years, to be achieved by installing scrubbers or using other pollution-prevention technology. This should reduce yearly toxic emissions of by-product volatile organic compounds (VOCs) by 1 million tons per year (the equivalent of 38 million cars, roughly 1/4 of all cars in the United States).
- In 1990 the EPA set guidelines to reduce VOCs 5% annually in car and light truck emissions, and 75% in hazardous waste operations.
- In October 1993, the EPA required all highway buses and trucks to

burn diesel fuel with 80% less sulfur in order to eliminate 90% of toxic exhaust particulates.

- The EPA has mandated stricter emission limits for cars and trucks in high carbon monoxide and smog areas.
- A growing number of states are mandating use of cleaner automobile fuel; research on cleaner fuels is continuing.
- California is purchasing electric cars and mandating clean fuel use. By 2003 10% of new cars sold in California must be zero-emission autos, most likely powered only by electricity. Ten other states have said they will follow California's lead.
- In California pollution from nonvehicular engines must be cut 46% by 1995 and another 55% by 1999; federal regulations for 1996 aim to reduce hydrocarbons by 32% and carbon monoxide by 14% by 2003.
- California is beginning to do more mass transit planning, focusing on light rail instead of highway planning.

What Needs to Be Done

- Raise fuel-efficiency standards; technology to do so already exists.
- Phase in zero-emission vehicles.
- Phase in zero discharge of VOCs and nitrogen oxides in industrial and utility air emissions; mandate that scrubbers be retrofitted; end emissions trading and focus on pollution prevention instead.
- Tax fossil fuels; end all subsidies to fossil fuel production.
- Fund and develop mass transit.
- Encourage use of catalytic converters and lead-free gasoline.
- Phase out the use of toxics in products and industries that contribute to toxic smog; phase in zero emissions of such toxics.
- Initiate worldwide efforts to limit and reduce biomass burning.

Also See

- Air pollution
- Fossil fuels
- Transportation

Soil degradation

Definition

Loss of soil to erosion or development of soil that is so nutrient poor or polluted that its ability to support crops, trees, or other vegetation is impaired.

Topsoil is formed from the breakdown and decay of organic matter (plants and animals), as well as from the erosion of rock over hundreds or even thousands of years. Soil that is nutrient rich cannot be easily, or quickly, replaced. In general soil is created in nature faster than it erodes; human activities, however, speed up the process of erosion.

Environmental Impact

Eroded soils: Plants cannot grow or reestablish themselves on eroded soils. The topsoil is washed or blown away or compacted into a hard surface, and the nutrients are lost. Soil that is washed into waterways creates turbidity, which causes fish populations to decline, destroys coral reefs, and fills in dam reservoirs, canals, and harbors that eventually have to be abandoned. Eroded land cannot hold water, leading to depleted groundwater supplies and contributing to flooding. As plants die off, wildlife loses habitat. Land eventually goes through a process of desertification, which can affect local weather. Degraded soils also lose held carbon dioxide and methane, which adds to the greenhouse effect.

Nutrient-poor soils: Some soils are nutrient poor to begin with, and agriculture or other pressures deplete their nutrients further until nothing can grow on them. Plants have trouble growing on nutrient-poor soils. Without plants to hold the soil, rain and wind erodes it. Eventually the soil goes through the process of desertification. If still used as farmland, it will require heavy fertilizer use (often artificial), and nitrates will be washed into rivers and ground-water. Fertilizer causes algae growth and can lead to eutrophication, which kills aquatic plants and fish. Nutrient-poor soils also cannot support natural microorganisms that break down organic matter, so nutrients are not replenished. As fertilizer use increases, the mining and production of fertilizer causes more air and water pollution and adds to smog and acid rain.

Polluted soils: Plants may take up toxics or other pollutants from the soil. If the plants are eaten by animals, the toxics continue up the food chain. If the plants die, the toxics remain in the soils. If plants cannot grow because of pollution or excessive salts from irrigation, the soil again erodes or turns hard.

Major Sources

Agriculture: The causes of erosion include failure to plant tree wind-breaks, failure to plant margin bushes or trees to prevent soil runoff at the edges of fields, failure after harvesting to leave crop residue to hold the soil, and failure to till and plant in such a way as to minimize water runoff. Agricultural soil becomes nutrient poor when the same crops are planted year after year, taking the same nutrients out of the soil. Agricultural pesticide use also depletes soil nutrients by killing microorganisms in the soil, which are responsible for decay of organic matter.

Deforestation: When forests or other vegetation are cut down, there are no longer tree roots to hold the soil or block the wind. Rain washes the topsoil away, and wind blows it away. Eventually this process can lead to desertification, in which all the topsoil has been eroded away; in some areas all that is left is a hard dry clay or sand. The topsoil in some tropical areas is very soft and weathered; without trees and vegetation to hold it, it can wash off with heavy rains within a few years. In some places rain can compact tropical forest soils until they become very hard.

Overgrazing: When cattle and livestock overgraze an area, they prevent grass and vegetation from growing and create conditions that favor erosion from wind and rain and eventual desertification. Livestock also trample and compact soil, so water does not soak in and instead runs off the surface, carrying the topsoil with it.

Irrigation: Arid regions lose large tracts of land to irrigation practices, which deposit salts from irrigated water on soils. Since arid areas have low rainfall, the salts are not washed away, and with rapid evaporation they accumulate in the soils, inhibiting plant growth and eventually leaving the soils rock hard. Too much irrigation can also raise the water table, dissolving salts that are added to the upper layers of the soil and speeding the process of salinization throughout the soil layers.

Dams: Reservoirs in tropical areas also create erosion problems downstream as silt is prevented from replacing lost soil along riverbanks and in river basins.

Mining: Strip-mining and pit-mining are responsible for erosion and topsoil losses in large areas.

Construction, roads, and airports: Construction, roads, and airports cause erosion and topsoil losses, or simply cover land with pavement or other materials.

Pollution: Eventually all air pollution falls from the sky in the rain or snow or as dry particles. Acid rain also leaches nutrients out of soils, making them infertile so plants cannot grow or grow poorly. Land-based pollution from oil, buried industrial chemicals, landfills, mining, nuclear wastes, pesticides, and discarded garbage is occurring to some degree in every country in the world.

Natural processes: Soil is constantly eroding naturally on seashores and coastlines, along river banks, and in deserts (from wind). Invading seashores deposit salts naturally on soils, and volcanoes spew ash over soils. Excessive flooding in areas can deposit salt water on soils and ruin them.

State of the Earth

The three biggest causes of all soil degradation worldwide are poor agricultural practices, deforestation, and overgrazing; each contributes roughly 30% to the problem. Soil degradation of arable land is estimated at 40% in India, 25% in Central America and Mexico, 15% in Africa, and less than 10% in South America. In Europe 20 million hectares have been damaged, the worst occurring in Germany, Hungary, Poland, and southern Sweden—acid rain and heavy metals are the main causes. The Brazilian Amazon has 12.5 to 25 million acres of severely degraded pastureland. Globally about 25 million acres of arable land are degraded yearly; nearly 3 billion acres of land were eroded in the past 50 years.

About 1/2 of U.S. farmland is losing topsoil faster than it is formed. In the United States the majority of soil loss is due to water erosion; the second largest cause is wind erosion. Most topsoil is lost as a result of livestock production and poor farming practices; the Dust Bowl of the 1930s was caused by poor farming practices, drought, and wind erosion. The U.S. Department of the Interior says 35% to 40% of federal lands are overgrazed and 15% to 25% are seriously overgrazed.

U.S. developers cover soil with millions of square feet of malls every year; roadways have paved over billions of square feet worldwide and 10% of all arable land in the United States—2 million acres of prime farmland yearly. Ninety-nine percent of U.S. tallgrass prairie—400,000 square miles—has been plowed under or paved over. Population increases everywhere are taking more arable soil out of production; China alone is losing 2 million acres a year.

Topsoil is being lost in the United States at an estimated rate of 5 billion tons a year—2/3 because of poor agricultural practices. The former Soviet Union loses 1.5 billion tons, and India, 6 billion tons. The United States is one of the few countries that is monitoring its soil losses. The Soil Conservation Service (of the Department of Agriculture) completes a National Resources Inventory every 5 years. The 1987 report estimated that 3 billion tons of soil eroded on 422 million acres of cropland, with 2 tons per acre lost to water erosion and 3.3 tons per acre lost to wind erosion. The 1992 report will be available in 1994 or 1995. Haiti, where most farming is done on mountain slopes, loses up to 15,000 acres yearly to erosion. Worldwide, agricultural topsoil losses are estimated at nearly 25 billion tons yearly. It is estimated that soils hold 1.5 trillion metric tons of carbon dioxide, which is lost when they are degraded.

More than 40% of the earth's lands are drylands (arid or semiarid), which degrade faster with poor or intensive agricultural use or over-grazing. It is estimated that more than 15% of all soils worldwide have been at least partially degraded; three billion acres of fertile land have been seriously degraded—the size of China and India combined. Land that is not severely degraded can be recovered with mixed crops, mulching, and proper land use.

Desertification is now an ongoing process in more than 100 countries, affecting 900 million people, especially in India. In West Africa (African Sahel) nearly 90% of productive drylands are already severely degraded. Desertification threatens 50% of Spain's farmland. Researchers are attempting to develop sustainable agricultural methods, using complex mixtures of prairie grasses, to replace topsoil at a rate of 4 to 10 inches over several years; one inch of topsoil can take anywhere from decades to hundreds of years to form through natural processes in most areas.

Dams have affected soils worldwide by preventing silt deposits down-stream and by increasing salinization problems if the reservoirs are used for irrigation. Because of the Aswan High Dam in Egypt, the Nile deposits only a few tons annually of rich organic sediment on shore banks, compared with more than 100 million tons annually before the dam was built; artificial fertilizers must make up the nutrient loss. Water behind the Aswan High Dam is estimated to hold as much as 10% more salt than water entering the reservoir; this water is used for irrigation and has led to massive salinization problems and soil degradation.

Coastal erosion is a problem worldwide because of dam reservoirs and irrigation; dams block the downstream passage of silt and thus prevent river silt from replenishing coastal silt lost to erosion. Loderup Beach in Sweden has lost 150 meters in the past 30 years; nearly all U.S. shorelines are eroding in all 30 coastal states.

Landfill leakage, pesticide use, industrial wastes, urban pollutants, and air pollution are affecting soil quality in nearly every country in the world. Toxics are still being buried underground, along with nuclear waste, and they are entering groundwater and soils in dozens of countries. The United States uses more than 1 billion pounds of pesticides each year.

Human Impact

Food: Soil degradation means less farmland is available to grow crops. Depleted soils also grow fewer crops and less nutritious food. Low crop yields in many poor countries cannot be increased through expensive fertilizer programs, and so they result in food shortages and even famine.

Toxics: Many toxic chemicals that end up in the soil are long lived and will slowly, continually, enter plant tissues that people eat. Thus over years it is possible to build up toxins in the body that will eventually lead to cancer and other diseases.

Fertilizer use: Depletion of soil nutrients necessitates more fertilizer use. Phosphate rock, which is the source of phosphorus in artificial fertilizer, releases radon when it is mined. Radon is a dangerous radioactive gas that causes cancer. Fertilizers leach from soils and end up as nitrates in drinking water; nitrates can lead to cancer. Fertilizer also causes eutrophication of waterways.

Weather: There is evidence that topsoil loss, and resulting desertification, may lead to local changes in weather, including drought, which accelerates crop losses and food shortages.

Flooding: Degraded soils that cannot hold water increase flooding.

Individual Solutions

- Buy produce at organic food cooperatives. Organic farmers grow their crops, in the main, through sustainable agriculture, which maintains topsoil and soil nutrients.
- Remember that air pollution is also soil pollution. Support strict clean air standards; drive less, carpool, walk, or use mass transit.
- Mix mulch or compost into your lawn and garden; plant trees.
- Avoid pesticide use; use organic fertilizers.
- Never throw oil, chemicals, or hazardous waste on land, down sewers, or down your sink; contact the U.S. Environmental Protection Agency (EPA) or local officials for hazardous waste pickup sites.
- Practice energy conservation, which reduces fossil fuel use and resultant pollution.
- Avoid disposable paper and wood products, such as napkins, paper

towels, or chopsticks; use cloth or reusable products instead. Avoid tropical rain forest woods such as ebony, teak, or mahogany.

- Eat less meat, especially red meat.

Industrial/Political Solutions

What's Being Done

- Farmers worldwide are being encouraged to use conservation tillage and to plant windbreaks and margin grasses and shrubs to prevent field erosion. The Conservation Reserve Program pays farmers rent or other aid to take highly erodable soils (36.5 million acres) out of production and plant grasses or trees on them; this subsidy program has prevented 700 million tons of erosion per year. Australian farmers have come under soil conservation mandates, which are replacing unsuccessful voluntary programs worldwide. The U.S. Agency for International Development is currently helping Haiti conserve its soils.
- Agroforestry, which emphasizes planting trees with crops to help prevent soil erosion, is being practiced in developing countries.
- A number of nonprofit and governmental groups are working to save rain forests through incentives, pressure, ecotourism, indigenous people's programs, and scientific assistance.
- Integrated pest management (IPM) and biological controls that emphasize more sustainable practices to control pests in farming are being used instead of high inputs of pesticides and fertilizers.
- Wetlands are being reclaimed in the United States and a number of other countries in Europe and Asia. Wetlands help prevent flooding, which leads to soil erosion.
- In 1991 EPA regulations were implemented to help control coastal erosion.
- The 1990 U.S. Clean Air Act set acid rain controls.
- Research is being conducted on prairie grasses and other plants that may be able to replace topsoil losses much faster than previously thought.
- Federal land livestock grazing fee incentives are being proposed to encourage better land use by ranchers.
- Mining fees for public lands are being raised.

- Institute soil reclamation through sustainable farming practices.
- Establish tax and loan incentives for sustainable agriculture; set up a national training program for sustainable farming.
- Increase wetland reclamation; pursue a no-net-loss policy on wetlands.
- End all burying or dumping of toxic by-products or nuclear waste in soils.
- Renew the Conservation Reserve Program in the 1995 U.S. Farm Bill; 22 million acres will end their contracts with the program in 1996 and 1997.
- Limit livestock grazing on arid and semiarid land; use conservative stocking rates (number of head of cattle allowed per acre) with less brush control (burning of brush to clear land).
- Make the U.S. Bureau of Land Management (BLM), the nation's largest landholder, with 266 million acres, more accountable for the caretaking of the nation's assets.
- Plant trees on marginal cropland and in any open acreage to limit erosion.
- Mandate testing of all chemicals applied to the land for long-term health effects before they are used.
- Help developing countries accomplish soil reclamation through sustainable practices.
- End timber company subsidies; fund the U.S. Forest Service independently of logged wood profits; mandate that the Forest Service aggressively prosecute timber theft in U.S. forests.
- Increase world pressure to end rain forest destruction; end all logging in old-growth forests; support only sustainable forestry and end clearcutting practices.

Also See

- Deforestation
- Livestock problems
- Sustainable agriculture

Solar Energy

Renewable energy from the sun, which is stored, collected, and then used as heat or electricity.

Solar energy is also being researched for creating hydrogen fuel from water, detoxifying hazardous waste, and purifying water.

Environmental Impact

Clean energy: Solar energy is considered one of the cleanest and most environmentally safe sources of energy. It creates no pollution in its collection or use. Using solar to replace fossil fuels or other energy sources would alleviate environmental concerns associated with each energy. For example, the acid rain, global warming, and air pollution associated with fossil fuels would greatly decrease; the generation of nuclear waste from nuclear power plants could be stopped; the destruction of fish populations and downstream ecologies associated with dams and hydroelectric energy could be eliminated.

Drawbacks: Some fossil fuel use is required to manufacture solar equipment, but this amount is still less than that required to manufacture conventional energy equipment. Large areas of land are needed for large solar energy power plants, but it is still less than the amount of land required to strip-mine the coal used in fossil fuel power plants. It is believed that rooftops of buildings and parking lots can offer the needed land space. Water used for cooling in solar power plants may be scarce or costly in desert areas.

Sun: The sun daily sends an inexhaustible supply of energy down to the earth—close to the energy of 200 million power stations yearly, 15,000 times the energy used by the whole world, 1,000 watts of sunlight per square meter. Solar technology now allows solar cells to absorb greater amounts of sun energy even on cloudy days; thus passive solar and photovoltaics can be used virtually anywhere. The Southwest desert area of the United States is best suited, however, for large solar power plants, and clear sunny days still provide the most energy.

Solar energy is currently collected by three main methods:

Photovoltaics (PVs): Photovoltaics use devices (called cells) made of semiconducting materials that absorb light (photons) and convert it directly into electricity (by emitting electrons). The waferlike cells have no moving parts, require no maintenance, and are long lived; they convert about 10% of the sunlight striking them into electricity. This efficiency will climb to 15% in the coming years. Photovoltaics are often used to generate electricity for rural or remote areas that normal power lines cannot reach; the electricity generated is used for water pumping, irrigation, desalinization, refrigeration, and lighting. Photovoltaics are also used in many consumer products and in other large projects such as central power stations. Use of photovoltaics is expected to expand greatly for home electricity, in mini-substations on commercial buildings, and in industry.

Active solar (solar thermal): Collectors track the sun and concentrate its heat and light, using it to heat water, oil, salt, air, or antifreeze, which in turn is used to create electricity. This method is usually used at a central power plant to generate electricity to be transported elsewhere.

Passive solar: Stationary panels collect heat and use it to heat rooms or water. This method is used mainly for houses, greenhouses, and water heaters. Salt ponds can also trap solar energy to be used for water and space heating or to generate electricity.

5

Using passive solar energy in building designs can save up to 70% of heating costs. In Israel, 65% of the homes (700,000) have solar water heaters; Japan uses more than 4 million; Cyprus uses them in more than 90% of houses, apartment buildings, and hotels. Solar cook boxes have been tested successfully in developing countries to eliminate the need for burning wood for cooking.

Photovoltaic programs could provide much of U.S. heating needs and all of U.S. electricity needs in 4 to 5 decades if fully supported. Studies have shown that reaching this latter goal would require only 0.3% to 0.4% of the land area of the continental United States (30,000 square kilometers); this amount is not exorbitant compared with the large amounts of land required by strip mines for coal and other fossil fuel uses. The U.S. Department of Energy showed that coal already uses an equivalent amount of land. Thin-film photovoltaic cells will increase efficiency and lower solar energy costs soon. Nine commercial solar power plants are now operating in the Mojave Desert in the United States.

Europe, Japan, and the United States are all currently trying to develop large-scale utility photovoltaic power plants. Sanyo of Japan has designed rooftop solar shingles that will supply all of a building's electrical needs, with spare energy to be used for recharging electric cars. Germany is constructing about 2,000 rooftop systems for lighting and heating in residential and commercial buildings; the governments of Austria, Italy, Japan, and Switzerland are also investing in residential rooftop solar development.

Mexico uses photovoltaics in residences, and its program is one of the fastest growing in the world. Many developing countries are encouraging PV use, especially Algeria, Brazil, China, India, and Venezuela. India has the most aggressive government PV program among developing countries. Colombia has 17,000 small systems in use.

Indonesia is undertaking a large-scale program to power remote villages with photovoltaics. For instance, PVs are the only electricity source for a remote mountain village there, and they cost less than the batteries and kerosene they replaced; they are effective, even though the village averages only 3 sunny days per week. Three remote villages in the desert of Saudi Arabia have used PV power since 1982; the Greek island of Kithnos produces 1/3 of its electricity with photovoltaics. PVs also power many mountaintop telecommunications stations worldwide; systems such as these can operate unattended for as many as 10 to 20 years. Rural PV electricity is replacing diesel in off-the-grid uses in many developing countries, including Brazil, Dominican Republic, Kenya, Sri Lanka, Tuvalu, and Zimbabwe, as well as rural New York State.

The U.S. Coast Guard already has more than 20,000 PV-powered navigation buoys. Cheaper than conventional electricity sources, PVs are being used in radios, televisions, watches, calculators, portable phones, and highway, billboard, and street lighting. PVs generate on-board electricity for all space shuttles. Photovoltaic "farms" can be run with little or no supervision.

Car models powered by solar energy and batteries have already been developed by a number of automobile manufacturers. Germany, Japan, Saudia Arabia, and the University of California at Riverside will all soon be demonstrating use of solar-produced hydrogen for vehicles and other applications. An ultralight plane recently completed the first solar-powered cross-country trip; solar-powered car races have been held in Australia and the United States.

Solar power plants in California are already generating cheaper electricity than new nuclear plants. Some hybrid power plants in southern California use 75% solar energy and 25% natural gas as a backup for peak power needs.

Human Impact

Health: Solar energy causes no health problems. Because it is a clean energy, using it wherever possible will lessen health problems related to fossil fuel and nuclear energy use.

Worker safety: Manufacturing photovoltaic cells often requires the use of heavy metals, such as arsenic, cadmium, and silicon. Heavy metals, which are very toxic, can cause a myriad of serious health problems, including neurological problems and cancer. Silicon dust can cause respiratory diseases. Thus workers involved with making solar energy equipment must be protected from exposure to these materials.

Individual Solutions

- Support solar energy research and incentives.
- Purchase solar-powered products; look into solar energy, coupled with small wind turbines, for supplying your home electricity needs.
- Build with solar heating; use solar energy for space heating, lighting, water heating, and ventilation.
- Practice energy conservation to reduce fossil fuel needs.
- Demand an end to fossil fuel subsidies and an equal amount of energy research dollars for solar energy.

Industrial/Political Solutions

What's Being Done

- The World Bank is supporting PV use for water pumps, irrigation, refrigeration, communications, battery charging, and lighting in rural areas.
- Research is being conducted on using solar energy to break apart water into hydrogen, which can then be stored and used as a clean energy in heating, transportation (in fuel cells), and electricity. International research is accelerating in PV applications.
- The U.S. Department of Energy is spending $20 million to study thin-film photovoltaics and utility-size applications at its National Renewable Energy Laboratory (NREL) and its Solar Energy Research Facility (SERF).
- The National Energy Policy Act of 1992 offers a 10% solar tax investment credit.
- The Public Utilities Regulatory Policy Act (PURPA) of 1978 required public utilities to purchase electricity from small renewable energy

producers at rates equivalent to the avoided cost of new or replacement generation.

- Electric utilities are studying the possibility of placing PV substations on commercial building rooftops; these substations are expected to be available in 6 to 7 years.
- California is requiring the use of electric cars; the state should have 200,000 by 2003. The South Coast Air Quality District in Diamond Bar, California, has a 3,000-square-foot solar cell array to recharge electric vehicles.

What Needs to Be Done

- Increase government investments in solar energy, especially in off-the-grid applications for residential, commercial, and industrial use.
- Tax fossil fuels and nuclear energy to reflect the environmental damage and health problems caused by fossil fuel pollution and nuclear wastes.
- End fossil fuel and nuclear subsidies.
- Use cost-effective passive and active solar technologies in new government and public buildings.
- Research ways to combine systems of solar and wind or other renewable energy sources.
- Strengthen local and state energy policies to reflect the environmental impact and sustainability of energy generation.
- Reform the electric utility industry; demand that it move in the direction of energy conservation and decentralized renewable energies that are environmentally sound and sustainable.

Also See

- Renewable energy

Solid waste

Definition

Anything that is dumped into the environment either as a finished or used product.

This section focuses on municipal solid waste—what is commonly known as garbage. Municipal solid waste is the waste collected by public or private haulers in communities from homes, institutions, commercial business, government, and industry. It does not include mining wastes, utility plant burned-fuel debris, construction debris, sewage sludge, incinerator ash, and some industrial wastes. Although wastewater and agricultural runoff are sometimes classified as solid waste because they contain particles and suspended matter, they are not included here.

Environmental Impact

The effects of solid waste on the environment depend on the method of disposal. Current methods include:

Landfills: For decades solid waste was simply put into big open pits; most current landfills are of this type. Newer, modern landfills are lined, sealed tightly, and monitored for pollution, methane emissions, and leakage. Garbage in landfills biodegrades slowly, since there often is not enough oxygen beneath the layers of garbage. Landfills leak leachate (rain runoff), which often is toxic and carries nutrients. Leachate pollutes local surface waters, soils, and groundwater. Leachate can also cause high growths of algae in water in a process called eutrophication, which can kill fish and plants. Anaerobic landfill decay produces methane, which is a serious greenhouse gas.

Incineration: Air emissions from the burning of solid waste are toxic and often contain mercury, lead, other heavy metals, sulfur dioxide, nitrogen oxide, particulates, and organic chemicals such as dioxins and furans. These toxics pollute lakes and soils and can be taken up by plant roots and animals in the food chain. Often, even if solid waste is in a nontoxic form, like plastic, when incinerated it can break down into chemicals that cause toxic air pollution. Incinerator emissions add to acid rain and greenhouse gases. Also, the leftover ash is highly toxic, yet it has been landfilled with municipal solid waste and presents risks with the leakage of landfill leachate. Incinerator ash used in road or other construction (mixed with cement or other materials) will eventually degrade, shedding its toxics into the local area where it is used.

Waste-to-energy plants: Some incinerators are designed to generate electricity or heat from burned garbage. The amount of energy generated is less than would be saved if the solid waste were simply recycled. Air emissions are toxic, as is the leftover landfilled ash.

Recycling: Recycling reduces mining and production pollution and energy and prevents solid waste from entering the environment.

Composting: Composting is the process of separating biodegradable from nonbiodegradable wastes and then exposing the biodegradable materials to oxygen so they will break down into small debris (called aerobic decay). Bacteria and small organisms break down the materials. This process, which is the way all organic materials are recycled in nature, can produce a rich fertilizer for soil. It is a clean, effective way of reusing organic material, unless the material also contains toxics such as pesticides, heavy metals, or synthetic organic chemicals; then returning the compost to soils pollutes the ground and the crops that grow on it.

Discarded waste: When solid waste is discarded into oceans, rivers, or waterways, it can kill marine animals such as seals, birds, turtles, and dolphins that mistake the garbage for food or simply do not see it. This danger is especially great with discarded plastic. Also, many materials might react with water or break down over time and become more toxic. Even if randomly discarded solid waste is not toxic, it is unsightly.

Production pollution: Manufacturing products that end up as solid waste requires raw materials such as trees for wood, oil for plastic and synthetics, and minerals. Coal and other fuels are needed for energy in the manufacturing process. Thus producing solid waste also means cutting down more forests, mining more land, and polluting more air and water during manufacturing.

Major Sources

Commercial: Restaurants and businesses discard used products, paper, cardboard, used materials, and containers. Their waste is a mixture of toxic and nontoxic materials.

Household: Household garbage includes food waste, lawn clippings, packaging, used or broken products, empty containers, furniture, and appliances; it is often a mixture of nontoxic and toxic materials.

Industrial: Industrial solid waste consists of manufacturing by-products, leftovers, and nuclear waste that is below regulatory concern and landfilled; it is often toxic or hazardous.

Government: The government throws away paper, cardboard, old furniture, used products, and containers; its waste is potentially hazardous and sometimes toxic.

Military: Military waste includes metal, plastic, and paper; this waste may contain potentially hazardous and toxic materials.

Medical: Hospitals, labs, and clinics discard toxic, hazardous, and plastic disposables.

State of the Earth

Each year the United States produces close to 200 million tons of municipal solid waste, roughly 3 1/2 pounds per person each day (nearly 4 1/2 pounds before recycling). Almost 80% of U.S. solid waste is recyclable, but the amount of garbage, even with recycling, is increasing yearly. The U.S. Environmental Protection Agency (EPA) states that 50% of most cities' garbage is business waste.

Municipal solid waste in the United States consists roughly of (by weight) 34% paper and paperboard, 20% yard trimmings, 9% plastic, 9% food waste, 8% metals, 7% glass, 4% wood, 3% rubber and leather, 2% textiles, and 4% other.

Packaging makes up more than 30% (by volume) of landfill solid waste; Americans throw out 10 to 20 billion disposable diapers yearly. In 1991, 45 billion pounds of plastic was produced; almost 60 billion pounds is projected for 1996—only about 4% is recycled. The United States exported 200 million pounds of plastic waste in 1991, mostly to Asia. In 1993 Germany collected 4 million tons of waste. Europe is aiming at 90% packaging recovery in 10 years; France has a program called Eco-Emballage aiming at reclaiming 75% of all solid waste for recycling, reuse, composting, or incineration energy recovery by 2003.

Many of the chemicals and additives used to manufacture or operate appliances or products are more toxic than the item that contains them. Examples are chemical waste from plastic production, glue and stains used in wood furniture, PCBs in electrical capacitators, inks and heavy metals used in label packaging, formaldehyde used in carpets and drapery, and chlorofluorocarbons (CFCs) left in an old refrigerator.

More than 1/2 of U.S. landfills have closed in the past 10 years, and nearly 1/2 of the remaining 5,800 landfills do not meet federal or state standards for human health and environmental protection. More landfills are being closed as they fail to meet 1993 and 1994 guidelines and as communities resist allowing new landfills in their area; 22 states will run out of landfill capacity within 10 years or less. The nation's 10 largest cities use a land area for their garbage that is larger than the state of Indiana. Landfills in Japan and the United Kingdom are almost filled.

All industrial countries have landfill problems. In the former West Germany thousands of landfill sites have been declared potentially dangerous because they threaten groundwater supplies. The EPA has stated that over time no landfill, even if sealed, is leakproof. Landfills handle 50% to 70% of hazardous wastes. Worldwide, landfills are responsible for 7% of human-caused global methane releases to the atmosphere; methane recovery from landfills is used mainly at large landfills; a large abandoned landfill can produce significants amounts of gas for up to 20 years.

About 2/3 of solid waste is organic and could be composted. The Netherlands composts more than 100,000 tons of material annually, and France has more than 100 facilities generating nearly 1 million tons of compost annually. The United States is beginning to research and develop extensive composting plans; there are more than 1,500 centralized composting operations across the country. The state of Washington has a compost facility that handles 30,000 tons of yard waste annually, which is sold as compost or soil mix.

The United States has about 150 waste-to-energy facilities (incinerators), which generate enough electricity for more than 1 million homes. They burn 100,00 tons of municipal solid waste daily, which is about 15% of the solid waste—1/3 of it ends up as toxic ash. Recycling recovers about 20% of solid waste, and landfills get about 65%. Japan recycles 40% to 60% of its waste. In a number of poor nations, people earn a living by sorting through the garbage at dumps and then recycling and selling it.

The United States is operating or has planned more than 200 material recovery facilities, which can recover over 50% of the garbage going to landfills. New Jersey recycles more than 50% of its discards; Seattle is already over 50% recycling of all its garbage; Florida recycles 25%. Most states have recycling goals of 25% or greater. When materials are recycled, anywhere from 20% to 95% of the production energy and air and water pollution are reduced. Recycling is increasing in all industrial countries; the United States recycled 38% of its paper in 1992. About 200 billion pounds of scrap material from industry were recycled in 1991.

Deposit laws for refillable containers are coming into effect in Canada, Denmark, England, Finland, France, Germany, Japan, the Netherlands, New Zealand, and Norway. Returnables are, ironically, the norm in India, South and Central America, and many other developing countries. Many countries are moving to limit plastic packaging, reduce the use of toxics in packaging, and end the use of nonrecyclable plastics; plastics don't biodegrade. Some plastic jugs and bottles can be refilled 20 to 75 times.

Fresh Kills in Staten Island, New York, the largest dump and human-made structure in the world, leaks 1 million gallons of toxic water each day to surrounding streams; this dump's wastes are being mined and recycled, and methane is being collected from it; methane production alone is estimated at 10 million cubic feet of gas per day.

The world's shipping industry dumps millions of plastic containers into the sea every year. Space junk, some 30,000 to 70,000 pieces of metal 1 centimeter or bigger, may soon prevent satellites and spacecraft from using low orbits—astronauts have already had to avoid chunks of debris when in orbit.

Human Impact

Landfills: Landfills can cause local groundwater contamination and pollute waterways; land used for landfills cannot be used for other purposes and may become polluted. Resources and energy are wasted, and production and mining pollution increases.

Incineration: Toxic air pollutants such as mercury, lead, and dioxin may cause cancer, neurological problems, skin problems, kidney and liver disorders, and respiratory problems. The landfilled ash may release toxics into local water supplies; ash used in construction will pollute the local area when it degrades. With incineration, resources and energy are wasted, and production and mining pollution increases.

Recycling: Recycling saves resources and energy and reduces pollution.

Composting: Composting, which generates fertilizer, is a positive use of materials (if refuse is nontoxic) that has minimal impact on people. Increases in paper or wood production to replace composted materials, however, lead to more pollution and the harvesting of more trees.

Source reduction: Source reduction (see glossary) is the waste strategy with the lowest impact by far. It uses fewer resources, creates less pollution, and therefore requires less energy, technology, and effort to manage.

Individual Solutions

- In general, buy durable goods that last or can be reused; even recycled goods continue the cycle of depleting resources, using more energy, and creating more pollution.

- Buy only reusable, recyclable, or refillable containers; avoid disposable products of any kind, including paper, aluminum, metal, wood, or plastic; reuse containers.
- Avoid buying products with packaging if possible.
- Recycle; more than 3,500 products are now being recycled.
- Sell or donate goods instead of throwing them away.
- Buy food in bulk; local food cooperatives often sell in bulk so you can reuse your own containers.
- Compost lawn clippings and food scraps.
- Oppose incineration.
- Keep toxics out of the solid waste stream; take them to local hazardous waste sites—call local government officials for information.
- Avoid plastics when possible; they produce toxics when incinerated, and they do not biodegrade. Most are not recycled, and if they are, they usually can be recycled only once. This is also true for most foam materials.
- To stop receiving junk mail, send your full name and address to Mail Preference Service, c/o Direct Marketing Association, P.O. Box 9008, Farmingdale, N.Y. 11735-9008. Ask to have your name put on the master "purge list."

Industrial/Political Solutions

What's Being Done

- The focus of solid waste management is shifting to source reduction; the best way to control the solid waste problem is to reduce the amount that is created. This means using little or no packaging, eliminating disposable products, and producing materials that can be reused, refilled, or returned for recycling. Primary objectives should include producing durable goods and avoiding planned obsolescence, or even recycling that continues patterns of raw material use.
- By 2003 products with packaging that can't be recovered will not be allowed in European Community countries; industry will be responsible for taking back and recycling durable goods.
- Programs have been initiated to collect household toxics and hazardous waste in Europe, Japan, and the United States.

- New 1993–1994 EPA landfill regulations seek to reduce pollution problems from older landfills.
- Most U.S. states have recycling goals of 25% of solid waste; 1,800 communities charge customers by the amount of garbage they discard.
- Companies are researching and using packaging—often made of cornstarch—that is either completely recyclable or totally biodegradable. Biopolymers are the newest subject of research. They mimic spider's silk and marine mussel adhesives and are much stronger than petroleum-based synthetics. Renewable and biodegradable, they will be used in packaging, food production, and medicine.
- Life cycle assessments are being used by companies more and more to evaluate a product from beginning (production) to end (disposal); waste audit processes are used to cut waste generation.
- U.S. states are passing minimum content laws, requiring some products to contain a certain percentage of recycled material. Florida requires legal documents to contain recycled paper. The federal government is purchasing writing and copy paper with a minimum of 20% recycled content, 30% by 1998.
- Companies and industry in the United States are forming coalitions to buy recycled products.
- In many areas of the United States composting is keeping organic materials out of solid waste landfills; Seattle has started a Master Composters program in an effort to educate the public. Many states have banned yard waste from landfills.
- A rapidly expanding program called Green Seal identifies environmentally safe products, which can then carry a Green Seal to alert customers. The European Commission, Canada, and Japan are also looking at eco-labeling.
- By July 1, 1994, Illinois, Maine, New Jersey, New York, and another dozen states will ban sales of products such as inks, dyes, pigments, adhesives, and stabilizers that intentionally have heavy metals (lead, cadmium, mercury, chromium) added; unintentional heavy metal content greater than 100 parts per million will be banned July 1, 1996.
- Paper mills in the United States are expanding to use waste paper; other companies are doing the same for other products.

- Vermont banned sales of disposable diapers in 1993; several children's hospitals in Minnesota and other states switched from disposable to cotton diapers. Disposable diapers are being recycled in Canada.
- Deposit legislation for can and bottle returns is in place in 9 U.S. states and a number of countries.
- Landfills are being "mined" to recover raw materials such as metals and are being tapped for methane production; other solid materials are being recycled to create other products.

What Needs to Be Done

- Increase recycling goals through incentives; charge customers for their garbage by weight, volume, or unit. Develop a national program of instruction and compliance for source reduction and recycling.
- Develop more markets for recycled goods; U.S. government purchasers should only buy paper with at least 50% recycled content and give favor to chlorine-free paper. Mandate minimum recycled content for products.
- Make manufacturers responsible for recycling their packaging and products. By mid-1995 German manufacturers will be responsible for collecting 80% of their packaging waste; fees are charged on packaging material in Germany, and it is illegal to discard packaging in household garbage.
- Pass deposit laws for bottles and cans; set higher taxes on nonreturnables.
- Reduce packaging; minimize use of plastic packaging.
- Phase out incineration.
- Packaging and products should not contain toxic materials, especially heavy metals such as lead, mercury, chromium, and cadmium.

Also See

- Hazardous waste
- Incineration

Sulfur dioxide (SO_2)

Colorless acidic gas with an intense odor.

Acid rain: Sulfur dioxide can change into sulfates (like sulfuric acid) or other sulfur oxides when it combines with water in the atmosphere. It then can be precipitated as acid rain or snow. Acid rain is responsible for acidifying lakes, rivers, and soils. It inhibits plant growth, destroys fish populations, and kills forests. It also leaches heavy metals into lakes and streams, further toxifying waters and fish living in those waters. Sulfur dioxide can settle as dry particles or soot onto soils and acidify them. It can combine with smog, making it more acidic.

Greenhouse effect: When sulfur dioxide changes into aerosol sulfates (through oxidation), it can remain in the atmosphere for years. The sulfates can block sunlight, producing a cooling effect, which can alter the greenhouse effect.

Ozone depletion: Sulfates in the upper atmosphere can increase the number of chemical reactions between chlorine and ozone molecules, speeding up ozone depletion. Thinned ozone increases the amount of ultraviolet radiation reaching the earth. Ultraviolet inhibits plant growth and can destroy plankton in the Antarctic, thereby affecting the ocean food chain. Sulfates may also block ultraviolet, producing uncertain net effects of ultraviolet radiation in some local areas. Particulates such as sulfates can also interfere with atmospheric methane, which converts reactive chlorine to nonreactive chlorine; this process also increases ozone depletion.

Major Sources

Utilities: The largest emitters of sulfur dioxide are electric utilities that burn high-sulfur coal. If fuel is refined and the sulfur content is reduced, less sulfur dioxide is emitted.

Burning of fuels and ores: Sulfur dioxide is produced by burning sulfur-containing fuel oil, natural gas, diesel, and gasoline and by smelting nonferrous ores.

Transported: The local areas around power plants without good pollution control equipment can have the highest concentrations of sulfur dioxide, but sulfates can be carried long distances by the wind. High stacks on power plants also allow sulfur to be carried farther on the wind.

Urban areas: Cities often have 10 times the amount of sulfur dioxide found in rural areas.

Mined sulfur: Sulfur is called brimstone in nature; it is mined in a number of ores. Oxygen is chemically added to sulfur (in a process called oxidation) to produce sulfur dioxide, which is used in bleaching, as a fumigant spray on some fruits, and as a preservative. Sulfur dioxide can be oxidized again to produce sulfuric acid. This compound is one of the most commonly used chemicals in industry, and when it vaporizes it can add to acid rain and air pollution problems.

Natural sulfur sources: Decaying vegetation, forest fires, bacterial decomposition, volcanoes, and lightning all produce sulfur. Normally human causes of sulfur dioxide emissions are large in comparison with natural causes. Volcanic eruptions, however, can emit massive amounts of sulfur dioxide in a very short period of time.

State of the Earth

The United States is the world leader for sulfur dioxide emissions, responsible for 20 million tons of sufur dioxide in 1990; the Soviet Union, 9 million tons; and West Germany, 5 million tons. Worldwide electric utilities generate 70% of all emissions; industry, 15%; transportation, 5%; and other fuel burning, 10%. One-fourth of U.S. SO_2

emissions are from power plants in Illinois, Indiana, Kentucky, and Ohio. Natural processes in the United States emit 1/4 of the amount generated by human activities.

All cuts in U.S. sulfur dioxide emissions have been dwarfed by the huge growth of energy production (mainly using coal) in other countries, which has resulted in increases in sulfur dioxide emissions. Many developing countries are using high-sulfur coal, often without any pollution control equipment; China is instituting a sulfur dioxide pollution tax because in 1991 it estimated that releases of 15 million tons of sulfur led to $2.8 billion in crop damage.

Lifeless lakes and rivers due to acid rain have been documented in the past decades in nearly every country in the world. Norwegian Atlantic salmon stocks are virtually extinct owing to acidified river breeding grounds. The National Acid Precipitation Project of the U.S. Environmental Protection Agency (EPA) concluded that 10% of Appalachian streams are acidic. Acid rain has also damaged trees along nearly the whole length of the Appalachian Mountains; spruce, maple beech, and oak have all been affected. Sulfate levels caused by acid rain have been decreasing in the eastern United States in the past decades as power plant emissions have been reduced. Fifteen percent of the forests in Europe and nearly 65% of those in the United Kingdom are damaged.

In June 1991, Mount Pinatubo in the Philippines erupted and spewed 15 to 30 million tons of sulfur dioxide into the upper atmosphere. The sulfur dioxide converted into sulfuric acid droplets, which will remain in the air for about 3 years; they absorb the sunlight in the upper atmosphere, warming it, but scatter light into space, thus cooling the lower atmosphere and planet surface. One-third of the sulfuric acid is washed out of the atmosphere each year as acid rain.

Human Impact

Sulfur dioxide, sulfates, and sulfuric acid all affect:

Respiratory problems: Sulfur dioxide triggers asthma attacks, increases respiratory illnesses such bronchitis, and diminishes the respiratory system's ability to deal with all other air pollutants.

Heavy metals: Heavy metals are leached by acid rain (in part originating from sulfur dioxide) from the soil into water and thus fish. Heavy metals cause a variety of serious diseases, including cancer.

Food resources: Fish populations and crops are damaged by acid rain, ozone depletion, and smog.

Timber resources: Forests are damaged by acid rain.

Buildings: Acid rain dissolves the outer faces of buildings and sculptures.

Ozone depletion: Ozone depletion increases skin cancer and cataracts.

Individual Solutions

- Practice energy conservation to reduce fossil fuel use.
- Buy energy-efficient appliances.
- Recycle; don't buy disposables; reuse containers; buy in bulk.
- Demand stringent acid rain pollution standards for industry.
- Investigate solar and wind energy for home use.

Industrial/Political Solutions

What's Being Done:

- More countries are adding scrubbers to electric utilities.
- Technology exists to reduce SO_2 emissions at coal utilities by 95% to 99%.
- The United States will reduce sulfur dioxide emissions by 40% by the year 2000 by imposing emission allowances under the 1990 Clean Air Act Amendments. This plan will cut about 10 million tons of sulfur dioxide annually from 1980 emission levels; 90% of the reductions will be from electric utilities.
- The EPA mandated an 80% reduction of sulfur in diesel fuel for highway trucks and buses by October 1993.
- Utilities are being forced to consider energy conservation and renewable energy, as the costs of new power plants and fossil fuels rise above those of renewable energy.
- More utilities are using prewashed coal, from which the sulfur has been taken out.

What Needs to Be Done

- Mandate wet scrubbers on smokestacks at coal- and oil-burning power plants, as well as at ore smelters. Scrubbers remove sulfur dioxide from smokestack emissions.
- Mandate the use of "clean" coal—coal that has had the sulfur washed out.
- Increase investments in solar and wind energy, especially in decentralized utility power generation, by using residential, commercial, and industrial solar and wind substations.
- End fossil fuel subsidies.
- Establish worldwide limits on sulfur dioxide emissions; help developing countries limit their sulfur dioxide emissions.
- Stress energy efficiency throughout the world.
- End emissions allowances trading—which allows polluters to buy the right to continue to pollute—and instead focus on pollution prevention.

Also See

- Acid rain
- Coal

Sustainable agriculture

Definition

System of farming that maintains agricultural land and its resources without depleting them for future generations. It uses the best of modern methods blended with time-honored farming traditions.

The main resources that sustainable farming tries to preserve are soil quantity and quality. It emphasizes using both renewable resources and management practices that allow repetitive use without resource depletion. Sustainable farming also considers social equity, not only by attempting to equalize conditions for current farmers, but also by assessing the cost of today's farming practices for future generations. They consider lost topsoil, which nonsustainable farmers might not factor into their costs. Another important consideration is the ecological impact of the farm. In a sense, sustainable farming is an attempt to model a farm after the local natural ecosystem, which perpetuates itself through natural, interdependent processes.

The goal of sustainable farming is to naturally balance soil fertility, humus content, and biological life. It often relies on crop rotation, orchards, agroforestry (planting trees with crops), farming several crops at once (strip farming, or intercropping), contour plowing (planting crops at right angles to sloped land, instead of up and down the slope), no-till plowing (planting without turning over the soil), and terracing (planting a sloped area with level terraces). All of these methods block pest migration, prevent pest population explosions, ensure nutrient levels, retain topsoil, and prevent runoff of pollution. Shelter belts and buffers of trees and shrubs are also used to protect wildlife and to help prevent erosion and the spreading of pests.

Organic farming eliminates all artificial chemicals such as pesticides, herbicides, and artificial fertilizers. Instead it uses natural pest control and natural, bulky fertilizers (compost, manure, mulch), which release nutrients at about the rate of plant uptake of those nutrients. This method allows plants to metabolize naturally, prevents excess water absorption that can occur with artificial fertilizers, and often yields crops that are nutritionally superior. Bulky fertilizers allow air into the soil, providing better decomposition of organic material, more complete nutrient release, and greater crop root uptake of those nutrients.

Organic farmers will leave land fallow (empty of crops) some seasons, so deep-rooted weeds can bring nutrients up from subsoil levels into the topsoil. Crop rotation ensures that all levels of the soil are exploited and none are overexploited. Green crops, like mustard, are plowed back under and increase nutrient levels further. Leguminous plants (clover and field beans) have nitrogen-fixing bacteria on their root nodules, and when used in crop rotations they ensure that nitrogen levels in the soil are maintained. Rotation prevents any one pest from proliferating. Natural crop varieties are often more resistant to local pests and more acclimated to local soil conditions than hybrids or nonlocal plants. All of the above help to eliminate the need for artificial fertilizers and toxic pesticides.

Grazing livestock on sloped hills, instead of planting crops, prevents erosion; livestock can be fed on crop residues, which reduces feed costs, and their manure becomes crop fertilizer. Livestock are raised with ethical considerations for their well-being and without feed additives, chemicals, and growth regulators; they are treated less as individual commodities to be maximized and more as a living part of a whole, interdependent system, which includes the soil, plants, animals, wildlife, and humans.

Often the terms sustainable, organic (used in the United Kingdom and United States), regenerative, and biological (used in Europe) farming are used to refer to the same type of farming. Biodynamic farming is a special type of organic farming.

S

Environmental Impact

Erosion: Sustainable farming methods of tilling and planting preserve topsoils; minimizing erosion keeps silt and nutrients out of rivers, preventing damage to fish and aquatic plant populations.

Pesticides and fertilizers: Eliminating chemical input of pesticides prevents these chemicals from leaching into lakes and rivers and entering the food chain; pesticides can cause neurological, reproductive, and behavioral problems in wildlife, as well as liver and kidney disease and cancer. By minimizing artificial fertilizer use, farmers can keep nitrates out of surface and groundwater. With sustainable farming, manure is not washed into local streams, but instead recycled; fertilizer and manure can cause eutrophication in water, which kills fish and aquatic plants. Soil organisms are harmed by use of both pesticides and artificial fertilizers. Air and water pollution from the manufacturing of agrochemicals is eliminated, as is the mining of phosphates and other chemicals for fertilizers, which releases radon.

Wildlife: Sustainable farming generally tries to achieve a harmony and balance with nature and the surrounding environment. Sustainable farming practices leave wetlands, crop residues, and ditch cover for wildlife.

Genetic diversity: Often sustainable farming ensures more genetic diversity by using open-pollinated seeds and natural seed varieties wherever possible, instead of hybrids, which often need many chemicals and which encourage the loss of diversity in the gene pool. The genetic diversity of wildlife and of wild plant populations is also given more protection.

Major Sources

Organic farming: Sustainable farming is often synonymous with smaller, family-run farms. Organic farms must often be state certified, and their soils must be free of any chemicals for a number of years (at least 3) before they can label their produce as certified organic.

Low-input sustainable agriculture (LISA): LISA is farming by using only minimal inputs of outside resources such as fertilizers, chemicals, fossil fuels, water, capital, and labor—it is not truly sustainable agriculture. To be considered sustainable, an agricultural system must rely solely on what is available in the immediate, local area.

Previous farming: Before chemical fertilizers and pesticides gained widespread use, sustainable practices were by and large the prevalent method of farming. When the United States was first settled, however, farmers often used land until its nutrients were exhausted and the soil was infertile, at which time they moved on to find new land.

Trends: Currently there is a trend in the United States and a number of other countries worldwide to lessen pesticide and chemical fertilizer use and to return to sustainable farming practices. In the long run, even though more labor intensive, these methods cost less, preserve the soil, and result in better crops, higher and more stable yields, and environmental preservation.

State of the Earth

Sustainable farms use 2/3 less energy than chemical farming and cost up to 1/3 less to run. They are cost effective and produce yields per acre that are as good as high-intensity chemical farming. They are more economically stable and less vulnerable to outside influences in the marketplace beyond their control. Nonsustainable farming uses up to 100 pounds per acre of fertilizer; worldwide estimates run as high as 150 million tons; as much as 50% of this fertilizer can be leached from fields by rain and irrigation water into rivers and groundwater.

The United States loses 5 billion tons of topsoil every year through water and wind erosion; the U.S. Department of Agriculture estimates that about 50% of Midwest farms have inadequate protection against erosion. One-third (100 million acres) of U.S. farmland uses conservation tillage. Worldwide, estimates are that 15 to 50 million acres of vegetative land turn to desert yearly through poor land management; 10% of all soils are already depleted. Forty percent of India's farmland is at risk from nonsustainable farm practices. Sustainable farms suffer 20% to 70% less erosion than conventional farms.

It is possible to renew lost topsoil with sustainable agriculture.

Researchers worldwide are experimenting with methods to rebuild soils; soil is basically a mixture of decay material, including minerals, nutrients, and living organisms. Prairie grasses are thought to be a very good soil-rebuilding source; they grow fast and hold a wide variety of species, and thus nutrients. When they are turned over and decay, they can rebuild lost soil and nutrients quickly.

Crop loss to insects has gone from 7% to 14% over the past 3 decades, even with the emphasis on pesticide use for monoculture farming; 50% of tropical country crops are lost to pests and disease. U.S. farmers use 1 billion pounds of pesticides each year; worldwide 4 billion pounds are used. Organic farmers use soap, oil and mineral, and botanical sprays to control pests; these solutions are nontoxic and biodegrade quickly.

Integrated pest management (IPM) uses natural barriers between crops to hinder insect movement (ditches, bushes); crop rotation to break the reproductive cycle of the insects; grazing animals to keep weeds down; biological controls like natural insect predators, parasites, pathogens, and pheromones; and plant varieties that are resistant to insects. IPM eliminates or minimizes pesticide use.

In China, for example, sorghum was planted between cotton plants to attract the natural enemies of cotton pests. Pesticide use fell 90%, and crop yield significantly increased. Polyculture, the practice of planting mixed crops, is being used today in combinations of corn-bean, wheat-flax, and flax-lentils. A South American wasp introduced in El Salvador to eat the eggs of pests eliminated the need for 10 pesticide applications every year; a wasp has also been used to control mealybugs on cassava in 25 African countries. The neem tree, which grows in Africa and Asia, repels many insects with its own chemicals and is under study by many scientists to see if extracts from it can be used as natural pest controls; the first neem tree–based pesticide plant is operating in India. Biotechnology, another tool of IPM, is a fast-developing field that presents some potentially high risks; it uses genetic engineering to develop new strains of plants to combat pests or to grow in arid regions, in acidic soils, or in other adverse situations. IPM techniques are considered truly sustainable, however, only if they use natural on-site and local farm resources.

Indonesia banned 57 of 63 pesticides used on rice and taught more than 200,000 farmers IPM techniques. It took this step after concluding that pesticides had created a pesticide-resistant planthopper, which caused a plague in 1986. Using less pesticide was cheaper, and yields were just as high as before.

Organic food sales were $1.25 billion in 1992 and are expected to triple by 1995. California winemakers are producing organic wines. Europe reduced its pesticide sales market 14% in 1992, and even further in 1993. Germany's Organic Aid Scheme subsidizes farmers switching from conventional to organic agriculture; some German states are up to 5% organic and are aiming for 20% by 2000.

In the United States certified organic farmers have been allowed to use pyrethrum (a natural insecticide that also kills beneficial insects) and strychnine (a natural poison that also kills raccoons and skunks), while importing bat guano (to the detriment of bat caves, and thus bats) and other oddities for fertilizer (creating mining problems). Thus the U.S. National Organic Standards Board is changing the definition of organic to include substances that are "safe for humans and the environment." Under this new definition, pheromones, which are not truly sustainable, will be allowed, since they are safe, while items such as pyrethrum, strychnine, and bat guano will not be allowed.

Human Impact

Food: Crops and animal products grown with organic agriculture are generally free of toxics and more nutritious. Commercially grown produce can have up to 75% fewer nutrients than organic produce. Health problems due to pesticide-contaminated food are eliminated; pesticides can cause birth defects, immune problems, kidney and liver problems, and cancer. Organic agriculture also lessens the threat of insect population explosions and one-disease crop wipeouts, and thus it also staves off the likelihood of famines related to crop failure. Productive soil and farmland are preserved for future generations. Fish and shellfish stocks are not harmed by erosion of soils or nutrients into waterways.

Water: Pesticides and nitrates from artificial fertilizers are kept out of drinking water; related cancer risks are eliminated. Sustainable farming preserves wetlands, which help replenish and filter groundwater supplies, as well as prevent flooding. River quality is maintained through minimal erosion.

Radon: Radon releases from phosphate mining for artificial fertilizers are eliminated, reducing related cancer risks.

Individual Solutions

- Buy locally grown organic foods, especially of native crop varieties instead of hybrids.
- Buy from farmers' markets and food cooperatives. Be aware that many small farmers also use pesticides; ask before you buy.
- Let large grocery chains know you won't buy chemically grown foods.
- Avoid commercially processed, canned, or frozen foods.
- Start your own organic garden; use natural fertilizers; avoid pesticide use.
- Compost your clippings in your lawn or garden.
- For a garden seed catalogue contact Seeds of Change, 621 Old Santa Fe Trail, Santa Fe, N.M. 87501: (505) 983-8956. The catalogue has an extensive list of edible plants, most of which are no longer used commercially, that are raised organically; recipes, numerous plant descriptions, and other information are also provided.

Industrial/Political Solutions

What's Being Done

- A number of countries are specializing in organic crops, which are gaining a stronger worldwide market every year.
- Some countries in the European Community provide subsidies for organic farmers or those farmers converting to sustainable agriculture.
- Pesticide use is being minimized in a number of countries, especially in Asia and Europe; the European Community has an organic food standard that requires no pesticide use and soil fertilized only with organic techniques.

- The U.S. Environmental Protection Agency (EPA) encourages integrated pest management practices; worldwide, organizations like the United Nations Environment Programme and World Bank are beginning to do the same. Some major growers are switching to IPM to save money, address consumer concerns about pesticides, and respond to insect immunity to available pesticides. However, in 1992 the U.S. government said it would allow sales of genetically altered foods. This will put more pressure on farmers to use hybrid seeds instead of natural varieties; hybrids often require more pesticides since they are less hardy than natural varieties.
- The Consultative Group on International Agricultural Research (CGIAR) and other organizations are conducting intensive research on sustainable, low-environmental-impact crops for developing countries; some methods, such as creating new hybrid crops, are thought to be the only way to increase crop production to keep up with population increases.
- World Neighbors, a Tulsa, Oklahoma, organization has more than 90 projects in 20 countries to teach sustainable agriculture to farmers.
- The U.S. National Organic Standards Board will finalize a new definition and new standards for what constitutes organic farming by 1995.
- The U.S. National Academy of Sciences has endorsed sustainable agriculture as more cost effective and ecologically sound than conventional farming.
- The U.S. Farm Bill contains the Conservation Reserve Program, which pays farmers to keep highly erodable soils (currently 36.5 million acres) as grasslands or forests; the Farm Bill also contains the Organic Food Production Act of 1990, which states farms must be free of chemicals for at least 3 years to be considered organic.
- The U.S. Swampbuster program withholds benefits to farmers who plow under wetlands to increase cropland.
- A number of U.S. states tax fertilizers and pesticides to fund sustainable farming research; Minnesota gives low-interest loans to farmers using low-input sustainable agriculture programs. The federal government is funding research on LISA.

6

What Needs to Be Done

- Back sustainable organic farming with incentives; end subsidies for farming practices that are not sustainable.
- Mandate independent testing for long-term health effects of all chemicals used on farms; require chemical companies to pay for such testing, and ban such chemicals until testing is completed.
- Force insurance companies, which are often large landowners, to have contracts with farmers that ensure sustainable practices.
- Support small, family-run farms.
- Tax use of pesticides and artificial fertilizers.
- Renew the Conservation Reserve Program in the 1995 U.S. Farm Bill; fund the Organic Food Production Act.
- Establish a national training program to teach sustainable farming practices.
- Ban genetically altered plant releases into the environment until studies show the crops to be benign environmentally.
- Ensure that international aid and lending institutions like the World Bank work directly with the farmers in a given country, and not just the government; women do 60 to 80% of the farming in many developing countries, and they should receive training and assistance.
- Prevent international seed corporations from taking seed rights from farmers who wish to save seeds from their crops. Current rules in the General Agreement on Tariffs and Trade (GATT) prevent farmers from saving seed from crops to grow new crops unless they pay a royalty; farmers in India rioted in 1993 to protest this measure.
- Land reform is essential in many countries so that local communities and farmers have ownership or long-term tenure, and therefore a stake in preserving land quality.

Also See

- Biotechnology
- Pesticides

Sustainable development

Definition

Developing and using renewable resources in a way that will not have a negative impact on the environment, so that future generations will have access to these resources in nearly the same abundance.

Included in sustainable development is improving the quality of life with social justice, equity, education, health care, and well-being for all people. The cost of any activity or product, industrial, agricultural, or otherwise, must include the cost of the impact on the environment and any loss to future generations. Sustainable resources include soil, water, air, fish and wildlife populations, forests, biodiversity, and renewable energy resources.

A sustainable economy is a renewable economy. For example, a logging town that runs out of timber is not renewable and has not managed its timber in a renewable manner; farming that erodes and pollutes the soil, reducing productivity, is not sustainable. Economic growth often means congestion, crowded conditions, concentrated pollution and waste, and wasted resources. Sustainable development can reach prosperity without such "growth."

People in many areas are no longer living in sustainable situations. An example is Japan, which imports up to 33% of its food and nearly 100% of key raw industrial materials; conversely, if Japan's population were sustainable, it would be small enough to be supported by food acquired from Japan's soils or waters and would make different choices for raw material usage or change the products it manufactures or consumes.

Industrial countries with nonsustainable lifestyles, such as Japan, the United States, and countries in Europe and parts of Asia, are importing huge amounts of raw materials and are responsible for many of the stresses to wilderness, land, and wildlife in many parts of the world. These countries are also responsible for the majority of toxic and hazardous wastes, greenhouse gases, ozone-depleting chemicals, and air and water pollution worldwide. Some developing countries are catching up in terms of pollution and emitted toxics, since they are industrializing with few pollution regulations.

Environmental Impact

Sustainable development would prevent or greatly lessen:

Deforestation: Sustainable development would end clear-cutting and destruction of rain forests and old-growth forests, the use of disposable wood products like chopsticks, and mining in rain forests.

Freshwater degradation: Sustainable development would end sewage, urban, and industrial discharges into water supplies and the agricultural use of pesticides and fertilizers.

Soil degradation: Sustainable development would preserve wetlands, use sustainable agriculture, and end deforestation and soil pollution.

Ocean degradation: Sustainable development would reduce fossil fuel use (resulting in less oil transported and thus fewer spills) and end coastal drilling, sewage dumping and runoff into marine areas, coral reef destruction, and production of disposable plastic products.

Air pollution: Sustainable development would lessen industrial emissions, end toxic emissions and incineration, use renewable energies, and switch transportation to electric or other low-emission cars and mass transit.

Greenhouse gases: Sustainable development would involve switching to renewable energy, conserving energy, producing less red meat, and ending deforestation.

Ozone depletion: Sustainable development would phase out CFC production, switch to safer alternatives than the planned HCFCs and HFCs, and destroy the CFCs in use.

Overfishing: Sustainable development would set fishing limits to regenerate and then sustain populations, end use of trawler nets, and lessen estuary and reef destruction and pollution.

Overhunting: Sustainable development would reduce the human population, lower meat consumption, preserve more habitat, and set hunting limits to sustain wildlife populations.

Extinction: Sustainable development would reduce deforestation, preserve rain forests, lower pollution levels, and preserve more habitat.

Wetland degradation: Sustainable development would use sustainable agriculture, reclaim wetlands, lower the human population, reduce pollution, and cut down on coastal development.

Toxic pollution: Sustainable development would prevent industrial discharges, involve a switch to different manufacturing methods, reduce use of toxics, lead to different choices about products manufactured, and end nuclear energy use.

Major Sources

The major areas to focus on for sustainable development are:

Population: As the human population increases, so do its needs; keeping the population at a sustainable number reduces environmental stress.

Poverty: Ending poverty would help decrease human population growth, since poor parents in a number of developing countries view having more children as a "social security measure"; they have more children so that someone will be available to take care of them when they are older. Thus, ending poverty would also ease stress on wilderness areas, land, and water resources.

Energy use: Switching to renewable energies and energy conservation would decrease major problems such as acid rain, air pollution, water pollution, mining, deforestation, soil degradation, and nuclear waste.

Agriculture: Switching to sustainable agriculture would end pesticide and artificial fertilizer use, resultant water pollution, and soil erosion.

Manufacturing: Using nontoxic methods and preventing toxic emissions would limit the toxicity of the environment. Choosing not to produce nonsustainable products such as disposable goods and many plastics would reduce pollution and raw material use.

Raw material use: Limiting mining and consumption of rain forest and old-growth forest wood, minerals, and fossil fuels would preserve these resources for future generations and result in less pollution and land degradation.

Transportation: Using even current technology to increase gas mileage, changing to electric zero-emission cars, and using mass transit would greatly lessen smog, air pollution, and acid rain.

Meat production: Choosing to eat less red meat puts less pressure on forests, frees up more land for agriculture, lessens methane production, and lessens the amount of grain, water, and fish needed to feed livestock.

Fisheries: Setting sustainable fishing limits allows fish populations to replenish themselves for continual harvesting.

Water use: Using water efficiently and reducing water pollution ensures clean available water resources.

Wilderness: Setting limits on the amount of wilderness that can be developed or used for any reason preserves biodiversity and wildlife populations.

Indigenous peoples: Indigenous peoples often live sustainably in their homelands; thus protecting their land rights would help to preserve wilderness areas and biodiversity and lessen extinctions.

Growth: Redefining a human goal as prosperity without "growth" allows the focus to shift from endless consumption of limited, finite resources to a high quality of life using renewable resources.

State of the Earth

Nonsustainable: Earth's farmers must feed 95 million more mouths annually, with 25 billion tons less topsoil; by 2035, to feed everyone, food production will have to double. Yet it is estimated that already more than 15% of all soils worldwide have been at least partially degraded.

Sustainable: A number of developing countries have ambitious family-planning programs that have reduced fertility rates. China's program of family planning, though criticized for being coercive, may reduce its estimated population growth by 1/2 billion by the year 2000. Sustainable farming yields are as high and cost effective per acre as high-input chemical farming, and erosion is 20% to 70% less than conventional farming; more countries are investing in sustainable farming methods.

Nonsustainable: Half of all tropical rain forests have already been cut down; 40 million acres of tropical forests are destroyed annually; worldwide 60% of all forested lands have been cleared; up to 5,000 species are going extinct each year.

Sustainable: Some U.S. companies have been using sustainable logging practices, such as selective logging instead of clear-cutting, for 70 years and are still logging the same temperate forests. A number of countries have areas of rain forest where nuts, rubber, and other renewable products are being harvested sustainably. Ecotourism in Costa Rica, the Galápagos Islands, and other places is financially supporting the protection of natural areas. Governments and nongovernmental organizations are offering to pay off a country's debt in exchange for guaranteed protection of tracts of rain forest. Instead of using fuelwood for cooking, some developing rural areas are using simply constructed solar cook boxes and ovens.

Nonsustainable: Livestock production consumes more than 1/2 of all water used in the United States; it accounts for 30% to 40% of world grain consumption and 70% of U.S. grain consumption.

Sustainable: Some people are choosing to eat less meat.

Nonsustainable: More than 90% of all fisheries are overfished; it may take 5 to 10 years for many fish populations to recover. Depending on the area of the world, 75% to 90% of all commercial marine fish caught are species that depend on coastal estuaries for reproduction, food, migration, or nurseries. Currently, 1/2 the population in the United States lives near coastlines; worldwide, 2/3 of all humans live near coastlines and rivers draining into coastal waters.

Sustainable: Driftnetting has been banned worldwide. There are attempts through the United Nations and governments to get world agreements on fish catch quotas on most stocks; countries have set 200-mile fishing limits off their shores to regulate their coastal stocks. In the United States the EPA assists states in managing and protecting coastal wetlands and estuaries.

Nonsustainable: Concentrations of all greenhouse gases are still increasing in the atmosphere; industrial nations create 80% of the greenhouse gases; developing nations will produce about 1/2 of all carbon emissions by 2050, because of their desire for quick economic growth.

The United States uses more than 30 times as much fossil fuel as some developing countries and 1/4 of all commercial energy worldwide; oil reserves could be gone in 30 to 50 years.

Sustainable: Wind energy production is expanding rapidly in over 100 countries. The U.S. Department of Energy (DOE) is funding major studies to make wind energy more efficient and affordable. Some electric utilities are investing in solar energy applications; a number of governments are funding decentralized solar rooftop "substations" to supply electicity. Electric utilities must purchase electricity from smaller producers at reasonable prices. Energy conservation and efficiency is being supported by governments in many countries.

Nonsustainable: Worldwide, landfills are filling up, space for garbage is limited, and incineration emits toxic pollutants; hazardous wastes are created daily.

Sustainable: Many countries in Europe are requiring industry to recover product packaging. In the United States, the EPA is focusing on source reduction (see Glossary) and incinerators are under stricter emission limits. There is intensive worldwide research to find ways to destroy hazardous waste and reduce the use of toxics in manufacturing.

Nonsustainable: Automobiles are one of the largest sources of air pollution; catalytic converters, which reduce air pollution from cars, are not used in a number of countries; leaded gas is still used in a number of countries.

Sustainable: California is requiring zero-emission cars and estimates that there will be 200,000 in use by 2003. Electric cars are already on the market in Europe, Japan, and the United States, and more are coming in the next few years; ultralights, cars made of composite material, are being researched and will get 150 to 300 miles per gallon. Trains and other mass transit options are gaining more favor in Europe and the United States.

The costs of achieving sustainable development are much smaller than the costs of trying to "fix" the problems created by our current nonsustainable lifestyles. For the United States alone, it is estimated that several hundred billion to trillions of dollars will be needed to clean up hazardous waste dump sites, dispose of radioactive waste, reduce acid rain, and reclaim degraded land from mining, pollution, and erosion; all of these problems are results of nonsustainable life-styles.

Human Impact

Health: Improved air, water, and food quality would lessen toxic diseases; respiratory, skin, kidney, and liver problems; and cancer. Health care would be provided for all people.

Food production: Sustainable development would help maintain fish populations, game fish and animal quality, land production, and soil quality. It would lessen erosion and reduce famine.

Medicinal: Plants in rain forests would be preserved and researched for medical cures.

Indigenous peoples: Indigenous people would be preserved, along with their knowledge and lifestyles.

Poverty: Poverty would be greatly reduced.

Violence: Less conflict over resources would occur.

Costs: Costs would fall thanks to reduced health problems, more efficient use of energy, less pollution and degradation of the environment, and fewer toxic cleanups.

Aesthetic: Wilderness areas and wildlife populations would be preserved for future generations; less crowding and noise pollution would occur in urban areas.

Individual Solutions

- Practice energy conservation at home and in transportation; water gardens and lawns sparingly; use low-flow shower and faucet heads and small toilet tanks.
- Recycle; don't purchase disposable products, especially plastic or paper, or items with a lot of packaging; avoid products with button batteries.
- Oppose incineration, which burns resources and pollutes lakes and fish with mercury.
- Eat less red meat.
- Buy organic foods at food cooperatives or farmers' markets.

- Use nontoxic household products; never pour toxics down the drain, in the sewer, or on the land; contact the U.S. Environmental Protection Agency (EPA) or local environmental groups about safe alternatives to toxics and hazardous waste pickup sites.
- Support family-planning availability.
- Don't buy tropical hardwoods such as teak, rosewood, mahogany, ramin, lauan, or maranti.
- Investigate solar or wind energy for your home; oppose nuclear energy and fossil fuel subsidies.
- Practice ecotourism—another name for sustainable tourism.

Industrial/Political Solutions

What's Being Done

- The 1992 Earth Summit in Rio de Janeiro produced the Rio Declaration and Agenda 21 (an 800-page action plan), both of which recommended sustainable development, eradication of poverty, reduction of unsustainable consumption and production, and protection of the global environment. They gave specific strategies and placed more responsibility on industrial countries to help the world achieve these goals. A new United Nations Commission on Sustainable Development will report on the progress of Agenda 21; its first meeting was held in June 1993.
- Nongovernmental organizations are promoting family planning, habitat preservation, ecotourism, and other sustainable practices worldwide.
- Research on renewable energy is being conducted worldwide.
- Efforts are being made through the United Nations to achieve world agreements on sustainable fish catches.
- Chlorofluorocarbons (CFCs) will be phased out worldwide in the mid-1990s.
- A worldwide ban on driftnets was established in 1993.
- Recycling is becoming widespread, in many industrial countries, and efforts are being made at source reduction.
- The EPA, World Bank, and other organizations are supporting inte-

grated pest management instead of high-input chemical farming with pesticides.

- Some countries are placing more emphasis on sustainable rain forest use.
- The United States and other countries have banned ocean dumping.
- The 1990 U.S. Clean Air Act regulates emissions that cause acid rain.
- Governments and nongovernmental organizations are conducting debt-for-nature swaps, indigenous people's aid, research, and many projects to help developing countries preserve their rain forests.
- The U.S. Council on Sustainable Development met in fall 1993 to develop recommendations for industrial, social, and economic policy based on the Earth Summit.
- Energy-efficient appliances, insulation, and other products already in the marketplace can reduce national energy consumption by the equivalent to the energy produced by several dozen power plants.
- The U.S. Resource Conservation and Recovery Act comes up for reauthorization in 1993.
- The U.S. Clean Water Act comes up for reauthorization in 1993.
- California is requiring use of zero-emission electric cars; the state should have 200,000 by 2003.
- People in some areas are using gray water for irrigation and using sewage to create fertilizer.
- Oil tankers and barges larger than 5,000 tons and using U.S. ports will be required to have double hulls by 2010.

What Needs to Be Done

- Make family planning available to all families worldwide.
- Offer education and equal rights to women and other oppressed people worldwide.
- End nuclear power; continue the ban on nuclear testing.
- Ban planned alternatives to CFCs, namely HCFCs and HFCs; both are greenhouse gases, and HCFCs are ozone depleters.
- Convert to renewable energies and energy conservation; end fossil fuel subsidies; increase investment in residential, commercial, and industrial solar and wind energy use.

- Set product, energy, and resource prices to reflect air, water, and land pollution.
- Phase out incineration.
- Phase in zero-emission transportation.
- Reduce use of toxics in manufacturing; shift to cleaner technologies.
- Reclaim wetlands; adopt a no-net-loss policy for wetlands.
- Institute land reform in developing countries so land is not controlled only by a few of the wealthy; recognize the land rights of indigenous peoples.
- Contain urban sprawl.
- Offer debt relief and aid only to developing countries that practice social justice for all their people; aid and assistance programs should involve local institutions and people and support environmentally sustainable projects.
- Increase world pressure to end warfare and regional conflicts.
- End toxic discharges into waterways and sewers and onto land; ban all hazardous waste transport to developing countries.
- Institute a worldwide ban on rain forest destruction; in the United States, end timber subsidies, and ban logging in old-growth forests.
- Pursue world agreements on limiting greenhouse gases.
- Reduce consumption of red meat.
- Convert to, and educate farmers about, sustainable agriculture; ban sales of banned pesticides to developing countries; end subsidies to nonsustainable farming; phase out pesticide use.
- Share clean technology, pollution technology, and renewable energy technology with developing countries.
- Evaluate all technology by criteria of minimal environmental stress, energy efficiency, and waste production.
- End the funding of nonsustainable energy and transportation projects by major lending institutions such as the World Bank.

Also See

- Energy conservation
- Renewable energy
- Sustainable agriculture

Toxic chemicals

Definition

Gas, solid, or liquid substances that are harmful to living organisms.

Most toxic chemicals are human-created (synthetic), but some occur naturally. In nature many plants create chemicals that are toxic to insects or animals that might eat them. Often chemicals that are toxic to animals are also toxic to humans.

Toxics are defined by their toxicity: a measure of how dangerous they are and how easily or how often they might cause health problems. Some chemicals are so toxic that just one molecule of exposure can cause serious health problems. Almost any chemical has some level of toxicity, which means if you take a large enough amount of any chemical into your body, it probably will cause health problems. The category of toxic chemicals, however, is usually restricted to those chemicals of which very small exposures can cause serious health problems.

Often toxic chemicals are regulated by weight per volume (say milligrams per liter) or parts per million (ppm) or billion (ppb); for example, when the U.S. Environmental Protection Agency (EPA) decides that a toxic chemical should never be higher than 5 ppm in drinking water, for safety, it means that for every million molecules of water, there should be no more than 5 molecules of the toxic chemical.

Examples of toxic chemicals are volatile organic chemicals (VOCs), benzene, pesticides, polychlorinated biphenyls (PCBs), herbicides, polycyclic aromatic hydrocarbons (PAHs), heavy metals, dioxins, DDT, mercury, lead, synthetic organic chemicals (SOCs), and formaldehyde. Toxic chemicals are often a part of hazardous waste.

T

Environmental Impact

Environmental buildup: Toxics enter the environment from air pollution, landfills, incineration, wastewater, sewage systems, and deliberate use, as in pesticides and herbicides. Often toxics do not biodegrade, or do so very slowly. Thus a large number of chemicals continue to accumulate in soils and water resources in greater and greater concentrations. Often toxics take years, decades, or even longer to be cleansed from water supplies. This is especially true for groundwater, where toxics may accumulate in surrounding rock and slowly leach into the water over a long period of time. Lakes, estuaries, marshes, bays, and inlets can also take a long time to cleanse themselves.

Food chain: From the soils and water, toxics can be taken up by plants and animals and bioaccumulate in the food chain. Toxics have been observed to cause breeding problems in predators at the top of the food chain—eggs break, mating behaviors are erratic, and young die. Shellfish and other animals can accumulate large amounts of toxics over time. Mammals and birds can develop cancer and neurological, respiratory, and circulation problems with exposure.

Population explosions: If top predators are killed, a pest population may explode, since it has no predator to control it. Some pests such as insects may develop resistance to certain toxics, such as pesticides, and may also undergo a population explosion when their predators are killed.

Soil: Toxics can kill microorganisms in the soil, which are resposible for the decay process that adds nutrients to the soil. When these microorganisms die, the soil becomes infertile and plant growth is inhibited. Toxics may migrate long distances through soils if there is moisture present.

Synergy: When some toxics are released into the environment they combine with chemicals in the air and water to form more dangerous chemicals.

Transported chemicals: Toxics can be carried by the wind or water; thus some have been spread over large areas when emitted into the environment.

Industrial processes: Industries produce toxic by-products, wastes, and manufactured chemicals; according to the EPA, photographic processes account for 40% to 60% of toxic chemical pollutants. A new photographic process has been developed that uses heat and magnetism instead of chemicals.

Petrochemical industry: The petrochemical industry is the source of 90% of manufactured synthetic organic chemicals and many by-product toxics.

The military, U.S. Department of Defense, and U.S. Department of Energy: The military generates chemical warfare toxics and by-products and waste from weapons production; the military is probably the largest single producer of toxic and hazardous wastes.

Mining: Mining produces toxic wastes, escaped heavy metals, and acidic chemicals.

Pesticides: Agriculture, landscaping, pest control for pets, and government spraying programs all use toxic pesticides.

Incineration: The by-products of burning solid waste include air emissions and toxic ash.

Hazardous waste sites: Dumped chemicals and other wastes can be toxic.

Medical waste: Medical waste includes radioactive material and chemicals.

Household products: Toxic chemicals are present in paints, detergents, cleansers, disinfectants, mothballs, appliance chemicals, lights, and numerous other products.

Landfills: Combinations of many different chemicals and products in landfills can lead to toxic formations and releases in leachate.

State of the Earth

Anywhere from 70,000 to 80,000 chemicals are commonly used in industry worldwide; 1,000 new ones are added each year. At least 80% have not been studied for health or environmental effects; the other 20%

have not been studied thoroughly. Up to 90% of industrial chemicals may be toxic to humans and the environment. U.S. industry emits 60% of its toxics into the air, 21% into wells (oil and other old wells), 12% onto land, and 7% into water (sewers, rivers, lakes). In 1991 U.S. industry emitted about 3.4 billion pounds of toxics; almost another 4 billion pounds were transferred to treatment facilities or recycled.

So many people have cancer along the Mississippi River from Baton Rouge to New Orleans that the area is called Cancer Alley or the Chemical Corridor; residents there are twice as likely to get colon or rectal cancer as those who live elsewhere. This is because more than 100 major chemical plants in the area discharge billions of pounds of toxic waste into the water yearly. Louisiana is 49th among states in the strictness of its environmental laws, first in total toxics released per capita, first in the number of high-risk facilities, first in oil spills, and second in the cancer death rate. Dow chemical company buried 273 million pounds of toxic liquid and sludge, which now threaten the Plaquemine Aquifer; Dow bought out a whole town, Morrisonville, which was near its Plaquemine plant.

In Minnesota, industry emitted 150 million pounds of toxics in 1988 to 1989 alone (roughly 40 pounds per person); 24 states use more than 1 billion pounds of toxics in industry yearly; Texas produces 100 billion pounds of toxics yearly. Houston industry released more than 1/4 billion pounds of toxics in 1990; New Orleans, 186 million pounds; and Chicago, 162 million pounds. In 1990 nearly 5 billion pounds of toxics were emitted and transferred in the United States. The majority of toxics in the United States still are often emitted or disposed of in economically poor areas. The EPA Toxic Release Inventory for 1991 shows that the top 10 toxic chemicals released to U.S. air were (in millions of pounds): methanol (199.7), toluene (198.6), ammonia (188.6), acetone (160.2), methylchloroform (137.5), xylenes (115.5), methylethylketone (103.4), carbon disulfide (89.3), hydrochloric acid (82.9), and methylene chloride (79.3).

Europe is recycling and detoxifying a larger percentage of its toxic wastes each year; Germany takes care of 1/3 of its toxic waste this way. Many developing countries, however, have few or no restrictions on toxic emissions or their disposal, and their use of toxic chemicals is increasing as they industrialize. Often they do not use up-to-date manufacturing pollution controls. Workers in these countries frequently have few or no safeguards against industrial toxins they are exposed to during manufacturing.

The EPA has estimated that 90% of Americans carry concentrations of dioxin, furans, chlorobenzene, dicholorobenzene, benzene, or styrene in their fatty tissues; all are human-created carcinogens. The EPA has also estimated that up to 100% of Americans have traces of PCBs in their bodies. Traces of PCBs, furans, and other toxic chemicals have been found worldwide in animals, soils, and aquatic sediments. Most toxics have not been followed in most animal populations; it is too difficult and costly, and the number of chemicals to track is too large.

Human breast milk has been found to contain more toxic chemicals than the U.S. Food and Drug Administration (FDA) allows in products we buy in the store; the concentrations of toxics in tested breast milk is 6 to 14 times higher than that allowed for adults. The largest study thus far found 192 chemicals (organic toxics) in breast milk. Organic chemicals contain carbon and thus can mimic many of the naturally occuring enzymes in living things. They can enter chemical reactions in the body and disrupt the normal biochemistry of cells, leading to cell mutation, cancer, and a host of other problems. Some estimates are that up to 80% of all breast cancers are directly related to environmental toxics like dioxins, PCBs, and DDT. Breast cancer kills 50,000 women yearly; the American Cancer Society says that the average woman has a 1-in-9 risk of cancer if she lives to be 85.

Most experts believe that increases in cancer are directly related to the amount of toxics in the environment; up to 80% of all cancer cases may be related to environmental toxics. The National Research Council has found that in the United States up to 7,500 infants each year have birth defects related directly to pollution, which may play a part in increasing birth defects by 150,000 annually. More than 10% of Russian infants have birth defects, and Russian life expectancy is falling— scientists believe these conditions are related to the extreme toxicity of Russia's environment.

T

Human Impact

Exposure: Toxic chemicals may be eaten in food, taken in with water or other liquids, breathed in, or absorbed through the skin or other membrane areas such as the eyes and nose. Regulations governing exposure to toxic chemicals for human health protection usually set a dosage based on the effects of one chemical acting alone. But humans are exposed to thousands of chemicals over their lifetimes, and no one understands the effect a number of different chemicals will have on the human body at the same time, over a long period of time. Also, some toxic chemicals unite synergistically to form other dangerous toxics that threaten our health.

Health: The kidneys and liver detoxify any toxics that enter the body; high levels of toxics strain both organ systems and lead to storage of some toxics that cannot be destroyed. Toxins often accumulate in fatty tissues or in the brain, liver, and kidney. As toxics concentrate in the body, other systems may break down, causing other diseases. In general toxics can cause cancer, organ diseases, endocrine breakdowns, immune system weaknesses, birth defects and miscarriages, neurological disorders, blood diseases, hyperactivity, asthma and allergies, skin problems, headaches, nausea, and death.

Children: Children are often much more susceptible to the effects of toxics than adults, and they react to much lower doses. This is because their nervous systems and other organ systems are still developing and will incorporate toxics more readily into their tissues.

Individual Solutions

- Buy nontoxic household and lawn-care products. If you are unsure about using certain chemicals or products, call the EPA or a local environmental group for a list of nontoxic alternatives.
- Never dispose of toxics down the drain, into your garbage, or onto the soil; call your local officials for toxic waste disposal sites.
- Buy carpet, drapery, or other products that do not use formaldehyde and that claim to be toxic free; use wooden floors instead of carpet.

- Keep toxics locked up, out of reach of children.
- Have your water tested, especially if your water is from a well.
- Demand toxic use reduction strategies for industry and large, prohibitive cleanup fees for polluting industries.
- If you are worried about emissions from a local industry, look up its emissions data in the yearly EPA report. Thanks to the Emergency Planning and Community Right-to-Know Act, industry must publish yearly Toxic Release Inventory (TRI) data on more than 300 chemicals or face substantial, cumulative fines. Private citizens also have a right to know of any stored toxic chemicals or hazardous wastes in their area. If you are concerned about a local industry, contact the EPA or state officials to request data on its chemicals and wastes.
- Report to the EPA if you think your company, or another company, is illegally dumping wastes, emissions, or other toxic or chemical substances; such actions are punishable by fines and prison sentences.
- Grow indoor plants—some remove toxics from the air. Some good examples are elephant-ear philodendrons, spider plants, chrysanthemum, and gerbera daisies.

Industrial/Political Solutions

What's Being Done

- Household toxics are being collected in the United States, Europe, and Japan.
- Toxics are being recycled by industries in the United States, Europe, and Japan.
- Researchers are working on ways to detoxify toxic chemicals before disposal; examples are solar energy, other chemicals, plasma arc torch furnaces, and bacteria (in a process known as bioremediation); new technologies and their use are rapidly expanding.
- Disposal methods for toxics include landfills, pits, ponds, deep well injection, incineration, vitrification, or encasement in cement or other substance.
- U.S. laws governing toxics include the Resource Conservation and Recovery Act (RCRA), the Toxic Substances Control Act (TSCA),

and the Emergency Planning and Community Right-To-Know Act. All govern the EPA's role in managing production, transportation, and disposal of toxics. Superfund, created by the Comprehensive Environmental Response, Compensation, and Liability Act, governs abandoned industrial hazardous and toxic wastes.

- The EPA publishes a toxic release inventory (TRI) for more than 300 chemicals at about 24,000 facilities; it plans to add 170 chemicals. The European Community is considering a similar program.
- The EPA has concluded the best way to manage toxic chemicals is to reduce the amounts and numbers of them that are created.
- The 1990 U.S. Clean Air Act Amendments require the EPA to regulate and limit emissions of air toxics. In February 1994, the EPA began enforcing the Chemical Manufacturing Rule of the 1990 Clean Air Act, which requires the EPA to regulate 189 toxic air pollutants by the year 2000. The first step by the EPA was to require chemical plants to cut toxic air emissions of 112 pollutants by 88% within 3 years by installing scrubbers or using other pollution-prevention technology; this should reduce yearly toxic emissions by 506,000 tons.
- A number of U.S. states have passed toxic use reduction measures; toxic use reduction focuses on incentives and mandated reduction of toxic use at production sites by replacing toxics with safe alternatives in manufacturing.
- Toxic chemicals, which before were treated as a waste product, are now being traded from the industry that created them to one that can use them.
- Fire departments and community agencies are aware of toxics used on company premises so that they will know how to respond in an emergency.

What Needs to Be Done
- Phase out incineration; ban the landfilling of toxics.
- Strengthen incentives and mandates for toxic use reduction.
- Redesign industrial processes to eliminate toxics or to use them more efficiently.

- Prevent manufacturers from putting toxics into household products.
- Tax toxic chemicals as an incentive to reduce use and pay for pollution cleanup costs; make polluters pay for all cleanup costs.
- Create uniform product labeling that is clear to consumers.
- Require testing of all chemicals for long-term health and environmental effects before they are allowed to be used in products or emitted into the environment; phase out pesticide use.
- Require companies to make available toxic release inventory data for their overseas companies.
- Demand that companies be environmentally responsible, especially in developing countries that have few or no pollution regulations.
- Make companies responsible for recovery of produced toxics.
- Initiate worldwide agreements to phase out toxic use and emissions.
- Ensure that the North American Free Trade Agreement (NAFTA) does not supersede national, state, local, or tribal regulations.
- Ban U.S. companies from selling toxic chemicals banned in the United States to developing countries; this ban should include pesticides, herbicides, and pharmaceuticals.
- Mandate prison sentences for industry executives who knowingly release toxics into the environment.

Also See

- Hazardous waste
- Organic chemicals, synthetic

Transportation

Definition

Vehicles of any type used to transport people or goods.

Transportation involves everything related to the production and function of vehicles. For cars, for example, this would include the mining, processing, and burning of gas and oil; the mining of all metals and other required minerals; and the production and disposal of cars, cleansers, knickknacks, air fresheners, tires, air conditioners, Freon, antifreeze, brake fluids, plastics, and batteries. Water is used to wash cars, and enormous land area is required for roads. Also many chemicals, such as sulfuric acid, are used in the production of motor vehicles and their parts.

Environmental Impact

Air pollution: Particulates, nitrous oxides, carbon monoxide, lead, synthetic organic chemicals (SOCs), and other pollutants come from processed, burned, and vaporized vehicle fuels. Mining wastes are carried by the wind. Used-tire dumps have caught on fire. Vehicle emissions of carbon monoxide diminish atmospheric hydroxyl radicals, which destroy ground-level ozone. All air pollutants inhibit plant growth and contribute to smog. Smog destroys vegetation and forests, pollutes soils, and is a greenhouse gas.

Acid rain: Vehicle emissions of nitrogen oxides and sulfur dioxide contribute to acid rain, which acidifies lakes, rivers, and soils. It inhibits plant growth, kills fish, damages forests, and leaches heavy metals into water; these heavy metals enter the food chain through fish.

Freshwater degradation: Fossil fuel mining wastes and wastewater emit salts, silt, minerals, and toxics such as heavy metals and organic chemicals into waterways; used fuel oils and other liquids are dumped into sewers; fuel-processing wastes are dumped into soil and water. All inhibit plant growth and kill plants and animals; these toxics enter the food chain and cause reproductive problems, cancer, birth defects, and death in animals. Water used to wash motor vehicles further depletes water resources.

Greenhouse effect: Vehicle emissions, fuel processing, and methane escaping from natural gas production lead to increased greenhouse gases. Vehicle emissions of carbon monoxide reduce atmospheric hydroxyl radicals, thus increasing methane. Greenhouses gases are responsible for trapping heat around the earth.

Soil degradation: Soil degradation results from mining for fuels, minerals, and construction materials and the resulting waste; road construction; disposed-of wastes such as used tires and vehicles; and disposed-of used oil and other fluids. These wastes are often toxic and inhibit plant growth; roads and mining also take land out of production.

Ocean degradation: Oil spills and used fuels and wastes from boats and ships destroy coastal estuaries, aquatic plants, and animals.

Ozone depletion: Leaking or vented air conditioners in cars add chlorofluorocarbons (CFCs) to the air; high-flying airplanes emit nitrous oxides, which play a role in ozone thinning. Ozone depletion destroys Antarctic plankton, inhibits plant growth, and causes harm to animals.

Wildlife habitat: Habitat is lost as a result of mining, pollution, roads, and airports; high-speed trains block migration routes. All force wildlife to crowd into other areas or leave an area, and thereby increase extinctions.

Major Sources

Automobiles, trucks, and buses: In industrialized countries automobiles, trucks, and buses are responsible for 20% to 25% of carbon dioxide emissions, 30% of CFC emissions, 50% to 75% of smog chemicals, 65% of carbon monoxide emissions, 50% of nitrous oxides emissions,

40% of hydrocarbon emissions, and emissions of toxics such as sulfur dioxide, formaldehyde, benzene, heavy metals, particulates, and SOCs. Used fuels and parts for these vehicles are discarded; fuels and raw minerals must be mined. These vehicles are large land users; they create heavy congestion; they are expensive and dangerous. In general, diesel fuels produce more particulates and other emissions than gasoline or other fuels. Biofuels made from biomass burn more cleanly and are being used in a number of countries.

Airplanes: Airplanes emit nitrous oxides and particulates; they are heavy users of fuel and land. They create noise pollution and congestion and are expensive. In 1990 airplanes were responsible for 15% of the world's annual consumption of transportation fuel.

Rail (light, heavy, intercity, and freight): Rail uses less oil, energy, and land than cars and trucks; it produces fewer air emissions (from an electric power source) and results in fewer accidents, less traffic congestion, and more efficient passenger and freight hauling. Light rail is cheaper than high-speed trains, but all are cheaper than cars or trucks. The cost of constructing rail for trains is 1/10 that of building highways. Rail can provide easier access to cities and multiple locations than airplanes, and light rail is quieter.

Ships and boats: Ships, boats, and barges are responsible for oil leaks, discarded garbage, used fuels, and river shoreline erosion.

Bicycles: Bicycles generate virtually no pollution; they are quiet and inexpensive.

Electric and solar electric vehicles: Electric and solar electric vehicles have zero emissions; They use less energy than vehicles with internal-combustion engines and no oil. They will, however, result in increased utility power needs to charge batteries unless solar-powered rechargers are used; their batteries will also need to be recycled. Electric vehicles are quiet. A number are currently on the market, and more are coming soon.

Ultralights: Ultralights are supercars using defense industry composites. Light, stronger than steel, with small engines and low friction, these vehicles can travel 150 to 300 miles per gallon of fuel. They are being researched and designed.

Hybrid vehicles (using electric and internal-combustion motors): Hybrid vehicles have the problems and benefits of both electric and internal-combustion engines; prototypes have been built and are under study.

Hydrogen-powered and fuel-cell-powered vehicles (using hydrogen or other fuel): Vehicles powered by hydrogen or fuel cells are still being researched; prototypes have been built and tested. They use clean fuel and generate few air emissions. The energy is renewable if water is the hydrogen source instead of fossil fuels.

State of the Earth

There are about 200 million motor vehicles in the United States—150 million cars. There are more than 1/2 billion motor vehicles worldwide, a number that could double in two decades; more than 90,000 are built each day and 35 million each year. In the United States cars are responsible for 4/5 of all miles driven. Roads and parking lots occupy 1/4 to 1/2 of all land space in large cities, 50% to 70% in Los Angeles. Transportation accounts for 2/3 of all oil used in the United States; worldwide, transportation accounts for 1/3 all oil usage and 1/5 of total energy usage. Trucks account for more than 90% of all road damage.

There are nearly 1 billion bicycles worldwide—China has 1/4 of them, followed by the United States (with 100 million), Japan, India, and Germany; 3 million people biked to work in 1992 in the United States. Bikes are still the main form of transport in many countries.

London, Moscow, New York, Sydney, Tokyo, Toronto, and many other cities have major rail systems; Los Angeles is starting rapid expansion of light rail. Amtrak is buying several dozen high-speed rail systems by 1997. Chicago is studying designs for personal rapid transit (PRT) systems, which involve small cars, carry 3 to 5 people, and may be more versatile than trains in destination and pickup spots. Rail is heavily used in countries in Africa, Asia, and Europe. Traveling by rail is 20 to 80 times safer than traveling by car or other vehicle; trains use 1/4 the energy of planes and 1/3 the energy of cars per passenger-mile.

The Paris-London route will be serviced mostly by the underwater

T

train tunnel beneath the English Channel. Superfast trains that travel at 155 to 300 miles per hour are making a comeback in Canada, Europe, South Korea, and elsewhere. Europe is planning an 18,000-mile network of tracks costing $76 billion. Germany wants to replace air and car travel with trains for distances shorter than 300 miles. France is the leading exporter of fast-train technology; Brazil, Pakistan, and Taiwan are all expanding rail service.

Electric cars, vans, and buses have already been designed, built, and tested by every major car company worldwide; several thousand are already on the road in the United States; 2,000 are on the road in Switzerland. More models will come on the market in the mid-1990s; some require outlet charging, some solar charging, and some both.

BMW, Mazda, and Mercedes-Benz have tested hydrogen-powered vehicles. Hydrogen is created by splitting apart water molecules by solar energy or in fuel cells; it can be created from renewable sources such as water or biomass and gives off virtually no pollutants. It is more efficient than gasoline, can be replaced more easily than electric vehicle battery charges, and provides more range than batteries. The British Columbia Transit System is testing hydrogen-powered buses.

The Rocky Mountain Institute believes ultralight cars, weighing 2/5 of current car weights and made of composite defense industry materials that are stronger than steel, will be able to get 150 to 300 miles per gallon with a 10 to 20 horsepower hybrid electric drive and will be on the market in 4 to 7 years. General Motors built a 4-passenger ultralight car in 1991 that got 100 mpg at 50 mph with a standard larger engine.

Internal-combustion engines turn 80% of fuel energy into waste heat; electric cars use 1/10 the energy of gasoline engines and use no engine oil or antifreeze. They are more reliable and cost less to maintain since no oil changes or tuneups are required. The total cost is the same over the life of the vehicle. Electric cars can travel at highway speeds (some can accelerate quickly), and their engines last 10 to 20 times longer than internal-combustion engines. Light electric vehicles (LEVs), 2-passenger cars for short-range urban use, are being developed in Switzerland.

In 1991 the U.S. Enviromental Protection Agency (EPA) found that nearly 100 major urban areas in the United States exceeded air pollution standards; cars contribute to more than 50% of air pollution in most cities; motor vehicles are the largest source of air pollution. About 15% of cars on the road in the United States are older models, responsible for 30% to 50% of all car emissions. Worldwide, 1/4 million people die and 10 million are injured in road accidents each year. Thirty thousand in the United States die from cancer and emphysema due to motor vehicle pollution. Cities throughout Asia and Latin America have serious motor vehicle pollution problems; many burn leaded gas. Even gains in worldwide automobile pollution controls would be offset by the increase in motor vehicles expected with current population growth. Brazil has used sugarcane-created ethanol in its vehicles for years, which has helped to reduce automotive pollution. Blends of diesel and plant oils are being tested in buses in Germany and South Dakota. A 50-50 blend of water and either gasoline or diesel fuel with an added emulsifier has been developed by a U.S. company; the Transportation Department, military, and postal service are already testing it with great interest. The fuel burns efficiently and cleanly, but needs ethanol added to it at low temperatures to keep it from freezing.

Volvo has a prototype car that gets 80 mpg on the highway, seats 4, and is safe in a 35-mph head-on collision—the U.S. crash standard for head-ons is at 30 mph. Nearly every major car company has prototypes that get 100 miles per gallon.

Millions of gallons of toxic oil-drilling wastes are dumped each year in unlined pits throughout U.S. oil-producing states; millions of gallons of water contaminated by the drilling processes are dumped yearly, untreated, into waterways and oceans. About 1 billion gallons of oil pollute the oceans each year; 1/2 come from the continents, the result of river and coastal runoff, and 1/2 come from deliberate dumping of fuels at sea, accidents, and production platforms.

One gallon of gas can contaminate 1/4 million gallons of drinking water. Every gallon of gas burned produces 20 pounds of carbon dioxide; a car emits its own weight in carbon every year. Several hundred billion gallons of gasoline are burned worldwide each year. In the United States about 1/4 million tires are discarded (2 to 3 billion tires from previous years' discards are stockpiled illegally around the country). Each year more than 10 million cars are discarded; 94% of cars are recycled. Car metal is recycled at a very high rate, but many of the other materials are incinerated or landfilled; more complete recycling is being developed.

Human Impact

Health: Smog, acid rain, and air pollution contribute to lowered resistance to colds and pneumonia, asthma, cancer, chest pains, death for people with cardiovascular problems, drowsiness, eye irritations, heart diseases, lead poisoning, and scarring of the lungs. Toxics in water or food can cause neurological problems, birth defects, skin problems, kidney and liver failure, cancer, and death. Carbon monoxide hinders oxygen transport in the blood. Ozone thinning increases incidences of skin cancer and cataracts. Airport noise causes hearing loss and stress.

Food: Cropland is ruined or paved over; smog, acid rain, and air pollution reduce crop yields. Water pollution harms finfish and shellfish stocks; ozone depletion may reduce fish stocks.

Aesthetics: Polluted or mined areas destroy wildlife habitat and natural areas. Airport noise disrupts daily life; traffic congestion increases stress.

Individual Solutions

- Drive less; carpool; walk; bicycle; use mass transit; combine errands.
- Buy fuel-efficient vehicles; keep your car tuned and tires inflated; wash it sparingly.
- Never discard any solution, additives, oils, or wastes into the sewer, street, or drains or onto the soil; contact the EPA or local officials for location of recycling centers.
- Make sure your car air conditioner's CFCs are recycled; have it checked for leaks; do not refill a leaking air conditioner.
- Write or call legislators to demand higher fuel efficiency in cars and zero tailpipe emissions.

Industrial/Political Solutions

What's Being Done

- More than 100 million bicycles were produced worldwide in 1993.
- Canada, Japan, the United States, and most of Europe are requiring catalytic converters on all new cars; catalytic converters reduce nitrogen oxides by 60% to 75%, and convert 85% of carbon monoxide and 90% of hydrocarbons to water and carbon dioxide.
- Leaded gas is no longer used in most industrial countries.
- A number of car companies have already designed, built, and tested electric and solar electric cars. Japan has guaranteed markets for electric cars since local governments must purchase them. Fast charging systems for electric cars have already been designed; France will install battery-charging devices in 22 French cities by 1995. Germany and Japan are close to debuting the first hydrogen-powered vehicles.
- CFCs are being phased out of automobile air conditioners worldwide.
- Research is being conducted on cleaner fuels such as ethanol, bio-fuels, synthetic fuels, and gasohol.
- Production of alternative fuels and vehicles is being encouraged in Europe, Japan, and the United States, with zero-emission standards, subsidies for the production of alternative vehicles, and tax breaks for buyers of alternative vehicles such as electric cars.
- Switzerland passed a 1994 resolution to ban all heavy trucks in 10 years from passing through their country enroute to and from other European countries; starting in 2004 all heavy trucks passing through Switzerland must be put on railroad flatbeds. Austria is considering a similar initiative.
- The 1990 U.S. Clean Air Act Amendments required reductions in car and truck tailpipe emissions of hydrocarbons, carbon monoxide, and various nitrogen oxides beginning in 1994. Auto manufacturers must also reduce vehicle emissions from gasoline during refueling. Nine cities with the worst ozone problems must use cleaner gasoline by 1995. A second-stage initiative will mandate that gas stations fit

pump nozzles with vapor collectors to capture fumes. The EPA mandated 80% reduction of sulfur in diesel fuel by 1993.

- Supertrains are being studied by the U.S. government to handle traveler flow on highways and airports; Amtrak is planning to purchase 26 new trains that will run at 100 to 200 mph; its order may double in the coming years. A long-term National Maglev Initiative is developing maglev train (speeds up to 300 mph) technology for the United States. Maglev trains move by magnetic levitation (see glossary).
- Car batteries are being recycled.
- The U.S. Department of Energy (DOE) is doing research on hybrid vehicles and has a program that will pay a selected car company $150 million to develop a hybrid over the next 10 years. A U.S. Advanced Battery Consortium was established in 1991; it is a joint venture between the government and battery and car companies to develop new batteries for high-performance electric vehicles. Westinghouse received $1.9 million from the DOE for a 2-year contract to develop zinc-air cells, an advanced battery. A new, powerful, lightweight, and leakproof lithium battery may make lead-acid and nickel-cadmium batteries obsolete.
- The 1991 $151 billion U.S. transportation bill allows states to decide where to allocate federal transportation funds.
- California is mandating that 2% of cars sold there must be zero-emission vehicles (electric vehicles) by 1998; by 2003, 10%, or about 200,000. Ten other states have indicated they will follow the lead of California; these East Coast states account for 30% to 40% of the U.S. car market. California is investing in mass transit planning in addition to highway planning.
- A growing list of states is mandating use of cleaner automobile fuels and better emissions testing.
- Car companies are using fewer kinds of materials in cars to make them readily recyclable.
- In some U.S. states used car tires are being burned as fuel chips or recycled into other products. Currently 80 million tires are reused each year, and predictions are for 250 million a year by 1998.

What Needs to Be Done

- Increase funding for mass transit.
- Increase construction of bike lanes on roadways; integrate bikes into transit planning; create incentives to use bicycles.
- Raise fuel efficiency for all cars.
- Reduce funding and building of rural roadways; limit urban sprawl.
- Make manufacturers responsible for recycling cars and their components.
- Achieve worldwide use of catalytic converters and unleaded gas.
- End fossil fuel subsidies; tax fossil fuels to reflect environmental damage and costs.
- Phase out all subsidies of cars.
- Phase in zero-emission standards on all cars.
- Convert freight shipping from trucks to rail.
- Design cities to accomodate mass transit systems and to make bicycling and walking convenient.
- Initiate a program to buy or somehow replace older vehicles, which are responsible for most pollution.

Also See

- Fossil fuels
- Renewable energy

W

Warfare effects on the environment

Definition

Environmental damage, pollution, or stress caused by weapons use; weapons production; armies during war or war games; terrorism; maintenance of armies, navies, or air forces; or displaced people as a result of war.

Warfare damage to the environment can be incidental to other objectives during the course of a war or intentional as a way of destroying an enemy's resources.

Environmental Impact

Wildlife: Animals can be used as a food source for an army or displaced refugees, killed for "fun," destroyed as innocent bystanders during a conflict, and frightened when migrating. Toxic pollutants from weaponry production or warfare can be introduced into the food chain. These toxics bioaccumulate and can cause cancer, reproductive failure, neurological disorders, behavior changes, and death in young and adult animals. Nuclear radiation causes cancer and death and also enters the food chain. Wildlife habitat destruction also occurs as a result of fighting, meeting the fuelwood needs of displaced people, using defoliant herbicides in chemical warfare, or moving soldiers and equipment, all of which can accelerate extinctions.

Air pollution: Air pollution results from industrial weapons production, nuclear weapons plant radioactive emissions, jet aircraft use during peacetime and warfare, burning oil wells, bombed chemical or industrial factories, and manufacturing plants.

Air pollution from all sources affects soil and water quality, plant growth, and thus animal populations.

Freshwater pollution: Freshwater pollution includes chemical runoff, used fuels and oils from army equipment, nuclear weapons testing, industrial weapons production, mining wastes, bombed industrial plants, and nuclear and toxic wastes. This pollution ruins shorelines, threatens aquatic environments, and kills fish.

Soil degradation: Abandoned equipment, buried toxic and nuclear wastes, used oils and fuels, war games, abandoned weapons that did not discharge, pollution from mining for weapons materials, and erosion from bombed and traveled land all lead to soil degradation. In degraded soil plant growth is inhibited, or plants take up toxics through their roots, which animals eat.

Ocean degradation: Ocean degradation results from navy garbage thrown overboard, used oils and fuels, discharged oil, discarded nuclear waste, and abandoned nuclear weapons or submarines. These materials threaten shorelines and aquatic plants and animals.

Major Sources

Wars: Wars have occurred on every continent in the world and in most countries in the past century. Areas with recent widespread warfare or resultant related problems include Africa, which for decades has had internal conflicts in a number of countries; Armenia and Azerbaijan; the Persian Gulf; and the former Yugoslavia. Even in countries with no current conflicts, major disposal and environmental problems exist because of warfare-related nuclear, chemical, and other toxic weapons and wastes.

Weapons: As weapons become more powerful, or sophisticated, more damage is done to the environment. The pinnacle of destruction is embodied by nuclear weapons, which can annihilate a given area, as well as contaminating it for decades or centuries, making it unusable and unlivable. Nuclear weapons could also threaten the whole world, since the products of an explosion could be carried on the wind and threaten the global climate. Current conventional weapons are also capable of causing widespread environmental destruction. More and more countries in recent decades have developed weapons production plants, which are often very toxic and polluting—this is especially true for nuclear plutonium production and reprocessing.

Terrorism: Environmental terrorism, in which extremist factions purposely cause damage by assaulting a country's resources such as water, food, dams, or a nuclear power plant, is a current risk.

Industrialized areas: Any conflict in an industrialized area risks large amounts of toxic chemical releases, since most industries, even peacetime manufacturing plants, use toxic chemicals in day-to-day operations.

State of the Earth

Internal strife in a number of African countries continues to stress its peoples and environment. Ethiopia, Mozambique, Somalia, and Sudan have suffered from civil conflict for almost two decades; the results have been crop wipeouts, soil erosion and desertification, the starvation of millions of people, and the creation of millions of refugees. Worldwide, there have been more than 100 regional conflicts in the past 4 decades.

Current sporadic fighting in Volcanoes National Park, Rwanda, still threatens the last few hundred mountain gorillas. Elephants, rhinos, and other wildlife in Africa have been killed by the hundreds and thousands by passing or invading armies. Fighting in Sudan and Zaire in 1993 forced elephants back into Uganda, where large herds had been decimated by poaching and warfare in past years.

In the Bosnian conflict, destruction of sewage and industrial sites have caused river pollution; one spill alone sent 50,000 tons of oil into a river feeding into the Danube. Polychlorinated biphenyls (PCBs) from destroyed electrical equipment have spilled onto soils and into water. Landfills are out of reach of many citizens.

Agent Orange, a defoliant used in Vietnam by the United States, contains dioxin, a toxic chemical. Millions of acres of forest were destroyed by this chemical, and local inhabitants suffered from birth defects, spontaneous abortions, and cancer; U.S. Vietnam vets are still suffering health problems from Agent Orange exposure. The Cambodian countryside contains booby-trapped land mines.

In 1989 the Pentagon closed off several hundred square kilometers of public land in Nevada, when it found more than 1,000 live bombs, more than 100,000 pounds of shrapnel, and nearly 30,000 rounds of ammunition accidentally dropped outside an Air Force bombing range. The U.S. military has 1/2 million M-55 rockets, armed with nerve gas, which are deteriorating, as well as thousands of other chemical weapons around the country that need to be destroyed. An incinerator on Johnston Atoll in the South Pacific was built in the early 1990s to incinerate old weapons, but it generates toxics in the process and has been broken half the time it has been in operation.

In the past four decades the U.S. Department of Defense (DOD) and Department of Energy (DOE) have generated millions of tons of toxic wastewater from their nuclear weapons complex. The DOE has 4,000 contaminated sites over tens of thousands of acres, and the DOD has 12,000 contaminated sites at active military installations. More than 100 serious Superfund sites have been created by weapons manufacturers. Chemicals from decades of nerve gas and other toxics production have been dumped at the Rocky Mountain Arsenal, outside of Denver, Colorado. This area was once called "the most contaminated square mile on earth"; plutonium is in the soils, radioactive materials are in the groundwater; it is now a wildlife refuge.

In 1949 in Washington State, at the Hanford nuclear reservation, which supplied plutonium for atomic bombs, operators secretly released radiation equivalent to a major nuclear accident into the air, the Columbia River, and nearby soils—more than 400 billion gallons of radioactive water and waste; harmful levels of radiation covered half of Oregon and Washington. Millions of gallons of highly concentrated radioactive waste and thousands of toxic chemicals, potentially explosive combinations, are now stored in 149 unstable tanks, and 66 are leaking; it will take at least 30 years to clean up and cost tens of billions of dollars—right now the government does not have adequate cleanup technology for the job. Rocky Flats nuclear plant in Denver, Colorado, also a plutonium producer, leaked plutonium and other radioactive wastes into the air and surrounding environment in the 1950s and 1960s—dried plutonium blows on the wind. The Department of Energy's Savannah

River plant site in South Carolina, working with Dupont, for 30 years buried radioactive waste in cardboard boxes and dumped millions of gallons of radioactive waste into open pits. This waste now threatens the surrounding Savannah tributaries, as well as the Tuscaloosa Aquifer, one of the largest in the country.

Russia had much the same past negligent policies as the United States at its Chelyabinsk site and exposed 1/2 million people to radiation through accidents. Over a 40-year period the population of Semipalatinsk, Kazakhstan was exposed to above-ground atomic bomb blasts. Many people died and cancer, weak immune systems, and genetic defects are common. The Russian military dumped 11,000 to 17,000 radioactive waste containers in the Arctic Ocean and large amounts in the Barents Sea; it dumped 400,000 tons of chemical weapons in the Baltic. Lake Karichai of Russia has been used as a nuclear waste dump for the Mayak nuclear weapons plant in Siberia for 30 years. The lake now contains 1 billion curies of radiation and has been covered with a thick layer of concrete. The radiation is still leaking into nearby ground-water and rivers. Russia dumped 900 metric tons of low-level nuclear subradioactive waste in the Sea of Japan in October 1993.

A typical nuclear reactor will produce 500 pounds of plutonium each year; fewer than 2 dozen pounds are needed for a small nuclear bomb. Nearly 10,000 pounds of plutonium have been released worldwide in weapons testing. The Nevada test sites for the atom bomb exposed up to 100,000 people to radiation without their knowledge; several hundred nuclear bomb tests were conducted in secret from 1963 to 1990. Many victims developed leukemia and cancer, but the U.S. Supreme Court ruled that the U.S. government was immune to any lawsuits. The U.S. government conducted human radiation tests, often without the subject knowing, on over 1,000 people from the 1940s to the 1970s. Trees east of Nevada testing sites have shown high radioactive content when burned—they have been taking up fallout from the soils. Past nuclear bomb testing in Maralinga, Australia, by the United Kingdom, in the Marshall Islands by the United States, and in Mururoa, French Polynesia, by the French have doubled the rate of cancer in exposed children and adults living in or near the areas. In the Marshall Islands and in

Maralinga, natives were returned to high-radiation areas after testing. In 1993 food grown in the Marshall Islands had up to 60 times normal levels of strontium and cesium. Between 1950 and 1965 atmospheric radiation equaled 40,000 Hiroshima bombs; improvement-of-mortality statistics showed declines worldwide during this period.

In past warfare, often the first targets bombed have been power plants, which makes nuclear power plants a high-risk target in future conflicts. In 1993 the United States was still one of the biggest conventional weapons producers and exporters. France, Germany, Russia, and some countries in Asia also had large export programs.

In bombing Iraq during the Persian Gulf War, the United States released hundreds of thousands of pounds of nuclear and chemical toxic waste in the sites it targeted. The bombing also devasted widespread areas of land, wildlife, and vegetation. About 7,000 U.S. soldiers are suffering from Desert Storm syndrome, believed to be caused by chemical releases, chemical weapons, released uranium from U.S.-fired tank shells, or experimental vaccines; soldiers suffer memory loss, skin rashes, digestion problems, fatigue, respiratory problems, and aching joints, and their children display rashes and thyroid defects. About 40 tons of depleted uranium now litter Iraq and Kuwait.

Iraq's damaging and burning of 700 oil wells and releasing of oil into the Red Sea may prove to be the two most devastating environmental oil catastrophes to date. The oil slick in the Red Sea killed tens of thousands of animals, birds, and fish, bleached and eroded coral, and ruined beaches and fisheries. The oil well fires, put out in November 1991, created oily rain deposits over much of the region; these deposits ruined cropland and killed wildlife and vegetation. The burning wells also created acid rain that is harming forests in other countries—as far away as India. The wells left several million tons of oil on the land, which killed thousands of migratory birds that mistook the lakes of oil for water. Some of the oil has sunk into the soil up to 8 inches and hardened. The desert is also booby-trapped with land mines. In late 1993 Iraq began draining swamps and marshes to force southern opponents who live in those and adjacent areas to abandon their homeland.

W

Human Impact

Health: Health problems stem from nuclear testing, nuclear weapons production wastes, nuclear weapons use, other weapons use, herbicide use, and air and water pollution. Exposure to these toxics can result in cancer, nervous system disorders, digestion problems, growth problems, birth defects, kidney and liver failures, hair loss, appetite loss, leukemia, immune system damage, genetic damage, and death. Nuclear weapons production, use, and testing have contributed to increased local and global levels of background radiation, which increases the risk of cancer and other diseases. In general, warfare also increases the risks of diseases because of less sanitation, fewer medical supplies, and stressful situations.

Fresh water: Water shortages occur when water supplies, such as wells, lakes, and rivers, are contaminated or poisoned.

Food: Food shortages can occur when people are forced to flee their land and exhaust local sources of food, when food is contaminated by radiation or other warfare chemicals, when crops are burned, or when farmland is ruined.

Costs: The costs are very high for warfare-related pollution cleanups or reclamation of damaged areas; also, warfare is so expensive that it diverts funds from other essential needs.

Individual Solutions

- Support renewable energy.
- Oppose nuclear testing.
- Oppose below-regulatory-concern (BRC) nuclear waste dumping.
- Call or write your legislator to demand that the U.S. Departments of Defense and Energy be brought into compliance for environmental pollution.
- Demand an end to nuclear weapons production and to sales of large conventional weapons by the United States.

W

Industrial/Political Solutions

What's Being Done

- The Non-Proliferation of Nuclear Weapons Treaty (NPT) was signed by more than 150 countries to limit the spread of uranium and plutonium, and thus nuclear weapons; a U.N.-affiliated organization, the International Atomic Energy Agency (IAEA), tracks uranium and plutonium worldwide. The United Nations pressures countries to allow inspections of suspected nuclear weapons plants; U.N. teams inspected and dismantled Iraq's nuclear weapons plants.

- There is a current global ban on nuclear testing; China, however, ran a nuclear test in the fall of 1993.

- By congressional order, the U.S. army must destroy old chemical weapons by 1997.

- International arms control standards have been set to limit the spread of certain weapons systems, such as missiles; the United States in 1992 was the world's leading arms merchant, supplying more than 1/2 of all arms sold to developing countries.

- The London Dumping Convention Treaty of 1972 has permanently banned nuclear waste dumping at sea.

- In the past decade U.N. peacekeeping forces have tried to stabilize regions with their presence in more than a dozen countries, using short-term government takeovers, establishing food routes, disarming warring factions, and aiding peace negotiations. In some places, such as the former Yugoslavia, U.N. managment has been called ineffective, and even detrimental.

- Treaties on limiting nuclear weapons, destroying chemical and biological weapons, limiting missiles, and limiting conventional weapons have been signed.

- The United Nations adopted a resolution banning purposeful environmental destruction during war; Iraq signed but never ratified it.

- A number of countries have declared themselves nuclear-free zones and refuse to dock ships and submarines with nuclear missiles.

- The former Soviet Union is receiving international assistance in dismantling its nuclear weapons systems; Russia, with financial assistance and incentives from the United States, is offering to convert Ukraine's nuclear warheads to nuclear fuel.

- U.S. nuclear testing and weapons production sites are being cleaned up; the United States and Russia have a joint research program to share cleanup technology.
- Thanks to the 1992 U.S. Federal Compliance Act, which makes government polluters subject to the same penalties as private companies, environmental impact statements are required for most activities of the Department of Defense and the army.
- The United States is developing smaller nuclear missiles.
- Some defense industries in the United States are shifting to peacetime technologies.
- The U.S. Department of Defense is using biotechnology to research detection of biological weapons.
- The U.S. government is seeking sites to store spent nuclear waste.
- The U.S. government is investigating the Gulf War syndrome cases.
- The U.S. government has set up a $200 million trust fund for victims of Nevada nuclear bomb testing. Hanford "downwinder" victims are suing the government and associated companies.

What Needs to Be Done

- Continue the world ban on nuclear testing; pressure countries, such as China, that continue testing, with economic sanctions and boycotts.
- Ban biological warfare weapons and research.
- Pursue a world agreement to begin unilateral disarmament of conventional weapons; focus on defensive strategies.
- Reduce military spending.
- Provide security to developing countries by promoting an end to poverty, sustainable development, and shared technology.
- Offer aid only to countries that treat all their citizens with justice.
- Destroy all nuclear weapons; phase out all nuclear weapons design labs, manufacturing plants, and delivery systems plants.

- Dispose of all plutonium and radioactive materials under international supervision.
- Mandate that navy ships end ocean dumping of garbage and used fuels.
- Force weapons contractors to assume 100% responsibility for cleaning up contaminated areas; prosecute them for negligence.
- Boycott or pressure aggressive countries that begin military conflicts.

Also See

- Nuclear energy

Wetlands degradation

Damage to wetlands through pollution, or destruction by filling in, bull-dozing, or cutting down vegetation.

Wetlands are soggy, low places where the land is perpetually or seasonally saturated, or even partly submerged, by either fresh or brackish water. They are often transitional areas between terrestrial and aquatic systems, and the water table is often covering, even with, or near the surface of the land. The soils are capable of holding large amounts of water, and the plants can withstand periodic flooding.

Wetlands can be either coastal, such as estuaries or mangroves, or inland. Examples are swamps, bogs, peat lands, fens, marshes, swamp forests, and prairie potholes—which are wet only at certain times of the year. Two well-known North American wetlands are the Florida Everglades and the Okefenokee Swamp. Over long periods of time, lakes may slowly fill in and become wetlands, which over very long periods of time may turn to forest.

Wetlands are some of the most productive ecosystems in the world. This is because they are often protected from the elements to some degree, provide needed moisture, have abundant nutrients, and contain a diverse amount of food for a wide variety of animals. They are the breeding and reproductive sites of thousands of species of fish, shellfish, microorganisms, amphibians, reptiles, insects, invertebrates, and birds. Without wetlands, many species would go extinct, and many are going extinct as wetlands disappear.

There are also human-created wetlands, which are currently being used either for aquaculture or for filtering and purifying sludge and sewage. These can support wide varieties of plant and animal life, too. A human-created wetland, however, is not the same as a natural wetland or a reclaimed or restored wetland, in which the damage done to a natural system has been repaired. Human-created wetlands cannot duplicate groundwater recharge abilities, endangered species habitat, and biodiversity.

Environmental Impact

Plants and animals: When wetlands are destroyed, breeding, nesting, reproductive, and feeding grounds are lost for populations of animals that live there. Birds that migrate might not have a stopping place; coastal fish might lose breeding grounds; inland species that lived near the wetlands, such as big cats or other predators, lose hunting habitat. In general wetlands often play a major role in replenishing the biological diversity of an area. Loss of wetlands forces birds and other migratory species to overcrowd other wetlands. This loss will reduce breeding sites and populations, stress crowded animals, and make it easier for disease to spread through a population. Wetlands also provide environmental buffers for animals and local vegetation. If there is a drought, wetlands usually still hold some amount of water. If there is a storm, coastal wetlands provide a buffer against the wind and water surges. When these buffers are taken away, such areas and their inhabitants are subject to the effects of severe weather such as storms and drought.

Toxics: Toxic chemicals and other pollutants can remain in wetlands for years or even decades. These toxics enter the food chain when animals drink the water or eat microorganisms or plants that have taken up the toxics. Toxics often bioaccumulate and biomagnify, so that predator species hold more and more of them over time. Often toxics such as heavy metals, polychlorinated biphenyls (PCBs), pesticides, and other common pollutants cause reproductive and behavioral problems, birth defects, liver and kidney failure, cancer, and death of young and adult animals. If pollution is severe enough, whole species can be lost.

Erosion: Wetlands prevent erosion of soil in coastal and inland areas. Wetlands also prevent flooding by holding excess rainfall. When wetlands are lost, rain simply runs off the land and fills the rivers, increasing both flooding potential and erosion along rivers and coastlines. Runoff sediments can carry nutrients and heavy metals with them. This type of runoff, often called nonpoint source pollution because it occurs over large areas, can threaten aquatic animals and plants by poisoning them, making the water too murky, or increasing algae growth through eutrophication.

Groundwater: Wetlands are the major way that water is filtered through the ground to replenish and purify groundwater, which is held in aquifers. As wetlands are lost, groundwater is replenished much more slowly, if at all, which can increase the chances of drought in spring-fed areas. Loss of wetlands also means that polluted water is not detoxified and more groundwater may be contaminated.

Major Sources

Wetlands are degraded by:

Agriculture: Wetlands are plowed under for farmland. Agricultural runoff such as pesticides, fertilizers, and manure pollute wetlands.

Urban development: Wetlands are destroyed to build coastal and inland cities, airports, roads, and resorts. Urban runoff and wastes such as salts, oils, sewage, and wastewater pollute wetlands; this pollution increases with population growth.

River pollution: Rivers carrying wastewater, spilled oil, used chemicals, and industrial wastes often feed directly into, or are adjacent to, coastal estuaries.

Canals, ports, and harbors: Wetlands are destroyed through dredging and widening of areas for ships.

Oil: Spilled oil and offshore drilling wastes often go directly into the ocean waters, which end up in coastal estuaries.

Aquaculture: Mangroves and other coastal wetlands are destroyed to grow shellfish, shrimp, or other fish.

Dams: Dams restrict freshwater, nutrient, and sediment flow to coastal estuaries, which become more saline (salty) from ocean water and nutrient deficient. Dams can also divert water away from wetlands for agricultural irrigation and thus dry them out.

Air pollution: Incineration, industrial emissions, and automobile emissions can pollute wetlands.

Exotic species: If foreign plant species, such as cattails, invade wetlands, they can intrude on native sawgrass and sedges, inhibit water flow, and suffocate freshwater ponds and marshes.

Wetlands can cover less than an acre or thousands of square miles. They are home to more than 600 animal species and 5,000 plant species; 40% to 50% of U.S. endangered animals and 1/4 of endangered plants live in or rely on wetlands. Half of U.S. migratory birds rely on wetlands. Two-thirds of major commerical fish rely on estuaries and salt marshes for reproduction, feeding, or nurseries; 90% of Gulf Coast fisheries rely on estuaries and wetlands. Depending on the area of the world, 75% to 90% of all commercial marine fish caught are species that depend on coastal estuaries for reproduction, food, migration, and nurseries.

Currently about 1/2 the population in the United States lives in coastal areas; worldwide, 2/3 of all humans live near coastlines and rivers draining into coastal waters. The United States has developed—in other words, lost—50% of its coastal wetlands; Italy, nearly 100%; tropical areas, 50%.

More than 1/2 of all original wetlands have been lost in the United States; including Alaska and Hawaii, 1/3 of total U.S. wetlands have been lost. Half the states have lost 50% or more; California has lost 90%. The U.S. Fish and Wildlife Service estimates annual losses of nearly 300,000 acres. In the United States, 75% of total existing wetlands are privately owned; 25% are publicly owned. One hundred million acres remain; 5% are coastal wetlands, and 95% are inland wetlands. Alaska has 2/3 of the nation's remaining wetlands. Agricultural development accounts for the majority of freshwater wetlands losses; development and expanded river mouths owing to erosion, canals, ports, and marinas account for the majority of coastal saltwater wetlands losses.

Worldwide, wetlands are being lost to increased population pressures, which result in destruction of wetlands for farming land and development. Wetlands also are under siege from pollution, especially nonpoint sources such as rivers, agricultural runoff, industrial waste emissions, and air pollution. Many wetlands in the past have been used as dumping grounds for toxic waste and garbage.

Millions of migratory birds rely on wetlands for rest, food, and shelter. Of the millions of waterfowl that nest in the lower 48 states of the United States, more than 1/2 nest in prairie potholes; 98% of Iowa's potholes have been plowed under by farmers. 80% of the U.S. breeding bird population requires bottomland wooded swamps, which are flooded part of the year, for survival.

Experts have stated that flooding throughout the United States is increasing because of the loss of 50% of wetlands and the diking and damming of rivers. In 1993 the great flood in the Midwest was greatly exacerbated in the principal flooding rivers by the previous loss of 40% of the wetlands in Minnesota, 90% in Iowa and Missouri, 70% in Arkansas, and 45% in Lousiana.

Louisiana contains about 40% of U.S. coastal wetlands—more than 2 1/2 million acres—and is losing 30,000 acres of coastal wetlands yearly. Much of the loss is due to Mississippi sediment loss; instead of being dumped on the sides of the river in floods and natural processes— and thus replenishing the estuaries—sediments are kept in the river by dikes, levees, and dredging until they are dumped into the ocean. Also, canals and channels for boats and oil drilling and exploration allow salt water to invade these coastal wetlands and kill native plants.

The Everglades has 1,400 miles of canals, levees, spillways, and pumping stations built throughout it; water shortages occur there because of development and agriculture, water polluted with nutrients leads to algal and cattail explosions, and melaleuca is invading. Plans are being implemented to reduce nutrient discharges and increase water flow. A 1990 report from the U.S. Environmental Protection Agency (EPA) listed nutrients as a major cause of impaired estuaries; the United Nations projects nutrients to be a worldwide problem in 20 years unless corrective measures are taken.

Canada, which contains 1/4 of the world's wetlands, has lost 15% of its original wetlands. Germany and the Netherlands have lost 50% of their wetlands; Finland, 20%. The Netherlands is trying to reclaim its wetlands, and France, Germany, and Switzerland are trying to rehabilitate the Rhine River—but salmon still have not returned to its polluted waters. Europe's largest wetland, the Danube Delta, stretches over 2,860 kilometers in 9 nations; it is fed by the Danube River, which has 30 dams and whose river basin is used by 86 million people. The delta has shoreline erosion of 17 meters a year; its waters are high in nutrients, nitrogen, phosphorus, pollutants, and pesticides such as DDT and lindane (hexachlorocyclohexane). Fish harvests have dropped by 1/2, and intense algal blooms have been common in the past 2 decades.

Mangroves in Africa, Asia, and Central and South America have been depleted for firewood, rice fields, and aquaculture. India, the Philippines, and Thailand, have lost more than 80% of their mangroves; Bangladesh, Ghana, Kenya, Mozambique, Pakistan, Somalia, and Tanzania, 60% to 70%; and Indonesia 50%.

Human Impact

Fresh water: Wetlands are "holding tanks" that allow excess water to filter down to aquifers; thus wetlands losses result in less groundwater and drinking water. Without wetlands, rain washes into rivers and increases flooding, and river quality deteriorates, which can affect fish populations and drinking water. Wetlands also filter pollutants out of water as it filters through them. Increased pollutants in drinking groundwater may increase serious diseases such as kidney and liver problems, reproductive problems, birth defects, and cancer.

Food: Deep ocean and coastal fish and shellfish populations are threatened and destroyed with wetland losses.

Erosion: Topsoils can erode when wetlands are lost; increasing erosion may also threaten coastal development.

Recreation and aesthetics: Sport hunting, fishing, canoeing, and bird-watching all are reduced with the loss of wetlands.

Native plant and timber losses: Wild rice, mangrove trees (valuable for their oils and as firewood), and other plants may be lost with wetland destruction.

Human-created wetlands: Human-created wetlands can detoxify sewage waste, converting it to usable soil and nutrients, and increase habitat for some wildlife.

Individual Solutions

- Never pour any toxics down the drain, in the street, or on the soil. Contact local officials or the EPA for hazardous waste collection sites.

- Support wetlands reclamation projects or groups such as the Nature Conservancy, which buy land and set it aside so it can't be developed.
- Buy and use nontoxic household products; contact the EPA for a list.
- Buy organic food, which eliminates pesticide and chemical fertilizer runoff into rivers and estuaries.
- If you eat shrimp, inquire where it was grown; shrimp aquaculture is responsible for mangrove destruction in many countries, especially Thailand.
- Oppose incineration; recycle; avoid disposable products made of plastic, paper, or wood.
- Practice energy conservation in your home, travel, and life-style; this reduces fossil fuel use and lessens air pollution.
- If your land has wetlands, and you have any questions regarding developing or altering them, contact the EPA wetlands division.

Industrial/Political Solutions

What's Being Done

- International agreements and programs have been established to protect major wetlands. Examples include the North American Wetlands Conservation Act and the North American Waterfowl Management Plan, which protect bird migration areas in Canada, Mexico, and the United States. The Convention of Wetlands of International Importance tries to protect waterfowl habitat worldwide.
- Wetland reclamation is occurring in the United States and some other parts of the world.
- The EPA's National Estuary Program chooses wetlands (in selected states) that are in need of management and protection; the EPA then works with the state to develop a management plan.
- States receive federal assistance to develop wetland and coastal protection programs.
- The Farm Bill contains the Conservation Reserve Program, which pays farmers to keep highly erodable soils as grasslands or forests; this program helps minimize nutrient flow into wetlands.
- The U.S. Clean Water Act comes up for reauthorization in 1993 and 1994. It protects wetlands from being used as dumps or fills.

- Coastal oil drilling has been temporarily halted in a number of areas, including the Florida coastline.
- The Swampbuster program withholds benefits to farmers who plow under wetlands to increase cropland.
- U.S. floodplain management is shifting from construction of levees to restoration of floodplains and wetlands.
- Oil tankers and barges larger than 5,000 tons and using U.S. ports will be required to have double hulls by 2010; smaller barges and tankers have until 2015.
- Private sporting and environmental groups buy wetlands to preserve them.

What Needs to Be Done

- Develop state and federal no-net-loss wetlands policies; reclaim lost wetlands; broaden protection. The National Academy of Sciences urges a gain of 10 million acres of wetlands over the next 2 decades to preserve water supplies and wildlife habitats.
- Renew the Conservation Reserve Program in the 1995 Farm Bill.
- Prohibit industrial chemical discharges into waterways and sewage systems.
- Support sustainable agriculture.
- Restrict coastal and inland development that threatens wetlands.
- End all offshore coastal drilling for oil.
- Mandate that offshore rigs not be allowed to dump drilling wastes and other wastewater into the ocean.
- Mandate that any inland river barges carrying hazardous or toxic materials be double hulled.
- Strengthen world agreements to preserve and protect wetlands; current agreements involve about 65 countries that have pledged to preserve at least 1 major national wetland.

Also See

- Freshwater degradation
- Ocean degradation

Wind energy

Definition

Renewable energy created from wind passing over rotating turbine blades that drive generators to produce electricity.

Small turbines can be used alone for small power needs. Large turbines are mounted on tall stands, often in clusters called wind farms; some of the stands are up to 10 stories high, and their turbines have blades several hundred feet long. The generated electricity is usually sold to a utility or electric network.

New variable-speed turbines, which can operate at different wind speeds, allow wind farms to gather electricity even when the wind is lower than average. There also are new designs, with vertical axis blades that are like large inverted eggbeaters, that can use wind from any direction. Conventional horizontal axis propeller blade designs can face downwind or upwind and must turn with the wind as it changes direction. Conventional blades are often higher, and thus sometimes more efficient than the vertical blade designs, since wind is often stronger higher off the ground. Many new wind turbine designs have two propeller blades, while older models have three.

Environmental Impact

Birds: At the Altamont Pass wind farm in California, a 2-year study found that 500 birds of prey were killed by wind turbines, through electrocution or collisions with the spinning rotors. Studies are being conducted to determine how to prevent the danger turbine props pose to raptors and other birds.

Clean energy: Wind technology is very clean and produces no pollutants or other adverse environmental effects. Using wind energy to minimize fossil fuel use would lessen acid rain, global warming, and air pollution caused by coal power plants.

Major Sources

Wind: When the sun heats the earth's atmosphere, different masses of air end up with differing air temperatures. The differences in temperature create wind currents. Some areas of the world produce fairly constant air currents year round, such as the equator, with its trade winds. But over land masses winds are usually less constant.

Best sites: In the United States the areas for wind generation that are being explored, or are already in use, are northern California, the Northeast, Texas, and the Great Plains. Often high hills or mountainous coasts have the best annual wind speeds. In temperate areas more energy may be produced during the windier winter months. Wind sites are graded as "good" if they have average wind speeds of 12 to 13 miles per hour, "excellent" at 16 mph, and "outstanding" at 19 mph. Often wind farms are ideally suited to farming areas, since the turbines take up little of the land they are sited on, they pose no threat to livestock, and planting can occur right up to the wind turbine. This makes the Great Plains an ideal area for vast wind farms. Turbines are also being placed on breakwaters in Los Angeles, Belgium, and the Netherlands, and on dikes in Denmark.

Residential: Residential wind power is as yet largely untapped, but small turbines are being used in European countries for residential home power needs; also, neighborhood cooperatives in some European countries are collectively buying one or more large turbines, which supply their electricity needs, and selling the excess energy to utilities.

State of the Earth

During the 1990s world wind energy is projected to exceed nuclear generating capacity. It is estimated England could supply 1/5 of its energy needs with wind-driven generators. The U.S. Department of Energy esti-mates that 36 states using wind energy could supply 20% to 25% of U.S. electricity needs by 2000; wind energy could supply up to 12% of the world's electricity needs. India has one of the fastest-growing wind energy programs. California currently has more than 17,000 operating turbines, generates 2/3 to 3/4 of world wind energy production, and meets 1% of its electricity needs with wind, supplying power to 1 mil-lion residents. It plans to supply 10% of its electricity needs with wind in the next decade.

Some 100 countries are expanding their investments in wind energy; the European Wind Energy Association aims at providing 10% of Europe's electricity by 2030. Denmark, the largest supplier of wind-mills, plans to supply 10% of its electricity needs with wind in the next decade; it has the first offshore wind farm. The Netherlands has the largest wind plant in the world; a 6-kilometer-long string of turbines. There are good wind sites in Northern Africa, Central and South Amer-ica, and South Asia. Ukraine hopes to have a 5,000-turbine wind farm on the Crimean Peninsula finished by 1998.

In some developing countries windmills are still being used as they originally were designed, to power farm machinery and water pumps and to grind wheat. Developing countries are using small windmills to recharge batteries, run television sets, power communication sets, run navigation aids, and perform essential services. Researchers estimate that residential units could supply large amounts of electricity in devel-oping countries, supplementing solar photovoltaic energy.

Studies have shown that on the high prairie of Buffalo Ridge in south-ern Minnesota, the average wind speed is 13 to 15 miles per hour, com-parable with California wind farms. Current studies show that wind energy would be as cheap as or cheaper than conventional power plants in southern Minnesota. Fourteen states have wind potential as great as California's; North Dakota alone could supply 1/3 of total U.S. electric-ity; Montana and Texas could supply all of U.S. electricity needs.

U.S. Windpower and Iowa-Illinois Gas and Electric have formed WindRiver Development Company; they are looking at 9 states in the

Midwest for wind farm expansion beginning in 1994. NSP of Minnesota has a contract to receive 25 megawatts in 1994 of an expected 100 megawatts of wind power to be generated in southern Minnesota by 1998; Iowa has begun a similar wind energy project. Other Midwest states are also studying wind energy. A number of companies are looking at sites in the Pacific Northwest, the Northeast, the Mid-Atlantic states, and Texas.

U.S. wind turbine manufacturers and electric utility companies are conducting a 5-year study to develop a variable-speed wind turbine that will be readily integrated into utility power systems worldwide.

Human Impact

Health: Wind technology produces no adverse health effects. Using wind energy will help lessen the adverse effects on health caused by fossil fuels.

Noise: People living near wind farms have complained about turbine noise. Research is being conducted to minimize the noise impact.

Individual Solutions

- Look into supplying your electricity needs with a small wind turbine (see *Wind Power for Home and Business,* by Paul Gipe).
- Demand that energy costs for fossil fuels and nuclear utilities reflect their environmental and health effects.
- Practice energy conservation.
- Investigate using solar energy for home heating and electrical needs; it can complement a small wind turbine to produce all of your energy needs.

Industrial/Political Solutions

What's Being Done

- The U.S. National Energy Policy Act of 1992 offers a production tax credit for wind energy.
- The U.S. Department of Energy (DOE) has an Advanced Wind

W

Turbine Program; it has awarded $7 million in research subcontracts develop more efficient production methods; it is especially active in the Great Plains region. Utilities and the DOE are evaluating new turbines through 1998 with a $40 million project; the DOE is testing 20 turbines at different sites.

- Public utilities are required to purchase electricity from small producers, and must do so at rates equivalent to avoided cost of replacement or new generation; this began with the Public Utilities Regulation Policies Act (PURPA) of 1978.
- Advanced computerized wind turbines will capture 3 to 4 times more energy than current turbines, be active from 95% to 98% of the time, and operate at wind speeds of 9 to 60 miles per hour.
- Electric utilities in the Midwest are purchasing wind energy.
- In Minnesota and a few other states, utility regulators must include environmental costs when deciding whether to approve new power plants and must approve the least expensive option presented.
- Wind turbines have been exempted from state property taxes in Minnesota.
- Researchers are studying how to minimize bird deaths due to turbines; some siting of wind turbines have been put on hold until this problem can be addressed. Screening turbines is costly and blocks wind; use of color and high-pitch frequencies are being examined.

What Needs to Be Done

- Strengthen incentives for wind energy projects, especially community and industrial use of small turbines, as well as larger turbines that can sell surplus electricity to power utilities.
- Tax fossil fuels to reflect the environmental and health problems caused by their pollution and to encourage renewable energy use.
- End subsidies for fossil fuels.
- Conduct research on integrating systems of wind and solar energy or other renewable energy sources.
- Strengthen local and state energy policies to reflect environmental impacts and sustainability of energy generation.

- Reform the electric utility industry; demand that it move in the direction of decentralized renewable energies that are environmentally sound and sustainable.

Also See

- Renewable energy

Glossary

absorb: to suck up, soak up, or take in.

acid: liquid or substance capable of releasing hydrogen ions (H+) with a pH of less than 7; the lower the pH number, the stronger the acid.

acidity: degree or quality of being acid.

adsorption: extraction of a material from one phase into another; for example, adsorbing radon from the atmosphere into activated charcoal to measure its concentration; used in water purification.

aeration: adding oxygen, especially to water; occurs naturally and helps to purify rivers and lakes.

aerobic: process that uses air or oxygen.

aerosol: small suspended solid or liquid particles in a gas, such as the atmosphere; in environmental pollution, aerosols often contain sulfur. Three major sources are volcanic eruptions, biomass burning, and the burning of fossil fuels.

afforestation: planting trees on land used previously for purposes other than forestry.

aflatoxin: toxic substance made by the mold *Aspergillus flavus* that can cause liver cancer and immune system damage; found in peanuts, grains, and nuts and showing worldwide increases. The use of artificial fertilizers, which increases the water content of foods, is suspected of increasing the presence of aflatoxins. The molds are more prevalent in the tropics and subtropics and are common in soils. Soggy soils due to flooding can increase aflatoxin problems. Heat increases, due to the greenhouse effect, may increase their presence in the food chains. They can cause chromosome damage in animals and are more dangerous to malnourished children, whose bodies are less capable of excreting them.

Agent Orange: herbicide that contains dioxin; used as a defoliant in Vietnam by the United States.

agrichemical: pesticide or fertilizer manufactured by chemical and petrochemical industries for agricultural purposes.

agricultural runoff: water that washes off farmland into adjoining land or waterways, often containing farm chemicals like pesticides, fertilizers, manure, or other nutrients; an example of nonpoint source pollution.

agroforestry: agricultural practice of interplanting trees and shrubs with food or other crops. Trees help prevent erosion and provide firewood, nuts, and fruit; their roots draw nutrients from the deep soil, which reaches and nourishes the topsoil when the trees lose their leaves.

air: composed on average of 78% nitrogen (N), 21% oxygen (O_2), 1% argon, 0.03% carbon dioxide (CO_2), and inert gases such as neon, helium, krypton, and xenon. Clean air would also have water vapor and traces of carbon monoxide (CO), ozone (O_3), nitrogen oxides (NO_X), and nitrogen dioxide (NO_2), as well as particulates and dust.

alkaline: liquid or substance (called a base) with a pH greater than 7; the larger the number, the stronger the base.

alpha particle: highly charged particle emitted from a radioactive atom, consisting of two protons and two neutrons. It is too weak to permeate human skin, but it can cause health problems if ingested, inhaled, or absorbed through a skin wound. Usually produced from radium, uranium, or plutonium, it is a form of ionizing radiation.

alternative energy: energy produced by means other than fossil fuels. It often refers to renewable energy such as wind, solar, hydroelectric, biomass, ocean, geothermal, or hydrogen energy.

amalgamation: combining, especially in reference to combining mercury with another metal. In gold pan mining, liquid mercury is poured over crushed ore and combines with gold, which is separated by hand and pressed into a cloth to remove excess mercury (which is reused); the gold is heated with a blowtorch to remove the remaining mercury, which evaporates and represents 70% of the mercury released into the environment in pan mining. Amalgams of mercury are used in dental fillings.

amphibian: cold-blooded vertebrates that have gilled aquatic larvae, are air-breathing adults, and that have other characteristics of both fishes and reptiles. Examples are frogs, toads, salamanders, and newts.

anaerobic: not using or requiring air or oxygen.

anthropogenic: of human origin; relating to the influence of humans on nature.

aquaculture: raising of algae, seaweed, fish, shellfish, shrimp, or other aquatic animals or plants, in fresh water or seawater, in pens, ponds, streams, or tanks, along shorelines, or in deep ocean waters.

aquifer: underground cavity or porous area in soil or rock that is capable of holding a large amount of water.

arable land: land suitable for farming and agricultural purposes.

arid: extremely dry; often referring to desertlike areas, regions, or lands with little yearly rainfall or with only short seasonal bursts of rain. With annual rainfall of 10 inches or less, such areas have too little rainfall to support agriculture.

artesian system: aquifers that are surrounded by impermeable rock and that often contain water under pressure.

arthropod: crustaceans, insects, arachnids (spiders), and centipedes; these make up 75% to 99% of all species of animals.

atmosphere: area surrounding the earth, including three chief layers: the troposphere, stratosphere, and mesosphere. The troposphere is 5 miles high at the poles and 10 miles high in the tropics. The higher in the troposphere you go the colder it gets; the troposphere holds 75% of the atmosphere's mass, most water vapors and clouds, and is where storms occur and airplanes fly. Above the troposphere is the stratosphere, up to 30 miles above the earth. The higher you go in this layer, the hotter it gets. The ozone layer, which absorbs the sun's ultraviolet radiation, is in the stratosphere. The mesosphere, which is above the stratosphere, goes up another 20 miles and is cooler.

atom: nucleus consisting of protons and neutrons and surrounded by orbiting electrons. The number of electrons always equals the number of neutrons. The number of protons determines what element it is. The number of neutrons determines what isotope it is.

atomic weight: mass of an atom of an element compared with the mass of the carbon-12 isotope, taken as the standard at 12.

autoclave: steam sterilization equipment, in which pressurized steam is held at over 250°F for at least 30 minutes; it is used to sterilize contaminated materials, often bioinfectious medical material.

background radiation: natural radiation found in the environment as a result of solar, earth, and cosmic sources. The mining of uranium and other minerals, microwaves, radio waves, biomass burning, and other human-created sources have increased the levels and types of background radiation.

bacteria: microscopic organism, usually single celled, that lacks a nuclear membrane; DNA is held in the cytoplasm. Bacteria are either spherical (coccus), rodlike (bacillus), or spiral (spirillum) and can occur singly or in pairs, chains, or clusters.

base: alkaline substance capable of releasing hydroxyl (OH-) ions with a pH greater than 7. The greater the number, the more alkaline the substance.

becquerel (bq): unit for measuring the speed of radioactive decay; 1 becquerel equals 1 disintegration of 1 atom per second. The becquerel is replacing the measure called the curie (Ci), which equals 37 billion bq.

benthic: referring to the sea bottom, from the shoreline to deep sea; benthic creatures, such as lobsters and octopuses, live on the sea bottom.

beta particle: electron that is emitted from a radioactive nucleus at a velocity approaching the speed of light. Beta particles can travel only several centimeters through human tissue before being stopped, so internal organs are generally protected, but eyes are not. These particles are found in low-level waste and are a form of ionizing radiation.

bioaccumulation: storage of substances such as toxics or pollutants in living organisms, often in fatty tissues.

bioavailability: measure of a substance's ability to be separated from whatever holds it (air, water, soil, food) and enter the human body and the circulation.

biochemical oxygen demand (BOD): index for the amount of oxygen needed by bacteria to decompose a given amount of organic waste; the higher the BOD, the greater the amount of waste or pollutant. The term is usually applied to waste and sewage water. Indirectly, BOD is a measure of eutrophication.

biocide: substance that is toxic or destructive to many different living organisms.

biodegradable: capable of biodegrading.

biodegrade: to break down biologically into basic elements through the action of microorganisms such as bacteria. Both human-created and natural materials can biodegrade. A tree or paper, for example, degrades into humus, which further breaks down into carbon and other elements.

biodiversity: measure of the number of different ecosystems, species, and individuals within a species—also called genetic diversity.

biodynamic farming: organic farming using special, sustainable techniques to fertilize soils and grow crops. Biodynamic farming treats the whole farm as a living organism that needs to be in balance for most productive results. It has been shown to yield higher crop yields than conventional farming.

biofuel: organic material used to create heat or other fuels that can be burned; it often consists of biomass energy sources such as plant crops, algae, or manure.

biogas digester: device that uses anaerobic bacterial digestion to convert manure and organic waste into a methane-rich gas, called biogas. The resulting decomposed wastes can be used as a rich fertilizer, better than manure.

biological controls (biocontrol): use of living organisms such as insects, parasites, viruses, and bacteria to control unwanted pests such as other insects or weeds.

biomagnification: concentration of heavy metals and other pollutants more heavily in those species at the top of the food chain. For example, a big fish or bird that eats many little fish will have concentrated in its muscle and fatty tissues all of the heavy metals and pollutants carried by all of the smaller fish; the same is true for humans.

biomarker: detectable change in cells, biochemistry, DNA, or other process or structure in a living organism that is outside the norm for the species; these changes occur in response to foreign toxins. Biomarkers are being investigated more and more frequently as possible early warnings of environmental pollution and toxics. For example, DNA may be bound to a pollutant molecule, or an enzyme that breaks down toxics may be present in the body.

biomass: weight of living material, expressed as dry weight, of a living organism, population of organisms, or community of organisms; the term is also used in reference to organic material, usually plants or manure, that have been produced through the process of photosynthesis.

biome: specific, mature, stable ecosystem with unique dominant plant life, weather patterns, and wildlife; examples are deserts, tropical rain forests, and Arctic tundra.

bioremediation: process of encouraging bacterial populations to flourish and consume, alter, or detoxify hazardous wastes. It is not always successful, cannot be used with all wastes, and is still being researched.

biosphere: whole world's ecosystem of living organisms and their inanimate environment.

biotransformation: transformation of chemical compounds within a living system.

boreal forests: dense coniferous forests of Asia, Europe, and North America, containing fir, spruce, hemlock, and pine.

bottom ash: ash remaining on the floor of an incinerator or coal plant after waste or coal is burned; it is often toxic.

bottomland: low, often flat-lying areas in river floodplains that experience seasonal flooding; in the southern United States this land often includes hardwood forest swamps.

brackish water: water with a salt content too high to be drinkable and too low to be seawater.

breakwater: human-made barrier to shelter coastal areas from wave activity.

brine: liquid, often water, with a very high dissolved mineral or salt content.

British thermal unit (Btu): energy necessary to raise the temperature of 1 pound of water 1°F. It is also roughly the heat produced from burning a kitchen match from end to end.

buffer: liquid whose pH is not greatly changed by small additions of a strong acid or base.

bycatch: nontarget fish species or other animals such as sharks, birds, or marine mammals that are captured through nets or other fishing practices and then discarded, often dead or injured.

carbon cycle: how carbon is recycled in the natural environment. Carbon is a basic building block in all living things, and is essential for life. Carbon is taken out of the atmosphere (as carbon dioxide) by rain, ocean absorption, and photosynthesis of plants. Carbon is released back into the atmosphere by volcanoes, natural fires, fossil fuel burning, tree destruction, organic decomposition, and breathing animals and humans. The burning of coal and forests is shifting the balance of the carbon cycle to create an overabundance of it in the atmosphere. Carbon is also held in environmental sinks in trees, phytoplankton, and the oceans.

carbon-14 dating: radiocarbon dating.

carbon dioxide (CO_2): odorless, colorless natural gas in the atmosphere and in water. Used by plants during photosynthesis to create simple sugars and oxygen, CO_2 is given off by cars, biomass burning, and fossil fuel burning. It is the most serious greenhouse gas, because of the large amounts created by human activities. With air moisture CO_2 forms carbonic acid, which can eat away at buildings.

carbon monoxide (CO): colorless, odorless toxic gas given off by car exhaust and fossil fuel burning; catalytic converters destroy 85% of CO emissions. CO can prevent the blood from carrying oxygen, causing weakness, fatigue, sleepiness, headaches, and death; pregnant women, infants, heart disease patients, and those with respiratory problems are at highest risk. The United States released 60 million tons of CO in 1990. CO depletes hydroxyl radicals, thus increasing tropospheric ozone and methane.

carcinogen: any agent that can cause cancer.

carnivore: animal or plant that eats only animals or meat.

carrying capacity: population that a given environment can sustain indefinitely without suffering damages to its basic structure.

catalyst: substance that speeds up or slows down a chemical reaction between 2 or more other substances; catalysts are recovered unchanged at the end of the reaction.

catalytic converter: device attached to internal-combustion engines to reduce emitted nitrogen oxides (by 60% to 75%), carbon monoxide (by 85%), and hydrocarbons (by 85% to 90%) in the exhaust. They still allow carbon dioxide emissions.

cesium-137: radioactive cesium. Cesium-137 is used in cancer therapy and food irradiation and is a common element of nuclear fallout, with a half-life of more than 30 years. It can accumulate in the kidney, liver, and reproductive organs and cause cancer.

Chipko movement: nonviolent protest begun in 1972 when tribal women in India started hugging trees to protect them from timber interests. The movement is modeled after that of 300 Hindus in 1730 who were killed as they tried to protect their trees. The movement has spread to a number of countries and is a revolt against viewing nature as a commodity.

chlorinated hydrocarbons: human-created compounds containing hydrogen, carbon, and chlorine. These compounds are very toxic and can bioaccumulate in the food chain; they include pesticides such as DDT, dieldrin, aldrin, heptachlor, lindane, endrin, mirex, and solvents including perchloroethylene, chloroform, and carbon tetrachloride. In humans, they can cause liver and kidney damage, birth defects, cancer, and death.

chlorofluorcarbons (CFCs): human-created compounds containing carbon, fluorine, and chlorine. No CFCs occur in nature. They are ozone-depleting chemicals (ODCs) that also add to the greenhouse effect.

chloroform: solvent formed by industry for a wide variety of uses, including refrigeration and the production of fluorocarbons, dyes, and drugs. Also formed when chlorine is added to drinking water and pools, and as a result of paper manufacturing, car exhaust, burning plastics, and evaporation from polluted rivers and lakes. Can cause dizziness, headaches, tremors, kidney damage, and cancer.

chlorophyll: green pigment of plants, which absorbs sunlight and enables them to use the process of photosynthesis.

chromosome: strand in the cell nucleus that carries DNA. Each species has a specific number of chromosomes; humans have 46 chromosomes in each cell.

circle of poison: process in which pollutants originate in one area, are sent to another, and end up returning to the source. An example is when U.S. companies export pesticides banned in the United States, only to see them return on produce imported from developing countries.

clear-cutting: practice of cutting all the trees in a stand, as opposed to selective logging, in which only mature, targeted trees are allowed to be harvested, allowing other trees to gain maturity. Clear-cutting causes erosion and extinction.

climax community: stable and dominant community of plants and animals unique to a particular area, soil, and weather conditions. Several communities may succeed each other in an area, finally being replaced by a climax community. For instance, a sandy area may yield to invading grasses, which give way to bushes, which are replaced by trees, which form a stable, dominant community.

cloning: taking a small piece of tissue or cells from one organism and growing other identical organisms from them; also called tissue culture.

coagulation: particles coming together and combining in clumps. Coagulation can clarify water: it occurs in lakes and rivers and is augmented in water purification plants by the use of alum.

cofiring: burning wood and coal in utilities.

cogeneration: simultaneous production of steam and electricity, often by using waste heat produced from burning fossil fuels. The steam is used to generate more electricity, heat nearby buildings, or run industrial processes. Cogeneration can cut carbon emissions and increase fossil fuel output by 30% to 60%; most cogeneration plants use gas.

coke: form of charcoal created by heating coal over 1,000°C in the absence of air to remove volatile constituents; it is used in steel making.

combustion: burning, usually with oxygen present, that forms oxides. Complete combustion of hydrocarbons yields carbon dioxide and water; incomplete combustion results in carbon monoxide and other products.

compost: organic material that has been biologically broken down by the process of composting.

composting: biological breakdown of organic materials, such as leaves, yard waste, and food waste into their basic nutrients, which can then be used as a rich fertilizer. Oxygen is necessary for the process to occur. Bacteria begin decomposing biodegradable material; fungi, protozoa, earthworms, beetles, centipedes, and millipedes aid the process. Composting works best if the items to be broken down are small to begin with and if the compost pile is turned occasionally to ensure the necessary access to oxygen.

condensation nuclei (CN): small particles or ions that serve as sites for vapor condensation; CNs such as sulfates can increase cloud formation.

conifer: a tree or shrub that bears cones, such as a pine tree. Most conifers are evergreens.

conservation biology: area of biology using many different disciplines, such as wildlife biology, ecology, and botany, to preserve, manage, protect, or restore specific endangered areas or species.

conservation tillage: leaving 30% or more of crop residue on the soil after planting to prevent erosion. There are numerous techniques, including chisel plow, ridge-till (creating ridges), and no-till (planting without turning over the soil).

continental shelf: underwater extension of the continents that slopes gently before reaching the continental slope, which drops sharply. Continental shelves at the slope reach depths of 50 to 1,500 feet and average about 400 feet.

coral: small sedentary marine animals that build up calcareous (limestone) external skeletons, which are added onto the colony with each successive generation. Coral reefs prevent erosion, protect shorelines from heavy waves, and are the most productive marine ecosystem. They support a high biodiversity of fish and marine life and serve as a major carbon sink.

coral bleaching: effect occurring in soft and hard corals, giant clams, and sea anemones, in which the colorful symbiotic algae associated with the organism loses its pigment or leaves the organism because of stress such as pollution, water temperature changes, or other causes.

cost-benefit analysis: weighing the cost or harm of a process or pollutant level against the benefit gained by the process.

criteria air pollutants: carbon monoxide, sulfur dioxide, nitrogen oxides, lead, particulates, and smog or ozone. These include the most highly regulated air pollutants under the 1990 U.S. Clean Air Act.

crustaceans: lobsters, crabs, shrimp, barnacles, krill, and other mainly marine creatures with hardened shells and highly developed pincers, legs, and other appendages.

curie (Ci): unit expressing the speed of radioactive decay, equivalent to 37 billion disintegrations per second; 1 picocurie (pCi) = 1 trillionth of a curie. The curie is being replaced by the becquerel (bq).

cyanide heap leaching: mining process in which huge mounds of ore are soaked with cyanide to leach out the gold. Heavy metals, nitrates, and other toxics are also leached out.

DDE (dichlorodiphenyldichlorotheylene): degredation product of DDT found as an impurity in DDT residues.

DDT (dichlorodiphenyltrichloroethane): pesticide (and biocide) first used in 1939 and banned in 1972. Extremely toxic to most living creatures, DDT bioaccumulates in the environment and takes decades to break down. It is still used in a number of developing countries.

debt-for-nature swap: agreement in which a government or organization will relieve or pay part of a country's debt if the country agrees to give a specified area of land protected status. Sometimes these "swaps" have helped developing countries; sometimes, however, paying off or reducing a debt has actually increased the debt load of the developing country when the lending bank revalues the debt.

decay: radioactive process in which substances give off particles, often electrons, and change into different isotopes; decay is also used to refer to the biological breakdown of organic matter, such as dead plants and animals.

deciduous: referring to trees that lose their leaves seasonally and become dormant, such as before drought to prevent water loss or before cold weather to prevent frost damage.

decommissioning: shutdown, dismantling, and disposal of nuclear reactors that are outdated, cannot be fixed, or are too radioactive.

decomposition: process of organic material decay, often by composting.

deep ecology: term used by Norwegian philosopher Arne Naess to signify that humans are a part of nature, a strand in the web of life; that nature is a cohesive whole rather than a sum of parts; and that economic goals should be subordinate to ecological concerns. This term is used in contrast to shallow ecology, which sees humans as ruling nature, analyzes and breaks nature into understandable parts, and puts economic goals ahead of environmental concerns.

deep well injection: injection of wastes under high pressure deep into the earth's porous rock layers.

deficit harvesting: fishing, hunting, or agricultural harvesting that depletes the resource being harvested beyond sustainable numbers. For example, cod can be overfished so that the population can no longer replenish itself and plummets in numbers.

defoliant: agent that prematurely removes leaves from a plant; examples are herbicides and insects such as caterpillars.

density: mass of a substance in a given volume; the higher the density of a substance, the more mass it has in a given volume. A gas has low density, while lead has very high density.

desalinization: removing dissolved or suspended salts, minerals, or solids from water (often seawater) to produce fresh water. It is often done through evaporation, solar power, or reverse osmosis—in which membranes separate the salts from the water.

desertification: natural or human-caused process in which there is progressive loss of plants and then topsoil of an area; this loss lessens the land's fertility and ability to hold water and ends in the creation of a desert. This is often a problem with marginal land in arid or semiarid areas.

detoxification: treating a substance so that its toxicity or harmful effects are removed.

diffusion: movement of atoms or particles through material via random collisions, called Brownian motion; for example, a gas entering a room will spread evenly throughout the whole room through diffusion.

distillation: evaporation and condensation of water; one of the most thorough methods of purifying water, though volatile organic chemicals may still remain.

DNA (deoxyribonucleic acid): carrier of genetic information in cells, composed of chains of phosphates, sugar molecules (deoxyribose), purines, and pyrimidines. DNA, shaped in a double helix, can self-replicate and determines RNA synthesis; genes form segments of DNA.

dredge spoil: layers of seabed (sand, gravel, mud, etc.) removed to deepen harbors, channels, and river mouths. Spoil, such as sand and gravel, is often used for construction purposes.

Earth Day: day in late April celebrated worldwide to honor, renew responsibility for, and educate people about the environment. First started in 1970, it is now celebrated by several hundred million people in some 150 countries.

Earth Summit: event sponsored by the United Nation's Environment Programme from June 3–14, 1992, attended by 178 countries. The result was Agenda 21, a blueprint for the 21st century for worldwide environmental and sustainable living goals.

ecological engineering: designing or reconstructing sustainable ecosystems that serve human needs; it includes restoration ecology, bio-engineering, habitat restoration, and reclamation ecology.

ecological niche: specific relationship to habitat and other animals and plants that only one particular species occupies by its presence and nature.

ecological succession: natural sequence whereby an initial given plant and animal community is replaced, often several times, by different plants and animals, until a stable and dominant climax community is reached. An example is a sandy beach invaded by grasses, later invaded by shrubs, then trees, and finally other dominant trees; changes in wildlife follow the plant changes.

ecological web: interconnecting relationships between all things in nature.

ecology: study of relationships between living organisms and their environments, both animate (living things such as plants and animals) and inanimate (nonliving things such as rocks, water, and air).

ecosystem: group of closely interrelated living things and a particular physical environment, all of which make up an ecological community. An example would be the animals, plants, and inanimate environment of a pine forest, a coral reef, or a prairie.

effluent: waste, usually in liquid or smoke form, that is discharged into the environment and often carries pollutants. Industrial waste, incinerator smokestacks, and sewage outputs are three of the largest sources of effluents.

electricity: flow of charges between positive and negative charges.

electrolysis: production of chemical changes by passing electric current through a substance.

electrolyte: electrically conductive medium.

electromagnetic spectrum: range of waves generated by electric and magnetic oscillations with wavelengths ranging from very short, high-frequency gamma rays to very long, low-frequency ELF waves.

electron: negatively charged particle that is in the orbit of atoms.

ELF: extremely low frequency electromagnetic wave less than 300 hertz (300Hz). ELFs are emitted by distribution lines, substations, and transmission lines operating at 60Hz in the United States and 50Hz in Europe; they are also emitted by video terminals, electronic office equipment, home appliances, and electric blankets. Project ELF is an extremely low frequency communications transmitter in northern Wisconsin that is used to signal nuclear submarines to launch a nuclear first strike; the project has been protested for years by area activists.

El Niño: poorly understood climatic phenomenon occurring off the Pacific coast of South America, near Peru, in which trade winds die down and thus stop churning the bottom ocean water upward toward the top of the sea. The warmer waters thus remain on the surface, and the deeper, colder, nutrient-rich waters cannot reach the surface. The resultant loss in nutrients at the surface results in plummeting fish and bird populations. The disturbance is believed responsible for global climatic changes like droughts, hurricanes, and excessive rainfall. Minor El Niños are thought to occur every 2 to 4 years, larger ones every 7 to 14 years.

endangered species: animals or plants that are at risk of going extinct because of habitat loss, food loss, or threatening toxics or pollutants.

energy audit: precise look at where energy is being used, wasted, and lost and where it can be saved or used better in a business, industrial process, or household.

environmental impact statement (EIS): study to determine exactly what effects a proposed project, waste material, or other item would have on the surrounding environment and human health, including suggestions to improve, replace, or disallow the proposed project.

environmental racism: siting of waste dumps, factories, road construction, or other undesirable projects in areas that are economically or politically poor.

environmental refugees: victims of unsustainable land use; people forced from their lands by governments or large farmers, or as a result of overgrazing, overcultivation, deforestation, or desertification.

environmental sink: area of the environment that acts as a temporary or semipermanent holding reservoir for a particular element, chemical, or compound, by keeping that material out of the environmental cycles. Examples are trees and coral, which act as sinks for carbon, storing it in tissues and keeping it out of the carbon cycle. Polar ice is a sink for fresh water. Frozen deep ocean sediments on the continental shelves and frozen tundras are both methane sinks.

epidemiology: area of medical science dealing with the incidence, distribution, and control of disease in a population.

erosion: process in which a substance, often soil or rock, is worn away over time, usually by water, wind, or glacial movement.

estuary: coastal area abutting the sea, often a river mouth, which is marshlike and brackish and supports one of the rishest supplies of life on the planet. Nesting birds, crustaceans, larvae, fish, and many plants use estuaries as nurseries and hatching grounds. They are rich in nutrients and play a vital role in maintaining sea life. Their salt content is diluted by fresh water from inland sources such as nearby or adjacent rivers.

eutrophication: process started when phosphorus or phosphates, nitrogen or nitrates, or organic nutrients such as sewage are added to a river or a lake, causing an algae bloom, a successive algae die-off, and then a decomposing-bacterial bloom, which uses enough water oxygen to suffocate native fish and plants. It is also a natural process that can occur in a lake over thousands of years or seasonally on a much smaller scale.

fallout: See nuclear fallout.

fallow: farming land left unplanted so the soil can replenish nutrients taken out by the previous crops.

fauna: animals living in an area.

feedback loop: process or chemical reaction that serves to increase itself. For example, if the earth is warmed too much, polar ice will melt and release locked-up, frozen carbon. The released carbon increases the greenhouse effect, and thus the earth's temperature, melting more ice and thus releasing more carbon.

feedlot: generally a fenced-in area with a concrete feeding trough along one side; often feedlots are mechanized and contain large numbers of animals, such as cows or pigs, in small areas.

feedstock: raw material supplied to a machine, a processing plant, or a particular production process.

fermentation: biological breakdown of a substance by yeast.

fertility rate: average number of children women bear in their lifetime.

fertilizer: natural or human-created substance that has the nutrients necessary for plant growth. Examples of organic fertilizers are manure and compost. Artificial fertilizer usually contains nitrogen (added industrially as ammonium nitrate), phosphates (from mined rock), and potassium (mined in potash deposits). Often artificial fertilizer is present in agricultural runoff and causes eutrophication of waterways and pollutes wells with nitrates.

fission: splitting the nucleus of an atom by neutron bombardment; the nucleus of the atom splits into two nuclei and releases neutrons, generating the great heat energy of fission.

flocculation: mixing that causes particles to collide and join in clumps; it occurs naturally in lakes and rivers and is used in water purification.

floodplain: area of land that is periodically flooded by an adjacent body of water; often a valley floor that a river will flood seasonally during heavy rains.

flora: plants living in an area.

fluorides: substance derived from a naturally mined mineral and added to water to prevent tooth decay. Excess fluoride causes blue skin, mottled teeth, bone cancer, kidney disease, and fluorosis—a disease that causes soft and deformed bones in infants. Some parts of the world, such as China and India, have water that naturally has excess fluoride that must be removed. In a 1993 study the National Academy of Sciences concluded that fluoride in drinking water poses no health risks; a number of experts, including senior toxicologists at the EPA, disagree with this conclusion, and it is still a controversy. Currently there is a movement nationwide to ban fluorides.

fly ash: ash that goes up the chimney with the smoke in an incinerator, industrial, or coal-burning smokestack; often toxic, it is made up of fine particles.

food chain: order in which energy or food is used by living things. For example, in the oceans the food/energy chain may run from the sun to plankton, to minnows, to bigger fish, to penguins, to seals, and to sharks. On land the food/energy chain may run from the sun to plants, to insects, to small birds, to snakes, and to larger predators.

fossil fuels: fuels such as coal, oil, and natural gas that were formed over millions of years through the decomposition and pressurized compaction of living things.

fuel cell: device that generates electricity electrochemically, like a battery. Hydrogen- or oxygen-bearing ions are produced chemically at one electrode (anode—negative electrode, or cathode—positive electrode) and travel through an electrolyte to another electrode and react with oxygen atoms, creating DC electricity, which is converted to AC by a conditioner.

fuel rod: uranium shaped into a rod and used as nuclear fission fuel in a nuclear power plant.

fuelwood: trees, shrubs, or other woody plants burned for heating and cooking needs. Fuelwood, still used by 2 billion people, is a major cause of deforestation. It is a renewable energy source (if plants or trees are regrown) as a biomass energy source.

fusion: joining of two nuclei of atoms under extreme temperature and pressure; still under research. Fusion is the process whereby the sun releases its energy.

Gaia: Greek goddess of Earth. The Gaia hypothesis was developed by Professor James Lovelock, who believes that living organisms and their physical environments evolve as a single unit and change each other; this hypothesis sees the planet as a living whole.

gamma radiation: ionizing/electromagnetic radiation composed of a wave of energy quantities that move at the speed of light and that have no mass or charge; they can pass through steel and concrete and can damage living organisms by disrupting cellular function.

gas exchange: adding oxygen to water to remove carbon dioxide and hydrogen sulfide; also called aeration. It is used to purify water and occurs in natural systems.

gene: unit of heredity; the smallest segment of DNA, containing instructions for the development of a particular inherited characteristic.

gene bank: storage facility for genetic material, plant or animal, to preserve a genetic line for future use; examples are wild, natural varieties of plant seeds no longer cultivated by farmers and sperm and ova of animals threatened with extinction.

gene pool: all of the genes of all of the individuals in a particular animal, plant, or human population. For instance, for a population of 500 whales, the gene pool would be all the genes of all 500 whales.

General Agreement on Tariffs and Trade (GATT): international agreement that stipulates world trade rules and arbitrates disputes over its terms. More than 100 countries belong. Begun in 1947, it is aimed at limiting restrictions on trade. GATT needs to be updated so that it recognizes the right of a country to set its own environmental laws, and GATT arbitrations should favor the side of environmental protection. According to 1993 interpretations, GATT arbitrators could force a country to throw out an environmental law if it were seen as restricting trade. In 1991, for example, Mexico argued that U.S. dolphin-safe tuna laws interfered with trade, and won; Mexico could have forced the United States to buy its tuna, even though Mexican fishermen were found to be using nets that kill dolphins. Mexico backed down, because of its interest in passing NAFTA. Many see current GATT interpretations as favoring large multinational corporate interests over the common good of countries, citizens, indigenous peoples, and the environment.

genetic damage: reproductive organ or tissue damage that can be passed on to offspring.

genetic diversity: variation among individuals of a specific population of the same species, determined by the range of characteristics coded in the DNA of each individual. Examples are different eye color and foot size.

genetic engineering: recombining or manipulating genetic material. It often involves inserting a piece of DNA from one species into the DNA strand of another species.

genetic erosion: loss of enough members of a population so that the variability of the DNA of the population is greatly reduced.

genetics: area of biology dealing with heredity and variation of organisms.

genome: one haploid set of chromosomes and their genes.

genotype: genes, or total hereditary constitution, an individual carries.

germination: sprouting of a seed, usually requiring warmth, water, and oxygen.

gram: metric unit of weight, equivalent to 0.035 ounces.

gray water: water recaptured from household sinks and pipes, except the toilet; it is treated and reused in irrigation, toilets, and industry.

groundwater: water that has percolated down through the soils or bodies of water and is held in the ground in pores, cracks, fissures, aquifers, and artesian systems.

habitat: area of wilderness that wildlife use as home or for survival.

half-life: period of time it takes for 1/2 the number of atoms in a given mass of a radioactive element to decay.

halogens: elements of fluorine, iodine, bromine, chlorine, and artificial astatine; they are similar in chemical structure and highly reactive. Compounds in which halogens are used are often highly toxic.

halons: group of chemicals with bromine that are adding to ozone depletion; they are used in fire extinguishers.

hectare: metric unit of land area roughly equal to 2 1/2 acres.

herbicides: chemicals used to kill unwanted weeds.

herbivore: animal that eats only plants.

hertz (Hz): frequency with which an alternating current changes direction; hertz is the number of complete cycles per second.

hormone: chemical substance secreted by cells of endocrine glands, which carry signals to influence the activity of other cells or organs.

Horn of Africa: west middle Africa, including Sudan, Ethiopia, Eritrea, Djibouti, and Somalia.

humus: organic material, such as leaves, grass, or dead animals and plants, that has decomposed; it adds nutrients to the soil and provides buffering capacity, metal binding capacity, soil stability, and water-holding ability.

hybrid: combination of two or more different types or varieties, usually in reference to combinations of genetically different individuals. Hybrid crops are often not as environmentally suited to a particular area as natural varieties, and they require more care, chemicals, and artificial nutrients to survive. A hybrid car could refer to an electric car with an internal-combustion backup engine; a hybrid power plant could refer to a solar power plant that uses natural gas fuel for peak power needs.

hydrate: compound formed by the union of water and another substance, such as methane held by water in the frozen arctic tundra.

hydrocarbon: organic compound made up of carbon and hydrogen. Most hydrocarbons occur in coal, oil, and natural gas; many are also called volatile organic compounds (VOCs).

hydroponics: growing plants using a nutrient-rich liquid instead of soil.

hydroxyl radicals (OH): Atmospheric molecule which reduces methane (CH_4), carbon monoxide (CO), and tropospheric ozone.

indicator species: species of plant or animal that is dependent on specific, critical factors in the environment for its survival and is often very sensitive to environmental changes. Its disappearance or reduction in numbers is used as a general sign of environmental degradation or destruction. Examples are songbirds, butterflies, and salmon.

inert: describes a stable chemical, gas, or compound that will not react with elements very easily, if at all. Glass is stable, nonreactive, and thus inert, whereas chlorine is reactive with a host of different chemicals.

infrared radiation: solar radiation, often called heat waves; the stronger the infrared, the warmer the temperature.

inorganic: composed of matter other than plant or animal material, such as rocks and minerals.

inorganic chemical: chemical with no carbon in its structure.

in situ: in the natural or original position or area.

integrated pest management (IPM): method of controlling pests that minimizes or eliminates pesticide use by using natural barriers between crops to hinder insect movement, crop rotation to break the reproductive cycle of the insects, biological controls like pheromones to confuse the insects, and plant varieties that are resistant to insects.

integrated waste management: multifaceted approach to solving waste management problems that uses source reduction, recycling, reprocessing, landfills, and incineration.

intercropping: growing two or more crops in the same field at the same time, or one after another. Often crops grown together will be harvested at different times.

invertebrate: living organism without a spinal column or backbone.

iodine-131: radioactive iodine. Common in nuclear fallout, iodine-131 has a half-life of 8 days; it accumulates in the thyroid and can cause cancer.

ion: atom that has lost or gained one or more electrons, giving it a positive or negative electric charge, respectively. A cation is an atom that has lost electrons and has a positive charge; an anion is an atom that has gained electrons and has a negative charge.

ion exchange: removal or addition of ions from or to a substance. It is used in water purification by adding resins to remove calcium and magnesium, which produce hard water.

ionization: forming ions through chemical reactions, radiation, or electrical discharge.

ionizing radiation: small, intense waves of radiation that can cause an atom to lose or gain one or more electrons and thus acquire an electrical charge. Ionized atoms are very unstable and can combine with other atoms and molecules, causing physical, chemical, and biological health problems in humans. Examples are alpha and beta particles, gamma rays, X rays, and ultraviolet rays.

irrigation: diverting water resources with channels, ditches, canals, or dams to agricultural areas that lack enough rainfall for crops.

isotope: atom with the same number of protons (but different number of neutrons) as another atom of the same element. Examples are naturally occurring uranium isotopes: U-234, U-235, and U-238.

joule: unit of energy, equal to 0.24 calories (1 calorie = 4.184 joules). One joule will raise the temperature of 1 gram of water by 0.24°C.

keystone species: animal or plant species whose actions or pattern of living determine the survival of other living creatures. Sea otters are keystone species because they eat sea urchins; if sea otters are killed, sea urchins destroy kelp beds, fish die off, and seals and eagles that eat the fish die off.

kilogram: metric unit of weight equal to about 2.2 pounds.

kilometer: metric unit of length equivalent to 0.62 mile.

kilowatt-hour (KHW): amount of energy that a 100-watt light bulb burns in 10 hours. One kilowatt (KW) = 1,000 watts; 1,000 kilowatts = 1 megawatt (MW); 1,000 MW = 1 gigawatt (GW) = the power of a large electric power plant.

krill: shrimplike crustacean living mainly in surface Antarctic waters, used as food by fish, birds, penguins, squid, seals, and whales. A keystone species in the Antarctic area, a krill is reddish-brown, about 3 inches long, and can occur in swarms of up to 1 billion.

landfill: often open pit where garbage is dumped. Rainwater runoff, or runoff that leaks into the ground, is called leachate and is often toxic. Modern landfills have liners, covers, treatment for leachate runoff, and monitoring systems for methane and for groundwater and soil contamination from landfill toxics.

larva: early form of animal or insect that hatches from eggs and then goes through an intermediate period of development before metamorphosizing into an adult. Examples are frog tadpoles and the aquatic larvae of dragonflies.

latitude: angular distance from the equator (0° latitude), with the poles being 90° latitude. Parallel circles measuring this distance and running east-west around the earth are called parallels of latitude. With latitude and longitude all exact points on the earth can be identified; this system of measurement is used in navigation.

leach/leaching: process by which a chemical or liquid dissolves or combines with a substance and removes it from another material that previously held it. For example, acid rain can leach heavy metals from soil or nutrients from leaves.

leachate: liquid resulting from water or other fluid percolating through soil or other substances and leaching out materials.

levee: natural or human-created bank, usually of dirt or silt, formed parallel to a river. Rivers that continually flood will eventually deposit enough silt to form levees, which help control further flooding.

life cycle assessment: evaluation of all impacts and aspects of a product, from procurement of needed raw materials to end product waste and disposal.

liming: addition of calcium, often calcium carbonate, to acidified waters or soils to restore pH.

longitude: angular distance from the prime meridian, at Greenwich, England, which is measured at 0° east or 180° west. Arcs running north-south from pole to pole measuring this distance are called meridians of longitude. Longitude can be expressed in either degrees or time. With latitude and longitude, all exact points on the earth can be identified; This system of measurement is used in navigation.

magma: mixture of liquid rock, gases, and mineral crystals formed as the extreme heat at the earth's core melts adjacent rock. Most magma is 50 to 100 miles beneath the earth's surface, but some is only 15 to 30 miles deep. Magma rises to the surface because it is lighter than surrounding rock. Magma that reaches the surface of the earth through volcanoes or other geological disturbances is called lava. When magma cools, it forms igneous rock. Magma ranges in temperature from 1,650° to 2,200°F.

magnetic levitation (maglev) trains: trains using magnetic forces to allow high-speed movement. The trains float above a fixed track, called a guideway, but do not touch it. The lack of friction allows speeds over 300 mph. Since the trains use electricity for power, they pollute less than other trains, and operate more quietly.

mangrove: forest along a shoreline in shallow brackish waters. Some of the tree roots project vertically above the water to obtain oxygen. Mangroves are among the most productive ecosystems in the world. They serve as fish, crab, and shrimp nurseries, prevent erosion, and extend shorelines by trapping mud with underwater roots.

marsh: treeless wetland with shallow water and floating or rooted plants; often found along bodies of water, which can be fresh water, brackish water, or salt water.

mass: quantity of matter in a given object.

megadiversity: extremely high biodiversity in plants or animals.

megawatt (MW): 1,000 kilowatts; 1,000,000 watts.

meltdown: process in which excessive heat in a nuclear reactor dissolves the reactor's protective case and housing, releasing large amounts of radiation.

meter: metric unit of length equivalent to 39.4 inches.

metric ton: metric unit of weight equivalent to 1.1 tons.

microbe: microorganism usually visible only through a microscope, such as bacteria or viruses.

microgram: metric unit of weight; there are 454 million micrograms in 1 pound.

micron: metric unit of length; 1 million microns = 1 meter.

mollusk: invertebrate animal with a soft, unsegmented body in a calcareous shell, such as a snail or clam.

monkeywrenching: term out of Edward Abbey's book *The Monkey Wrench Gang,* used to describe environmental sabotage techniques that are directed at property and equipment that threaten the environment; for example, pouring sand into the gas tank of timber machinery.

monoculture: farming method that involves repeatedly growing a large acreage of only one crop that is harvested all at the same time.

mudflat: bare, flat bottom of any body of water whose level has recently been lowered.

municipal solid waste: waste collected by public or private haulers in communities from homes, institutions, commercial businesses, and industry.

mutagen: chemical or substance that can cause changes in the genetic material of a living organism and that can thus cause mutations in the offspring of these same organisms. Carcinogens are often mutagens.

mutation: change in genetic material leading to changes in physical characteristics, which can be lethal.

nautical mile: unit expressing distance on sea or in the air; a British unit of 6,080 feet; an international unit of 6,076 feet.

negawatt: "negative watts"; refers to electricity saved (through conservation or efficiency) instead of used or wasted.

neutriceutical: genetically engineered food product used for a specific health effect, such as psyllium fiber taken to reduce cancer risks and garlic extact to reduce blood cholesterol.

neutron: particle with no charge that is in the nucleus of an atom.

niche: ecological niche.

nitrate: compound, often a salt, containing nitrogen and oxygen (NO_3^-). It is found naturally in soils and created by soil bacteria that convert nitrites to nitrates, which are used by plants to build amino acids. Nitrates are found in abundance in manure and added industrially as ammonium nitrate to artificial fertilizers; when present in agricultural runoff, both can cause eutrophication of waterways. Nitrates also can end up in drinking wells and when taken with water can be changed back into nitrites in the human body.

nitrite: compound, often a salt, containing nitrogen and oxygen (NO_2^-). It is found naturally in soils and created by soil bacteria from atmospheric nitrogen. If humans drink nitrates in contaminated water, their bodies can change nitrate to nitrite, which interferes with the blood's ability to carry oxygen and is called methemoglobinemia (in babies this is referred to as blue-baby syndrome), which can be fatal. Nitrites can also lower blood pressure and cause headaches, nausea, and diarrhea. Nitrites can combine with organic nitrogen compounds in the body to form nitrosamines, which are suspected of causing stomach cancer.

nitrogen cycle: natural process that recycles nitrogen in the environment. Nitrogen makes up 78% of the atmosphere. Certain soil bacteria and blue-green algae can take nitrogen out of the air, through nitrogen fixation, and convert it to nitrites and nitrates. The symbiotic bacteria rhizobium is the most common, and it is on the roots of legumes such as clover, beans, peas, alfalfa, and vetches. Decomposing animals and plants also release nitrites and nitrates to the soil. Plants use nitrates to build amino acids, which animals use when they eat the plants. When plants and organisms die, the nitrogen in their bodies is decomposed and returned to the soil, and eventually to the atmosphere again. There it can be reconverted, used again by the plants and eventually by plant-eating animals. Human activities are adding more nitrogen to the atmosphere and soils in the form of nitrogen oxides from burning fossil fuels and biomass and nitrogen in fertilizers, respectively.

nonattainment: failure to meet pollutant emission guidelines in specific sites or areas.

Nongovernmental organizations (NGOs): small organizations, minority groups, indigenous peoples, neighborhood associations, and small communities worldwide, NGOs are rapidly becoming one of the most important and influential components of public policy. About 10,000 exist worldwide and are linked by computer by ECONET.

nonionizing radiation: long, low-frequency waves of radiation such as infrared, ultraviolet, visible light, and that emitted by microwaves, radios, appliances, computers, water bed heaters, electric blankets, and power lines.

nonpoint source pollution: pollution emitted along a wide boundary, such as from landfills, streets, mining sites, building sites, and farmland.

nonrenewable resource: resource that cannot be replenished once it is used up. Fossil fuels, minerals, ecosystems, and species are examples.

nontarget species: species that is accidentally selected. Examples are insects that are killed by a pesticide that was aimed at another species.

North American Free Trade Agreement (NAFTA): international agreement among Canada, Mexico, and the United States to end any trade barriers and government influence between the 3 countries. Critics say side agreements covering worker wages and rights and environmental pollution issues are not strong enough and may jeopardize U.S. environmental policies as well as fail to improve Mexico's.

nuclear fallout: highly radioactive dust created during a nuclear explosion when surrounding soil, dust, or other materials are exploded up into the air and then precipitate as dust or in moisture. It can be carried long distances by the wind.

nuclear proliferation: acquisition of nuclear weapons by countries that previously did not have nuclear weapons capacity.

nuclear reactor: device in which a nuclear-fission chain reaction can be started, maintained, and controlled to generate energy.

nuclear winter: theoretically expected environmental changes with even a small exchange of nuclear warheads. Ash and soot from burning cities would rise into the atmosphere and block 80% of the sun's heat. This would lower temperatures as much as 20° in a few weeks, producing dryness, cold, and no sunlight: thus a nuclear winter. Food sources would be destroyed, and extinctions would occur on a mass scale.

nucleotide: building block of a DNA molecule made up of a phosphate, a sugar, and an organic base; there are 4 principal nucleotides in DNA.

Oceania: lands of the Central and South Pacific, including Melanesia, Micronesia, Polynesia, and sometimes Australia, Malaysia, and New Zealand.

off-the-grid power: small-scale electricity or energy generation independent of a major power plant; often done by photovoltaics, solar, wind, or microhydroelectric power. Off-the-grid power is currently used in residential, commercial, industrial, and rural sectors.

old-growth forest: high-biodiversity ecosystem that has many trees 200 to 300 or even 1,000 years old and that has not been harvested or logged.

omnivore: animal that eats both plants and animals.

ore: mineral mined for a valuable component. Examples are shale mined for the oil or coal component or various rocks mined for metal components, such as rock mined for copper.

organic chemical: chemical that contains the element of carbon. All living organisms are made of natural organic chemicals. Human-created, or synthetic, organic chemicals can mimic chemicals in the human body and thus gain entrance to the nervous system and brain.

organic food: crops or other food grown without the addition of any chemicals such as artificial fertilizers, pesticides, herbicides, or growth chemicals.

organochlorines: chlorinated hydrocarbons.

organophosphates: compounds containing carbon, hydrogen, and phosphorus; used in pesticides that disrupt the nervous system of pests. They are biodegradable in the environment and in living organisms. Examples are parathion and malathion, which was used in California to combat medflies. They are toxic to humans and can cause headaches, nausea, convulsions, paralysis, and death.

oxidant: substance that facilitates oxidation of another substance in a chemical reaction.

oxidation: addition of oxygen, or the removal or loss of hydrogen or electrons, occurring in a given compound or element. Oxidation reactions usually occur with reduction.

oxygen cycle: natural process of recycling oxygen in the environment. Plants create oxygen during photosynthesis by using carbon dioxide, water, and sunlight; animals and humans breathe in oxygen and give off carbon dioxide, which is again recycled into oxygen. Some oxygen is used to create ozone or held in sinks, like rivers, lakes, and the oceans, where fish filter it out with their gills.

ozone: molecule made of three oxygen atoms (O_3), created by oxygen reacting in the presence of ultraviolet light. Most ozone is in the stratosphere and is called the ozone layer; ozone created at ground level is a major component of smog. Ozone is very toxic to living organisms.

ozone layer: area of the stratosphere 10 to 25 miles high that contains most of the naturally occurring ozone. It protects the earth by absorbing ultraviolet-B (UV-B) and helps to hold heat around the planet.

Pacific Rim: rim of active volcanoes around the edge of the Pacific Ocean, including Japan, the Philippines, the western coast of the United States, Central and South America, and New Zealand.

particulates: liquid or solid particles in air or gas, also sometimes called aerosols.

parts per million (ppm): measurement of a substance, such as a toxic or pollutant. Saying that 1 ppm of a toxic chemical in water or air is toxic is equivalent to saying a concentration of 1 part of that toxin (by weight per volume) in every 1 million parts (by weight per volume) of water or air is toxic.

pathogen: organism capable of causing disease.

peat: partially decomposed black or dark-brown organic matter, mainly plants, formed anaerobically in a cool, humid, temperate waterlogged area. Peat is the first stage of coal formation. It is found near the surface and can be cut into bricks and burned as fuel.

ped: natural aggregate of soil particles; the smallest structural unit of soil. It is sometimes used to determine the health of a soil.

pedology: study of soils.

pelagic: related to the surface water of the oceans. Pelagic creatures, such as herring, krill, and plankton, live in the surface waters.

permafrost: permanently frozen area of soil beneath the earth's surface in cold regions. The depth of the permafrost varies from region to region.

pesticide resistance: ability of a rodent, insect, or plant to resist a pesticide. When a pest population is sprayed with a chemical, a few individuals will have genes that allow them to survive the chemical. Those few are the ones that will reproduce and create the next generation of immune insects or plants. Since insects and some plants reproduce in short time spans, an immune generation can be created very quickly. If the spray concentration is increased, again some of the new generation will be immune to the increased concentrations, and they will reproduce to create even more immune insects. If multiple chemicals are used, insects and weeds can eventually become immune to many chemicals.

petrochemicals: chemicals made from coal, petroleum, or oil.

petroleum: oil and oil products before or after refining.

pH: logarithmic scale of 0 to 14 that measures acidity and alkalinity, with 0 being the most acidic, 14 the most alkaline, and 7 neutral. A substance capable of releasing hydrogen ions (H+) is an acid; a substance releasing hydroxyl (OH-) ions is a base (alkaline). Every step increase or decrease in pH is a 10-fold difference in acidity or alkalinity. Thus a pH of 4 is 10 times more acidic than a pH of 5. Conversely, a pH of 9 is 10 times more alkaline than a pH of 8. Pure water has a pH of 7, with equal amounts of H+ and OH- ions.

phenotype: characteristics of an individual produced by the interaction of the individual's genotype and the environment; the total physical, biochemical, and physiological makeup of an individual.

pheromone: chemical substance produced by an animal or insect that stimulates other individuals of the same species to act in a certain way. For example, a female moth releases a pheromone so that the male moth can follow her trail and find her to mate; mammals use pheromones for mating and territory.

phosphate: naturally occurring salt mined in limestone deposits or other mineral deposits. Used in fertilizers or detergents, phosphates cause eutrophication in waterways through agricultural runoff and sewage. Mining phosphates can release radon.

photodegradable: ability of an object to be broken down by ultraviolet light, which causes oxidation of the substance. The substance becomes brittle and then breaks down under the elements of wind and rain.

photosynthesis: process by which the chlorophyll molecules of green plants use sunlight to convert carbon dioxide, salts, and water into simple sugars like glucose; oxygen is a by-product.

phytoplankton: free-floating, mostly microscopic, aquatic organisms that use photosynthesis to create their food from minerals and salts in the sea. Phytoplankton are at the bottom of all food chains in the oceans, and they are extremely important as one of the most productive food sources for other plankton, krill, fish, and thus marine mammals and many other aquatic creatures.

phytotoxic: toxic to plants.

plankton: all the small microscopic animals (zooplankton) and plants (phytoplankton) living in the surface waters of the oceans.

plasma arc torch furnace: furnace that generates heat as high as 10,000°F in a process called pyrolisis. Instead of burning material with oxygen, it forces the molecules of the waste to come apart without burning. Thus there is no ash by-product, and no toxic fumes are given off; by-products are metal slag and a hydrogen-rich gas. A plasma arc torch is a portable version of the furnace for use at hazardous waste sites.

plastic: any of a number of petroleum-based products that do not biodegrade, are responsible for large amounts of toxic wastes when manufactured, often can be recycled only once, and if burned can produce toxics such as dioxins. Plastics include PET (polyethelene terephthalate—soft drink bottles, peanut butter jars), HDPE (high-density polyethylene—milk, water, and liquid detergent jugs), PVC (polyvinyl chloride—blister packs, cooking oil bottles, liquid detergent containers), LDPE (low-density polyethylene—lids, squeeze bottles, bread bags), PP (polypropylene—syrup and ketchup bottles, yogurt containers, bottle caps), and PS (polystyrene—coffee cups, meat trays, packing peanuts).

plutonium (Pu): human-created substance produced in nuclear reactors by bombarding uranium-238 (U-238) with neutrons, which are absorbed into the uranium nucleus, converting it to plutonium. The plutonium can be recovered by reprocessing and used in bombs or other nuclear reactors. Plutonium is one of the most toxic substances known, with a half-life of 24,000 years. Once in the body, it tends to settle near the bone marrow, and will irradiate the surrounding tissues. It can cause cancer, birth defects, a host of other diseases, and death.

point source pollution: pollution from a specific point, such as a pipe.

polyculture: practice of planting mixed crops used by early farmers, many small farmers in developing countries today, and organic farmers.

polychlorinated biphenyls (PCBs): stable toxic, organic chemicals used in dyes, paints, fluorescent light bulbs, adhesives, and electrical transformers and capacitators. They are also created through incomplete burning of plastics. When PCBs are incinerated, dioxins are created. PCBs have been banned in the United States since 1976 in most items, except in totally enclosed uses. PCBs have been found worldwide in ocean sediments, wildlife, and other living organisms.

polyvinyl chloride (PVC): petrochemical formed from the toxic gas vinyl chloride (VC). It is used as the base to create plastic. The production of PVC and burning it can lead to formation of toxic chemicals such as dioxins.

potable: drinkable; used to describe water.

prairie pothole: wetland area that goes through seasonal periods of dryness and wetness. Used for breeding by many birds, it is often a shallow or bowl-like depression on a flat prairie. Prairie potholes are found in the northern Great Plains.

precycling: eliminating garbage before purchasing products by making environmentally aware consumer choices. For example, precycling might be buying an item without packaging rather than buying the same item packaged.

predator: animal, plant, insect, or any organism that kills other organisms for food.

prey: plant, animal, or other organism hunted and eaten by others for food.

proton: positively charged particle in the nucleus of an atom.

pyrolysis: heating a substance in the absence of oxygen, thus breaking it into simpler components.

qanat: underground irrigation conduit used in ancient times in the Middle East, North Africa, Spain, and elsewhere. The conduit was placed under the ground, in a hill, and connected to an aquifer. Rainwater collected in the conduit and ran downhill by gravity to a dug well or canal. The water was then used for irrigation and drinking water.

radiation: energy emitted by atoms or molecules in the form of particles or waves. Everything, including the cells in the human body, has electric and magnetic fields and therefore can produce radiation. Examples are ultraviolet, X rays, cosmic rays, gamma rays, and beta and alpha particles.

radiation absorbed dose (rad): amount of radiation energy absorbed in 1 gram of human tissue. The effects of a rad varies with the type of radiation involved. Rad is being replaced with the unit gray; 100 rads equals 1 gray (Gy).

radioactive: describing the decay of an unstable substance in which the atomic nuclei give off electrons (or other particles) and energy and become more stable.

radiocarbon dating: determining the age of a substance by measuring the amount of the isotope, carbon-14, in the substance. The bones or teeth are used for dead animals. Carbon-14 is an isotope of carbon that is taken in by living plants and animals until they die; C-14 decays with a half-life of 5,730 years.

radioisotopes: radioactive isotopes.

radionuclides: radioactive nuclei; nuclei undergoing decay.

reclamation: restoring habitat to its natural state.

recombinant DNA (rDNA): new combination of genes spliced together on a single strand of DNA. The technique involves using restriction enzymes to cut and paste DNA fragments; it is also called biotechnology.

recyclable: capable of being used in the same capacity over and over again.

red tide: naturally toxic algae (dinoflagellate) bloom that is red, green, or brown in appearance.

reduction: adding electrons to an atom, molecule, or ion.

reforestation: reclaiming forested land through tree planting or other means; also called reafforestation.

regenerative agriculture: sustainable agriculture; building up soil productivity with high levels of production and little or no impact on the environment.

rem (R—roentgen equivalent man): biological effectiveness of a given radiation; 1,000 millirems (mR) = 1 rem. Normal background radiation at sea level is 100 mR per year and increases above sea level because of solar radiation. The sievert is replacing the rem; 100 rems = 1 sievert (Sv).

renewable resource: resource that, if managed properly, can be renewed over a given time to yield the same amount of product over and over. Examples are managed forests, fish populations, and solar and wind energy.

reprocessing: reclaiming plutonium and uranium from spent nuclear fuel by dissolving the fuel in an acid bath. The fission products left over in the acid bath are highly radioactive.

restoration ecology: efforts to restore damaged ecosystems to their original state.

Richter scale: logarithmic earthquake scale, ranging from 0 to 9, devised by C. F. Richter in 1935 to measure the strength of an earthquake; the higher the number, the stronger the earthquake.

rift valley: trough in the earth's crust between normal faults; an area of tension in the earth's crust as a result of crustal plates moving apart. The Great Rift Valley of Africa is a major geological feature, the longest crack in the earth's surface at about 2,000 miles in length.

riparian: referring to something living or located on the bank of a river, lake, or sea.

risk assessment: attempt to estimate the damage that may be caused to the environment or human health by a given pollutant or toxicity problem.

RNA (ribonucleic acid): nucleic acid formed on chromosomal DNA that is involved with protein synthesis. It is composed of chains of phosphate, sugar molecules (ribose), purines, and pyrimidines.

salinity: amount of dissolved salts or minerals in a liquid; often used when talking about lakes, oceans, rivers, or soils. A saline solution is one containing salt.

salinization: deposition and accumulation of salts in soils, often through irrigation of arid and semiarid lands. When land is too salty to support plant life, it is said to be salinized.

savannah: grassland area between tropical rain forests and deserts, located in tropics and subtropics. Savannahs have few trees and a long dry season, and fires are not uncommon.

scrubber: equipment designed to prevent toxic emissions from leaving smokestacks. It is used to remove sulfur dioxide and nitrogen oxides in power plants, pollutants in industry, and fly ash in incinerators. One type of scrubber blasts wet streams of lime at the fumes rising out of the smokestack; it removes anywhere from 70% to 95% of the sulfur dioxide.

sea level: average level of the sea when there are no tides and waves.

sediment: silt, sand, soils, rocks, organic matter, or other substance that is carried by natural processes from one place to another.

sedimentation: natural process in which silt, sand, rocks, organic matter, and soils are carried by running water, wind, glaciers, or seas and deposited elsewhere. Rivers and running water deposit silt on the floors of harbors, lakes, and oceans. It is estimated that about 25 billion tons of sediment is carried to the sea each year by rivers and that about 15 billion tons of that is caused by humans through deforestation, dams, and construction. Also, sedimentation means the settling of particles by gravity to the bottom of a water source; it is a naturally occurring process and is used in water purification.

seed bank: place where different varieties of plant seeds are preserved so that their genetic material is not lost. Seed banks preserve mainly wild, natural varieties that are not farmed anymore.

seismic: related to vibration of the earth, often referring to earthquake-related disturbances.

selective logging: practice of cutting only mature, targeted trees, allowing other trees to gain maturity. Selective logging can allow a forest to be harvested sustainably indefinitely and does much less harm to the forest as a whole than clear-cutting (cutting all trees).

semiarid: dry, with a limited rainfall and sparse vegetation. Semiarid areas are often transition areas between arid land, such as deserts, and grasslands; they have an annual rainfall of 10 to 20 inches.

shale: rock formed by the compaction of sediments such as mud, clay, or silt that often has deposits of one main substance, such as coal, oil, or calcium.

shifting agriculture: See swidden agriculture.

sick-building syndrome: indoor illness that includes headaches, nausea, beathing difficulties, fatigue, difficulty concentrating, dry cough, eye irritations, and skin rashes. It is caused by bad ventilation, poor building maintenance, and indoor toxics.

silicates: complex metal salts containing silicon and oxygen. Silicates are the largest class of minerals in the earth's crust and are often used in building materials.

silviculture: branch of forestry concerned with the development and care of forests, often used for growing trees as commercial crops.

sink: See environmental sink.

sinkhole: deep hole formed by flooding or weathering of rock, usually limestone, through which water can run down to the water table.

slash-and-burn agriculture: See swidden agriculture.

smelting: process using heat and other substances, such as limestone or other chemicals, to separate a metal from its ore; also known generally as refining or processing.

soluble: substance that can be dissolved in a liquid base. Water soluble refers to a substance that can be dissolved in water; fat soluble refers to a substance that can be dissolved in fat. Chemicals that are fat soluble can gain access to nerves, the central nervous system, and the brain, making them much more toxic and dangerous.

solvent: liquid capable of dissolving or dispersing one or more substances.

source reduction: reducing the amount of garbage produced by reducing it at the source of production through recycling feedstock materials, changing manufacturing methods, using different materials, or eliminating processes.

species: group of animals, plants, or other organisms that can only reproduce among themselves.

stratosphere: layer of atmosphere just above the troposhere, between 10 and 30 miles high. The higher you go in the stratosphere, the warmer it gets; these temperature differences tend to keep different layers within the stratosphere separate, and they mix very slowly.

strontium-90: radioactive strontium; common in nuclear fallout. With a half-life of 26 years, it accumulates in bones and can cause cancer.

subtropical: bordering the tropical zones.

sulfate: oxidized sulfur.

supercritical fluid: fluid at high temperature and pressure that exhibits characteristics of both a liquid and a gas. Water is supercritical at 374°C and 221 bars pressure.

sustainable: describing a process or activity that does not deplete resources, replenishes them, or allows them to replenish themselves as they are used.

sustainable yield: the number or quantity of fish, animals, plants, or other resource that can be harvested without depleting the population or that allows the population or resource to maintain (or replenish) itself at its former numbers or quantity.

swidden agriculture: agricultural system used in the tropics by traditional tribes and villagers. The method is to cut down trees and brush in a small plot and burn them to release the nutrients to the soil; also called slash-and-burn or shifting agriculture. In 2 to 3 years the plot is no longer fertile, and the farmer moves to the next plot, leaving the first one to revert to rain forest over a period of 15 to 30 years. The cultivated plots grow multilayered gardens of many different edible plants, mimicking the layered growth of the rain forest.

synergy: process that occurs when two different substances unite to produce an effect that neither could produce alone. For instance, DDT's solubility in water is increased thousands of times if oil is present. Many insecticides, which may be relatively safe to use alone, together may destroy the enzymes in the body that would normally neutralize them, rendering them more toxic.

systemic: acting through the systems of an organism. In plants, an herbicide acts systemically by being absorbed through the roots and entering plant tissues; pesticide residues may be systemic, entering the fruit, stems, or other organs of a food crop.

taking: confiscating private land for public use.

target species: a species that is purposefully selected by a process. Examples are an insect for which a pesticide is developed to kill and a specific fish species that is caught using particular methods.

temperate: having a moderate climate; describing the regions between the tropics and polar circles.

temperature inversion: in air, a layer of warm air lying atop a layer of cold air (usually as you go higher in the sky the air is cooler); in water, a layer of warmer water below the cooler surface water (usually water is cooler as you go deeper).

teratogenic: tending to cause birth defects with exposure to an expecting mother.

thermal pollution: release of a substance, often liquid or air, that increases the heat in the area where it is released. For example, industrial wastewater dumped into a river or lake can raise the water temperature.

thermocline: often permanent water layer in lakes and oceans that separates warmer water layers from colder water with a sharp drop in temperature.

Third World: countries still developing their industrial complex and often economically weak; also called developing countries; includes countries in Africa, Asia, Latin America, and the Pacific.

tolerance level: legal residual amount of a substance or chemical that is allowed to be in food.

toxicity: measure of how dangerous a substance is to living organisms or to human health.

toxic use reduction: reducing the use of toxics in industry by substituting nontoxic alternatives.

trace substance: substance present in very small quantities, such as the trace gas argon in the atmosphere or the trace metal zinc in the body.

Often trace substances in living creatures are essential for physiological and biochemical processes to occur.

transgenic manipulation: genetic engineering. A transgenic animal has the genes of another organism spliced into its genetic code.

transmutation: changing one element to another; occurs naturally with radioactive decay and artificially with bombardment of an element with other nuclei or particles. This process is being studied to render nuclear waste less harmful by changing it to another substance.

transpiration: the loss of water vapor through plant leaf pores; often used by plants for cooling.

tropical: describing the region near the equator between the latitudes of the tropic of Cancer (at latitude 23 1/2° north, the northernmost latitude reached by the overhead sun) and the tropic of Capricorn (at latitude 23 1/2° south, the southernmost latitude reached by the overhead sun). The average monthly temperatures are greater than 68°F (20°C); rain forest and savannah are common vegetation types here; rainfall varies.

tropopause: boundary layer between the troposphere and stratosphere that is important in confining water to the lower atmosphere and troposphere.

troposphere: layer of atmosphere from the earth's surface to about 5 to 10 miles high. The higher you go in the troposphere the colder it gets; this tends to keep gases and particles in this layer well mixed. The troposphere holds 75% of the atmosphere's mass and most water vapor, clouds, and pollutants; it is where storms occur and airplanes fly.

tsunami: large, rapidly moving ocean wave created by an earthquake, landslide, or volcanic eruption.

tundra: cold, treeless plain in northern Arctic regions consisting mainly of lichens, mosses, shrubs, and grasses that support caribou, moose, elk, polar bear, insects, and a number of birds and rodents.

turbidity: cloudy condition of a liquid due to suspended matter such as clay, silt, or algae.

turbine: set of blades rotating around a hub, turned by wind or water, often used to drive generators to create electricity; used in wind and hydroelectric energy.

typhoon: tropical cyclone, or hurricane, in the western Pacific Ocean.

ultraviolet radiation: radiation from the sun.

umbrella species: often a large predator that requires a large range of land for survival; saving it also ensures the continued survival of all the smaller species in the same area.

uranium (U): naturally occurring radioactive element that is a silvery white metal. Uranium is a trace element in the earth's crust and is widely distributed in most soils and rocks. It serves as fuel in nuclear power plants and is used to create plutonium, which is used in nuclear bombs. Natural uranium is the heaviest natural element and consists of several isotopes. Uranium is enriched for use as nuclear fuel by increasing the concentration of the isotope U-235. Uranium is the largest source of radon, mining uranium releases radon.

vertebrate: organism with a spinal column or backbone, usually with an internal cartilaginous or bony skeleton. Vertebrates include mammals, birds, fish, and reptiles.

virus: smallest infectious particle; replicates only in a host's cells by directing a host cell to create material required to produce more virus.

vitrification: heating a material into a glasslike substance so that it is largely inert.

vitrified: describes a material already heated into a glasslike substance so it is largely inert.

volatile organic compounds (VOCs): compounds containing carbon, which evaporate easily at low temperatures. Many VOCs are hydrocarbons. They are emitted in industry and car exhaust and used in a number of household products. Large contributors to smog, tropospheric ozone, and indoor air pollution, they can cause eye, nose, and throat irritation; headaches; nausea; kidney and liver damage; and cancer. Examples are benzene and carbon tetrachloride; there are several hundred others.

water cycle: natural process in which water is recycled in the environment. Water evaporates from the oceans, is moved by wind over land masses, and then is precipitated out in the form of snow or rain. It is collected in watersheds, which are areas that funnel the flowing water into rivers, lakes, and eventually oceans. Other rain or snow is absorbed into the soil or held in marshes, bogs, and wetlands, where it trickles down into the aquifer system. Living organisms take in water and give it off during respiration. Much water evaporates from land and again is carried on the wind to repeat the cycle.

watershed: natural area that funnels water into rivers and lakes.

water table: upper surface of underground water, which is the upper surface of groundwater saturation.

watt (W): unit of power; the production of 1 joule of energy per second equals 1 watt.

weight: measure of gravity on an object.

wildlife trade: legal and illegal international trade in animals and plants, products, or parts thereof. About 1/3 of wildlife trade is illegal and is a major reason that endangered animals could go extinct.

World Bank: bank owned by governments of 160 nations that finances lending operations primarily from its own borrowings in world capital markets. The bank funds no nuclear energy, encourages ecotourism, and since 1970 has funded 400 hydroelectric and irrigation projects. Germany, Japan, and the United States hold 73% voting rights on a 24-member bank board. The World Bank has come under decades of criticism for funding environmentally degrading projects, especially hydroelectric, which have displaced hundreds of thousands of indigenous peoples and have often benefited wealthy politicians and businessmen more than the citizens of the countries involved.

World Heritage Site: natural or cultural site recognized internationally as possessing universal value and being deserving of collective responsibility. Countries nominate sites to the World Heritage Convention, and an international committee decides.

xerophyte: plant that can grow in arid or semiarid climates.

xeriscape: using local native plants and vegetation in gardens and lawns.

zooplankton: free-floating small animals, mainly crustaceans, in the surface waters of the oceans that feed primarily on phytoplankton. Zooplankton are in turn fed on by many fish and other animals.

APPENDIX A
Environmental Groups and Agencies

Acid Rain Foundation, 1410 Varsity Drive, Raleigh, NC 27606; (919) 828-9443. Primarily air-quality issues: acid rain, air pollution, global climate.

African Wildlife Foundation, 1717 Massachusetts Avenue, NW, #602, Washington, DC 20036; (202) 265-8393. Wildlife conservation projects in Africa.

American Cetacean Society, P.O. Box 2639, San Pedro, CA 90731; (310) 548-6279. Protection of marine mammals, especially whales and dolphins.

American Council for an Energy-Efficient Economy, 1001 Connecticut Avenue, NW, Suite 801, Washington, DC 20036; (202) 429-8873. The most energy-efficient appliances.

American Oceans Campaign, 725 Arizona Avenue, Suite 102, Santa Monica, CA 90401; (310) 576-6162. Concerned with offshore and inland pollution of oceans.

American Rivers, 801 Pennsylvania Avenue, SE, Suite 400, Washington, DC 20003; (202) 547-6900. Protection of U.S. rivers.

American Wind Energy Association, 122 "C" Street, NW, 4th Floor, Washington, DC 20001; (202) 408-8988. Wind industry association.

Antarctic Project, 424 "C" Street, NE, Washington, DC 20002; (202) 544-0236. Protection of Antarctica.

Bat Conservation International, P.O. Box 162603, Austin, TX 78716; (512) 327-9721. Worldwide bat conservation.

Center for Environmental Information, Inc., 50 W. Main Street, Rochester, NY 14614; (716) 262-2870. Special library and information assistance.

Center for Marine Conservation, 1725 DeSales Street, NW, Suite 500, Washington, DC 20036; (202) 429-5609. Protection of coastal and ocean wildlife and resources.

Center for Responsible Tourism, P.O. Box 827, San Anselmo, CA 94979; (510) 843-5506. Promoting ecotourism and indigenous rights worldwide.

CERES-Coalition for Environmentally Responsible Economies, 711 Atlantic

Avenue, Boston, MA 02111; (617) 451-0927. Promotion of investing money in environmentally responsible companies agreeing to the CERES Principles.

Citizen's Clearinghouse for Hazardous Waste, P.O. Box 6806, Falls Church, VA 22040; (703) 237-2249. Provides citizen assistance for environmental justice and to fight for environmental cleanups.

Clean Water Action, 1320 18th Street, NW, Suite 300, Washington, DC 20036; (202) 457-1286. Grass-roots group working on water quality, incineration, global climate, and toxics.

Congressional Legislation Hotline, (202) 225-1772. Information on congressional bills.

Conservation International, 1015 18th Street, NW, Suite 1000, Washington, DC 20036; (202) 429-5660. Active in preserving temperate and tropical ecosystems.

Consultative Group on International Agricultural Research (CGIAR), 1818 H Street, NW, Washington, DC 20433; (202) 477-1234. World Bank group helping developing countries expand agricultural production, with a focus on sustainable agriculture.

Council for Responsible Genetics, 5 Upland Road, Cambridge, MA 02140; (617) 868-0870. Organization working to safeguard the environment and public health from biotechnology applications that may pose a threat.

Cultural Survival, 215 First Street, Cambridge, MA 02142; (617) 621-3818. Promoting ecotourism and indigenous rights worldwide.

Defenders of Wildlife, 1101 14th Street, NW, #1400, Washington, DC 20005; (202) 659-9510. Active in legal action to protect wildlife and habitat.

Ducks Unlimited, One Waterfowl Way, Long Grove, IL 60047; (708) 438-4300. Works to preserve wetlands for waterfowl populations.

Earth First!, P.O. Box 5176, Missoula, MT 59806. Activist group supporting nonviolent opposition to activities that threaten nature, such as clear-cutting old-growth forests.

Earth Kids Organization, P.O. Box 3847, Salem, OR 97302; (503) 363-1896. Children connected worldwide by computer, sharing ideas to protect the environment.

Earth Island Institute, 300 Broadway, Suite 28, San Francisco, CA 94133; (415) 788-3666. Diverse projects such as marine mammals, sea turtles, and Lake Baikal.

Earthwatch, P.O. Box 403, Mt. Auburn Street, Watertown, MA 02272; (800) 776-0188. Scientific field research done with unskilled volunteers.

Ecotourism Society, P.O. Box 755, North Bennington, VT 05257-0755; (802) 447-2121. Helps indigenous peoples promote ecotourism.

Environmental Defense Fund, 257 Park Avenue South, New York, NY 10010; (212) 505-2100. Scientists, economists, and lawyers address global issues.

Food & Water, Inc., (800) EAT-SAFE. Information on food quality and hazards in our foods.

Freshwater Foundation, Spring Hill Center, 725 County Road 6, Wayzata, MN 55391; (612) 449-0092. Surface and groundwater protection with scientific work and education.

Friends of the Earth, 1025 Vermont Avenue, NW, Third Floor, Washington, DC 20005; (202) 783-7400. Worldwide work on numerous issues working with citizens to influence public policy.

Global Green U.S.A., P.O. Box 21451, Columbus, OH 43221-0451; (805) 565-3485. American partner of Green Cross International, a group aimed at organizing grass-roots pressure to change governments so they will work with their citizens to protect the environment.

Global Response: Environmental Action Network, P.O. Box 7490, Boulder, CO 80306-7490; (303) 444-0306. Organizes monthly global letter writing campaigns for adults and elementary and junior-high students with newsletters on current environmental problems to mount pressure on decision makers.

Greenhouse Crisis Foundation, 1130 17th Street, NW, Suite 630, Washington, DC 20036; (202) 466-2823. Addresses global climate issues.

Greenpeace, 1436 U Street, NW, Washington, DC 20009; (202) 462-1177. Active and political work to protect the environment worldwide.

Human Ecology Action League, P.O. Box 49126, Atlanta, GA 30359-1126; (404) 248-1898. Information clearinghouse on hazardous and toxic threats.

Humane Society of the United States, 2100 L Street, NW, Washington, DC 20037; (202) 452-1100. Protection of wild and domestic animals.

Institute for Consumer Responsibility, 6506 28th Avenue, NE, Seattle, WA 98115; (206) 523-0421. Information on current boycotts.

Institute for Local Self-Reliance, 2425 18th Street, NW, Washington, DC 20009; (202) 232-4108. Promotes local action and participation by citizens.

International Alliance for Sustainable Agriculture, 1701 University Avenue,

SE, Minneapolis, MN 55414; (612) 331-1099. Promotes sustainable agriculture practices.

International Eco-Agriculture Technology Association, Inc. (I.E.A.T.A.), P.O. Box 998, Welches, OR 97067; (800) 798-5543. Promotes environmentally sound agricultural practices worldwide.

International Federation of Organic Agriculture Movements, c/o Berne Ward Geier, Okozentrum, Imsbach W-6695, Tholey-Theley, Germany; 6853 5190. Promotes organic agriculture worldwide.

International Institute for Energy Conservation, 750 First Street, NE, Suite 940, Washington, DC 20002; (202) 842-3388. Active in energy efficiency projects in developing countries and Eastern and Central Europe.

International Society for Animal Rights (ISAR), 421 South State Street, Clark Summit, PA 18411; (717) 586-2200. Actively involved with major boycotts worldwide to defend animal rights.

International Whaling Commission (IWC), The Red House, 135 Station Road, Histon, Cambridge CB4 4NP England; (0223) 233971. World commission to protect whales.

Institute for Food and Development Policy, 398 60th Street, Oakland, CA 94618; (510) 654-4400. Studies food, water, and health issues in developing countries.

Izaak Walton League of America, 1401 Wilson Boulevard, Level B, Arlington, VA 22209-2318; (703) 528-1818. Promotes land acquisition for protection.

Jobs and the Environment Campaign, 1168 Commonwealth Avenue, Third Floor, Boston, MA 02134; (617) 232-5833. Works to create jobs that are good for both people and the environment.

Kids for Saving Earth, 620 Mendelssohn, Suite 130, Golden Valley, MN 55427; (612) 525-0002. Educates and empowers children to protect the environment.

Land Stewardship Project, 14758 Ostlund Trail North, Marine on the St. Croix, MN 55047; (612) 433-2770. Promotes environmentally sound agricultural policies.

National Audubon Society, 700 Broadway, New York, NY 10003; (212) 979-3000. Works to preserve threatened ecosystems.

National Coalition against the Misuse of Pesticides, 701 E Street, SE, Suite 200, Washington, DC 20003; (202) 543-5450. Education and action on pesticide use in urban and farm settings.

National Parks and Conservation Association, 1776 Massachusetts Avenue, #200, Washington, DC 20036; (202) 223-6722. Dedicated to preserving and improving the national park system.

National Recycling Coalition, 1101 30th Street, NW, Suite 305, Washington, DC 20007; (202) 625-6406. Nonprofit organization committed to increasing recycling efforts by providing information and shaping public and private policy.

Natural Renewable Energy Laboratory, 1617 Cole Boulevard, Golden, CO 80401; (303) 275-4065. U.S. government lab working on renewable energy.

National Seed Storage Laboratory, U.S. Department of Agriculture, Fort Collins, Colorado. National storage of natural varieties of plant seeds for public and future use.

National Wildlife Federation, 1400 Sixteenth Street, NW, Washington, DC 20036; (202) 797-6800. Protects fish, wildlife, and habitat.

Natural Resources Defense Council, 40 West Twentieth Street, New York, NY 10011; (212) 727-2700. Protects endangered wildlife and habitat and natural resources.

Nature Conservancy, 1815 Lynn Street, Arlington, VA 22209; (703) 841-5300. Worldwide efforts to acquire land for preservation purposes.

Nuclear Regulatory Commission, Washington, DC 20555; (301) 492-7000. Oversees nuclear power industry in the United States.

Oceanic Society Expeditions, Fort Mason Center, Building E, San Francisco, CA 94123; (800) 326-7491. Offers environmentally sound trips.

People for the Ethical Treatment of Animals (PETA), P.O. Box 42516, Washington, DC 20015; (301) 770-7444. Protection of animals in all settings.

Population Council, One Dag Hammarskjold Plaza, New York, NY 10017; (212) 339-0500. Works on population issues in developing countries.

Rainforest Action Network, 450 Sansome, Suite 700, San Francisco, CA 94111; (415) 398-4404. Direct action to protect rain forests internationally.

Rainforest Alliance, 65 Bleeker Street, New York, NY 10012-2420; (212) 677-1900. Protects rain forests worldwide.

Rocky Mountain Institute, 1739 Snowmass Creek Road, Snowmass, CO 81654; (303) 927-3851. Promotes energy efficiency.

Save the Redwoods League, 114 Sansome Street, Room 605, San Francisco, CA 94104; (415) 362-2352. Works to save forests, especially redwood trees.

Seed Savers Exchange, 3076 North Winn Road, Decorah, Iowa 52101; (319) 382-5990. Seed bank for out-of-use natural varieties of plants.

Seeds of Change, 621 Old Santa Fe Trail No 10, Santa Fe, NM 87501; (505) 983-8956. Garden seed catalogue company that specializes in varieties (grown organically) often no longer grown commercially.

Sierra Club, 730 Polk Street, San Francisco, CA 94109; (415) 776-2211. Promotes responsible use of earth's ecosystems.

Soil and Water Conservation Society, 7515 N.E. Ankeny Road, Ankeny, IA 50021; (800) THE-SOIL. Promotes worldwide sound soil and water use.

Solar Energy Industries Association, 122 "C" Street, NW, 4th Floor, Washington, DC 20001; (202) 408-0660. Information on current solar industry efforts.

Union of Concerned Scientists, 2 Brattle Street, Cambridge, Massachusetts 02238; (617) 547-5552. Information and action on technology's impact on society and the environment.

United Nations Environment Programme, 2 UN Plaza, Room 803, New York, NY 10017; (212) 963-8138. Environmental organization focusing on diverse issues worldwide.

U.S. Agency for International Development, 2201 "C" Street, NW, Washington, DC 20520; (202) 647-4000. Administers economic and humanitarian assistance programs in developing countries.

U.S. Bureau of Land Management, 1849 C Street, NW, Rm 5600, Washington, DC 20240; (202) 208-5717. Largest land holder in country; manages and oversees uses of public lands.

U.S. Bureau of Mines, 810 Seventh Street, NW, Washington, DC 20241; (202) 501-9649. Oversees mining industry in the United States.

U.S. Department of Agriculture, Fourteenth Street and Independence Avenue, SW, Washington, DC 20250; (202) 720-8732. Responsible for inspecting, grading, and certifying all agricultural products; oversees and assists U.S. farmers, especially with soil quality and conservation.

U.S. Department of Energy, Forrestal Building, 1000 Independence Avenue, SW, Washington, DC 20585; (202) 586-5000. Oversees U.S. energy policies; involvement with nuclear weapons production and nuclear energy. Renewable and energy efficiency hotline: Conservation and Renewable Energy Inquiry and Referral Service (CAREIRS), (800) 523-2929.

U.S. Department of Health, Food and Drug Administration (FDA), 5600 Fishers Lane, Rockville, MD 20857; (301) 443-1544. Oversees food and drug quality in the United States.

U.S. Department of the Interior, 1849 C Street, NW, Washington, DC 20240; (202) 208-3100. Oversees U.S. responsibility to Native American tribes and natural resources preservation.

U.S. Department of Transportation, Nassif Building, 400 Seventh Street, SW, Washington, DC 20590; (202) 366-4000. Oversees national transportation networks.

U.S. Environmental Protection Agency, 401 M Street, SW, Washington DC, 20460; (202) 260-2080 (information services). Charged with protecting U.S. environmental resources such as soil, water, and air quality.

U.S. Fish and Wildlife Service, 1849 C Street, NW, Washington, DC 20240; (202) 208-5634. Oversees preservation of fish and wildlife populations and habitat.

U.S. Forest Service, Fourteenth Street and Independence Avenue, SW, P.O. Box 96090, Washington, DC 20250; (202) 205-1760. Oversees the preservation of U.S. forests.

U.S. Public Interest Research Group, 215 Pennsylvania Avenue, SE, Washington, DC 20003; (202) 546-9707. Conducts independent research and lobbies for national environmental and consumer protection.

U.S. Senate Document Room, (202) 225-3456 (to check on legislation)

White House Office of Environmental Policy, Old Executive Office Building, Room 360, 1600 Pennsylvania Avenue, Washington, DC 20501; (202) 456-6224. Information on the government's environmental policy.

Wilderness Society, 900 Seventeenth Street, NW, Washington, DC 20006; (202) 833-2300. Protects wildlife and wilderness.

Wildlife Conservation Society, New York Zoological Society, 185th Street and Southern Boulevard, Bronx, NY 10460; (718) 220-5100. Conducts research and projects for wildlife and habitat worldwide.

World Bank (International Bank for Reconstruction and Development), 1818 H Street, NW, Washington, DC 20433; (202) 477-1234. Offers international loans and aid for a variety of projects in developing countries (see Glossary).

World Council of Indigenous Peoples, 555 King Edward Avenue, Ottawa, Ontario, Canada KIN 6NS. Global federation interested in representing the causes of all indigenous peoples.

World Health Organization (WHO), Avenue APPIA, CH-1211 Geneva 27, Switzerland; (22) 7913105. Research and projects in developing countries related to health issues.

World Neighbors, 4127 NW 122nd Street, Oklahoma City, Oklahoma, 73120; (405) 752-9700. International efforts to teach sustainable farming techniques, especially in developing countries, by farmers training other farmers.

World Wildlife Fund, 1250 24th Street, NW, Washington, DC 20037; (202) 293-4800. Worldwide efforts to protect wildlife and habitat.

Worldwatch Institute, 1776 Massachusetts Avenue, NW, Washington, DC 20036; (202) 452-1999. Researches and informs the public and governments about global problems and trends.

World Resources Institute, 1709 New York Avenue, NW, Suite 700, Washington, DC 20006; (202) 638-6300. Research on policies worldwide concerning the environment, resource management, economic growth, and international security.

Zero Population Growth, 1400 16th Street, NW, Suite 320, Washington, DC 20036; (202) 332-2200. Educates the public about population growth effects on world and advocates programs to end global population growth.

APPENDIX B
Further Reading

Books

A Consumer's Dictionary of Household, Yard and Office Chemicals, Ruth Winter, Crown, 1992. Easy-to-read guide and listing of toxics people are often exposed to.

Acidic Deposition: Sulphur and Nitrogen Oxides, Allan H. Legge and Sagar Krupa, Lewis Publishers, 1990. Collection of scientific studies on acid rain.

Acid Rain Rhetoric and Reality, Chris C. Park, Methuen, 1987. Thorough discussion of acid rain studies and information, particularly in Europe and the United States.

Agenda 21: The Earth Summit Strategy to Save Our Planet, edited by Daniel Sitarz, EarthPress, 1993. Reviews Agenda 21 in detail.

Air Pollution, Kathlyn Gay, Franklin Watts, 1991. Primer on air pollution.

Air Pollution's Toll on Forests and Crops, edited by James J. MacKenzie and Mohamed T. El-Ashry, Yale University Press, 1989. Global, thorough discussion and studies of the effect of air pollution on plants.

Air Pollution, the Automobile, and Public Health, edited by Ann Y. Watson, National Academy Press, 1988. Technical book examining public health issues and car pollution.

The Almanac of Renewable Energy, Eric Golob and Richard Brus, Henry Holt & Company, 1993. Overview of renewable energy resources and their impacts on society.

Alternative Energy Sourcebook, edited by John Schaeffer, Real Goods Trading Corporation, 1991. Comprehensive collection of energy-sensible appliances, technologies, and goods.

Alternative Transportation Fuels, edited by Daniel Sperling, Quorum Books, 1989. Essays on current use and environmental advantages of alternative fuel use.

Amazon Conservation in the Age of Development, Ronald A. Foresta, University of Florida Press, 1991. How development and politics have affected the Amazon and a look at the processes that spurred the development.

Ancient Forests of the Pacific Northwest, Elliott A. Norse, Island Press, 1990. Thorough discussion about old-growth forests, their role in global ecology, and the effects of the timber industry with recommendations for future preservation.

Annual Review of Energy and the Environment, edited by Jack M. Hollander, Annual Reviews, Inc., 1992. Global overview of selected energy policies, resources, impacts, and conservation.

Asbestos in Public and Commercial Buildings: A Literature Review and Synthesis of Current Knowledge, Health Effects Institute—Asbestos Research, 1991. Thorough review of literature on asbestos.

Asbestos: The Hazardous Fiber, Melvin A. Benarde, CRC Press, 1990. Thorough discussion of all aspects of asbestos.

Backyard Composting: Your Complete Guide to Recycling Yard Clippings, Harmonious Press, 1992. Easy-to-read guide to composting.

Beyond Beef: The Rise and Fall of the Cattle Culture, Jeremy Rifkin, Dutton, 1992. History, legacy, and current global problems associated with raising cattle.

Beyond 40 Percent, Institute for Local Self-Reliance, Island Press, 1991. Examination of record-setting recycling and composting programs.

Biochemistry of Trace Metals, edited by Domy C. Adriano, Lewis, 1992. Technical book on trace metals.

The Biodynamic Farm, Herbert H. Koepf, Anthroposophic Press, 1989. How-to book on biodynamic organic farming.

The Bio Revolution: Cornucopia or Pandora's Box? edited by Peter Wheale and Ruth McNally, Pluto Press, 1990. Essays on concerns and problems regarding biotechnology applications.

Biotechnology, Agriculture and Food, Organisation for Economic Co-operation and Development, OECD, 1992. Evaluation of biotechnology applications, primarily in agriculture, and public and economic policies.

Blue Skies, Green Politics: The Clean Air Act of 1990, Gary C. Bryner, CQ Press, 1993. Easy-to-read discussion of the Clean Air Act of 1990.

The Burning Season: The Murder of Chico Mendez and the Fight for the Amazon Rainforest, Andrew Revkin, Houghton Mifflin, 1990. The story of activist Chico Mendez's fight to save the rainforest and his resulting death.

But Not a Drop to Drink, Steve Coffel, Rawson Associates, 1989. Thorough discussion of pollutants in tap water, their sources, health effects, and solutions.

Caring for the Earth: A Strategy for Sustainable Living, World Wildlife Fund,

1991. Practical guide for individuals, communities, governments, and countries on sustainable living.

Car Trouble, Steve Nadis and James J. MacKenzie, Beacon Press, 1993. Problems cars pose and new technology, fuels, and different transportation options presented as solutions.

The Challenge of Extinction, Dorothy Hinshaw Patent, Enslow, 1991. Overview of extinctions and solutions to prevent them.

Chemical Deception, Marc Lappé, Sierra Club Books, 1991. Thorough discussion of chemicals in our everyday life and how they affect health.

Chemistry of Atmospheres, Richard P. Wayne, Clarendon Press, 1991. Textbook of atmosphere chemistry and chemical reactions.

Coal: The Energy Source of the Past and Future, Harold H. Schobert, American Chemical Society, 1987. Thorough discussion about coal and how it is used.

Coastal Alert, Dwight Holing, Natural Resources Defense Council, 1990. Easy-to-read discussion of coastal offshore oil production, citizen action, and alternatives.

Community Recycling, Nyles V. Reinfeld, Prentice-Hall, 1992. Detailed examination of how to plan, operate, and manage a recycling system.

Complex Cleanup: The Environmental Legacy of Nuclear Weapons Production, U.S. Congress, Office of Technology Assessment, 1991. Congressional examination of legacy, sites, and problems of cleanup of U.S. nuclear production facilities.

Composting Municipal Sludge, Arthur H. Benedict, Noyes Data Corporation, 1988. Technical studies of different techniques of composting sludge.

Confessions of an Eco-Warrior, Dave Foreman, Harmony Books, 1991. A founder of Earth First! discusses his views and actions.

Conservation Biology, edited by Peggy L. Fiedler and Subodh K. Jain, Routledge, Chapman & Hall, 1992. Collection of scientific studies on theory and practice of nature conservation, preservation, and management.

1994 Conservation Directory, National Wildlife Federation, 1994. Listing of environmental, governmental, and international agencies.

Conserving Biodiversity: A Research Agenda for Development Agencies, U.S. National Research Council, National Academy Press, 1992. Research methodology and policies to preserve biodiversity.

Constructed Wetlands for Wastewater Treatment, Donald A. Hammer, Lewis Publishers, 1989. Technical discussion of treating wastewater with artificial wetlands.

Consumer Guide to Solar Energy, Scott Sklar and Kenneth Sheinkopf, Bonus Books, 1991. Guide to easy and inexpensive residential solar applications.

Costing the Earth: The Challenge for Governments, the Opportunities for Business, Frances Cairncross, Harvard Business School Press, 1991. How business can thrive using environmentally sound principles and how governments can aid this process.

Crisis in the Atmosphere: The Greenhouse Factor, Ed Phillips and D. B. Clark, 1990. Discussion of the greenhouse effect.

Crisis in the World's Fisheries, James, R. McGoodwin, Stanford University Press, 1990. Discussion of the state of world fisheries, related policies, and the causes of the crisis.

Deadly Deceit: Low-Level Radiation, High-Level Cover-up, Jay M. Gould and Benjamin A. Goldman, Four Walls Eight Windows, 1990. Discussion of low-level waste, its dangers, and its sources.

Deeper Shades of Green: The Rise of Blue-Collar and Minority, Jim Schwab, Sierra Club Books, 1994. Examination of the role of minorities and blue-collar workers in the environmental movement, focusing on 8 case histories.

Dioxin Treatment Technologies, Office of Technology Assessment, 1991. Background paper to Congress on dioxin cleanup technology.

Dolphins and the Tuna Industry, National Research Council, National Academy Press, 1992. NRC study of the impact the tuna industry has on dolphin deaths.

Dying Planet: The Extinction of Species, Jon Erickson, Tab Books, 1991. Basic, easy-to-read text on past and present extinctions.

Eco-Heroes: 12 Tales of Environmental Victory, Aubrey Wallace, Mercury House, 1993. Highlights ordinary individuals who have created extraordinary environmental change.

Ecolinking: Everyone's Guide to Online Environmental Information, Don Rittner, PeachPit Press, 1992. How-to book on gaining access to environmental computer sources, databases, and global networks.

The Ecology of War: Environmental Impacts of Weaponry and Warfare, Susan D. Lanier-Graham, Walker, 1993. Easy-to-read survey of past and present effects of warfare on the environment.

The Eco Wars, David Day and Kay Porter, 1989. Discussion of environmental victims created by outright murder, pollution, and other conflicts.

The Endangered Species Act, Daniel J. Rohlf, Stanford Environmental Law

Society, 1989. A guide to protection and implementation of the Endangered Species Act.

The End of Nature, Bill McKibben, Random House, 1989. Discussion of the greenhouse effect and how humans have irrevocably changed nature.

Energy Efficient Motor Systems, Steven Nadel, American Council for an Energy-Efficient Economy, 1991. Handbook on efficient motor technology, programs, and opportunities for their use.

The Energy-Environment Connection, edited by Jack M. Hollander, Island Press, 1992. In-depth discussion of energy impact on the environment and ways to minimize it.

Energy Policy in the Greenhouse, Florentin Krause, John Wiley and Sons, 1992. How energy policies and use impact global climate.

Energy: Principles, Problems, Alternatives, Joseph Priest, Addison-Wesley, 1991. Basic textbook on most energy sources, including renewables and conservation.

Energy Resources: Towards a Renewable Future, D. J. Herda and Margaret L. Madden, Franklin Watts, 1991. Basic history and current uses of fossil fuels, nuclear energy, and renewable energies.

Energy Unbound: A Fable for America's Future, Amory and Hunter Lovins, Sierra Club Books, 1985. Fable of energy efficiency as a solution to energy needs.

Energy Use and the Environment, F. Peter W. Winteringham, Lewis Publishers, 1992. Global look at how energy use affects the environment, with reference to future energy sources.

Environmental Hazards of War: Releasing Dangerous Forces in an Industrialized World, edited by Arthur H. Westing, Sage Publicatios, 1990. Dangers current industries pose in the event of a war and methods to minimize these dangers.

Environmental Hazards: Assessing Risks and Reducing Disaster, Keith Smith, Routledge, 1992. Thorough discussion of natural disaster planning, causes, and solutions.

Environmental Hazards: Toxic Waste and Hazardous Material, E. Willard Miller and Ruby M. Miller, ABC-CLIO, 1991. Basic textbook on toxic and hazardous waste.

Environmental Law in a Nutshell, Roger W. Findley and Daniel A. Farber, West, 1991. Case studies that illustrate environmental laws.

Environmental Radon, edited by C. Richard Cothern, Plenum Press, 1987. Technical discussion of radon in the environment.

Enzyme Technology, M. F. Chaplin and C. Bucke, Cambridge University Press, 1990. Textbook on biotechnology applications and uses of enzymes.

Every Person's Little Book of P-L-U-T-O-N-I-U-M, Stanley Berne, Rising Tide Press, 1992. Basic, easy-to-read book about nuclear energy development and problems.

The Expendable Future: U.S. Politics and the Protection of Biological Diversity, Richard Tobin, Duke University Press, 1990. Evaluation of past and current policy decisions in the United States related to protecting biological diversity.

Extinction A–Z, Erich Hoyt, Enslow, 1991. Basic dictionary of terms, some species, and processes related to extinction.

Extinction: The Causes and Consequences of the Disappearance of a Species, Paul and Anne Ehrlich, Random House, 1981. Thorough discussion of causes of and problems resulting from extinctions.

The Fate of the Forest: Developers, Destroyers, and Defenders of the Amazon, Susanna Hecht and Alexander Cockburn, Verso, 1989. History and conflicts shaping the fate of the Amazon rain forest.

Field Testing Genetically Modified Organisms, National Research Council, National Academy Press, 1989. Scientific evaluation of introducing genetically modified organisms and plants into the environment.

Fragile Majesty: The Battle for North America's Last Great Forest, Keith Ervin, The Mountaineers, 1989. The efforts to save old-growth forests.

Fuel Science and Technology Handbook, edited by James G. Speight, Marcel Dekker, 1990. Thorough reference book on fossil fuels.

The Gaia Atlas of First Peoples: A Future for the Indigenous World, Julian Burger, Anchor Books, 1990. A global sourcebook on indigenous peoples' lives, their visions, and the crises they face.

Geothermal Energy in the Western United States and Hawaii: Resources and Projected Electricity Generation Supplies, Energy Information Administration, 1991. Assessment of U.S. geothermal resources and potential.

Getting at the Source, World Wildlife Fund, Island Press, 1991. Strategies for reducing waste at the source.

Ghost Bears, R. Edward Grumbine, Island Press, 1992. Endangered animals and extinction.

Global Biomass Burning, edited by Joel S. Levine, MIT Press, 1991. Thorough collection of scientific studies on the effects of biomass burning worldwide from the Conference on Global Biomass Burning.

Global Forest Resources, Alexander S. Mather, Timber Press, 1990. Thorough book on forests worldwide and sustainable resource use.

Global Warming, Burkhard Bilger, Chelsea House, 1992. Discussion of global warming.

Global Warming Unchecked: Signs to Watch For, Harold W. Bernard, Jr., Indiana University Press, 1993. Thorough look at greenhouse theories and warning indicators.

The Green Cathedral: Sustainable Development of Amazonia, Juan De Onis, 1992, Oxford University Press. Discussion of why indigenous peoples must be saved to preserve the rain forests.

The Green Commuter, Joel Makower, Tilden Press, 1992. Environmental transportation with and without cars.

The Green Consumer, John Elkington, Penguin Books, 1990. How to be an environmentally aware shopper.

Green Earth Resource Guide, Cheryl Gorder, Blue Bird, 1991. Thorough listing of environmentally friendly products and services.

The Greenhouse Effect, edited by Mathew A. Kraljic, H. W. Wilson Company, 1992. Primer on causes, policies, possible problems, and solutions to the greenhouse effect.

1991–1992 Green Index: A State-by-State Guide to the Nation's Environmental Health, Bob Hall and Mary Lee Kerr, Island Press, 1991. State-by-state listing of environmental factors, such as water and air quality.

The Green Reader: Essays toward a Sustainable Society, edited by Andrew Dobson, Mercury House, 1992. A thorough international anthology that serves as an introduction to different approaches to environmental change.

Groundwater Treatment Technology, Evan K. Nyer, Van Nostrand Reinhold, 1992. Technology for cleaning up contaminated groundwater.

Growing Power: Bioenergy for Development and Energy, Alan S. Miller, World Resources Institute, 1986. Review of bioenergy uses worldwide, with suggestions for improving bioenergy development.

Guardians of the Land: Indigenous Peoples and the Health of the Earth, Alan Thein Durning, Worldwatch, 1992. Worldwatch paper on global overview of indigenous peoples, how they live, and the problems and crises they currently face.

Guide to Wood Heat, Dirk Thomas, Camden House, 1992. The basics of obtaining, cutting, and using wood in home heating.

Handle with Care: A Guide to Responsible Travel in Developing Countries, Scott Graham, Noble Press, 1991. Discussion of ecotourism by a traveling author.

Hazardous Waste: Confronting the Challenge, Christopher Harris, Quorum, 1987. Analysis of the 1984 Hazardous and Solid Waste Amendments.

The Independent Home: Living Well with Power from the Sun, Wind, and Water, Michael Potts, Chelsea Green, 1993. Practical guide to an energy-independent home.

Indigenous Views of Land and the Environment, Shelton Davis, World Bank, 1993. World Bank report on indigenous struggles and issues.

Indoor Pollution, Steve Coffel and Karyn Feiden, Fawcett Columbine, 1990. Overview of indoor pollution and precautions to take at home and work.

Infectious and Medical Waste Management, Peter A. Reinhardt and Judith G. Gordon, Lewis Publishers, 1991. Current methods of managing medical waste.

The International Politics of Nuclear Waste, Andrew Blowers, St. Martin's Press, 1991. Global nuclear waste legacy, problems, and suggested action.

International Wildlife Trade: Whose Business Is It? Sarah Fitzgerald, World Wildlife Fund, 1989. Thorough book about wildlife trade and factors that make it an ongoing problem.

Introduction to Genetic Engineering, William H. Sofer, Butterworth-Heinemann, 1991. Basic textbook on genetic engineering.

It's All in Your Head: Diseases Caused by Silver-Mercury Fillings, Hal A. Huggins, Life Sciences Press, 1989. Easy reading summarizing Dr. Huggins's decades of experience and expertise with problems caused by mercury fillings.

Keeping Options Alive: The Scientific Basis for Conserving Biodiversity, Walter V. Reid and Kenton R. Miller, World Resources Institute, 1989. Worldwatch paper presenting an overview on protecting global biodiversity.

Land Degradation: Development and Breakdown of Terrestrial Environments, C. J. Barrow, Cambridge University Press, 1991. Thorough global discussion of land degradation worldwide.

Land Planner's Environmental Handbook, William B. Honachefsky, Noyes, 1991. Thorough handbook stressing environmentally sound land use.

Landscaping with Nature, Jeff Cox, Rodale Press, 1991. How-to book on natural landscaping.

Land Use and Abuse, D. J. Herda and Margaret L. Madden, Franklin Watts, 1990. Primer on land use and degradation.

The Last Panda, George B. Schaller, University of Chicago Press, 1993. An examination of the politics and corruption in China that are ensuring the panda's demise despite international efforts to save it.

The Lead Clean-up Book, Lead Free Kids, 1992. How to safely take care of household sources of toxics, lead, and hazardous materials.

Life Support: Conserving Biological Diversity, John C. Ryan, Worldwatch Institute, 1992. Worldwatch paper on global efforts, reasons, and strategies to conserve biological diversity.

Low-Level Radioactive Waste: From Cradle to Grave, Edward L. Gershey, Van Nostrand Reinhold, 1990. Comprehensive reference on all aspects of low-level radioactive waste.

Making Peace with the Planet, Barry Commoner, Pantheon Books, 1990. Discussion of how modern technology is at war with the planet and natural ecosystems.

Managing Global Genetic Resources, National Research Council, National Academy Press, 1991. NRC discussion on preserving forest trees as a global resource.

Militarism and Global Ecology, Gary E. McCuen, Gem, 1993. Easy-to-read global overview of the environmental effects of the military.

Millennium: Tribal Wisdom and the Modern World, David Maybury-Lewis, Viking, 1992. Discussion of a number of indigenous peoples and their lifestyles and values.

Mineral Resources A–Z, Robert L. Bates, Enslow, 1991. Easy-to-read listing of mining terminology and minerals.

Mining America: The Industry and the Environment, 1800–1980, Duane A. Smith, University Press of Kansas, 1987. Discussion of mining history and impacts up to 1980.

Mining and the Freshwater Environment, Martyn Kelly, Elsevier Science, 1988. Effects of mining wastes and wastewater on freshwater sources.

The Monkey Wrench Gang, Edward Abbey, Lippincott, 1975. Fictional book depicting environmentalists plotting to blow up Glen Canyon Dam to preserve the river and downstream wildlife.

Multiple Exposures: Chronicles of the Radiation Age, Catherine Caufield, Harper & Row, 1989. Thorough discussion of the history and current legacy of nuclear energy and radiation.

National Acid Precipitation Assessment Program, NAPAP, 1991. Program report.

The Nature of Development: A Report from the Rural Tropics on the Quest for Sustainable Economic Growth, Roger D. Stone, Knopf, 1992. Sustainable development is examined in developing countries.

Nature Tourism: Managing for the Environment, edited by Tensie Whelan, Island Press, 1991. Basic concepts and policies for environmentally sound tourism development.

Nuclear Accidents, Joel Helgerson, Franklin Watts, 1988. Chronicle of major nuclear accidents worldwide.

Organic Farming, Nicolas Lampkin, Farming Press, 1990. Thorough explanation of all aspects of organic farming.

The Organic Gardener's Handbook of Natural Insect and Disease Control, edited by Barbara W. Ellis, Rodale, 1992. Thorough problem-solving guide to eliminating garden pests without chemicals.

Out of the Channel: The Exxon Valdex Oil Spill in Prince William Sound, John Keeble, Harper Collins, 1991. Account of the oil spill, aftereffects, and principal characters involved.

Over-Population: Crisis or Challenge? Nathan Aaseng, Franklin Watts, 1991. Easy-to-read primer about population problems and issues.

Ozone, Kathlyn Gay, Franklin Watts, 1989. Easy-to-read discussion of ozone in the ozone layer and as a pollutant in smog.

Ozone Crisis, Sharon L. Roan, John Wiley and Sons, 1989. The history of the ozone crisis.

Packaging and the Environment: Alternatives, Trends, and Solutions, Susan E. M. Selke, Technomic Publshing, 1990. Review of packaging energy use, pollution, and methods of disposal and recovery.

Plastics: America's Packaging Dilemma, Nancy Wolf and Ellen Feldman, Island Press, 1991. Easy-to-read, thorough discussion of the use of plastics in packaging, associated problems, and related legislation.

The Poisoned Well, edited by Eric P. Jorgensen, Island Press, 1989. Thorough review of groundwater pollution, cleanup, and protective legislation and programs.

Poisoners of the Sea, K. A. Gourlay, Zed Books, 1988. Discussion of worldwide ocean pollution with oil, toxics, sewage, and other pollutants.

Poisoning Our Children, Nancy Sokol Green, Noble Press, 1991. Thorough discussion of toxic threats facing our children.

The Population Explosion, Paul R. Ehrlich and Anne H. Ehrlich, Simon and Schuster, 1990. Predicted population growth and its effects on the world.

Primer on Greenhouse Gases, Donald J. Wuebbles and Jae Edmonds, Lewis Publishers, 1991. Somewhat technical book on greenhouse gases.

Providing Information to Groundwater Managers to Help Them Allocate Resources and Improve Their Programs, Harry P. Hatry, EPA, 1992. Specific management suggestions for groundwater protection.

Public Health and Preventive Medicine, John M. Last, 13th ed., Appleton & Lange, 1992. Thorough, technical textbook on all areas of public health, with a large section devoted to environmental health issues.

Radon, Radium and Uranium in Drinking Water, edited by C. Richard Cothern, Lewis Publishers, 1990. Covers all aspects of radionuclides in drinking water.

The Rainforest Book, Scott Lewis, Living Planet Press, 1990. Easy-to-read book on how you can save the rain forests.

The Recycler's Manual for Business, Government, and the Environmental Community, David R. Powelson and Melinda A. Powelson, Van Nostrand Reinhold, 1992. Thorough recycling manual for all materials, as well as ways to reduce waste and legislation and programs affecting waste.

Recycling in America: A Reference Handbook, Debi Kimball, ABC-CLIO, 1992. State of recycling in the United States.

Renewable Energy, edited by Thomas B. Johansson, Island Press, 1993. In-depth, and somewhat technical, examination of worldwide renewable sources of energy.

Research Priorities for Conservation Biology, edited by Michael E. Soulé and Kathryn A. Kohm, Island Press, 1989. Priorities of preserving wildlife and habitat.

The Residential Energy Audit Manual, edited by Dale Schueman, Fairmount Press, 1992. Detailed, practical, hands-on manual on energy loss and conservation practices for homes.

Rethinking the Ozone Problem in Urban and Regional Air Pollution, National Research Council, National Academy Press, 1991. Thorough, technical study of causes, explanations, sources, and remedies by the National Research Council for the problem of ground-level ozone.

Right-To-Know and Emergency Planning, Georgy G. Lowry, Lewis, 1988. Handbook of compliance for workers, communities, and states.

Rivers at Risk: The Concerned Citizen's Guide to Hydropower, John D. Echeverria, Island Press, 1989. Guide to understanding hydropower impact on rivers and actively participating in the renewal of hydropower permits.

Running a Biogas Programme: A Handbook, David Fulford, Intermediate Technology, 1988. How-to manual on biogas and biogas systems.

Rush to Burn: Solving America's Garbage Crisis? Newsday, Island Press, 1989. Easy-to-read primer on incineration and the waste crisis.

Safe Food: Eating Wisely in a Risky World, Michael F. Jacobson, Living Planet Press, 1991. How to find and choose healthy, toxic-free food and water.

Salmon and Steelhead: The Struggle to Restore an Imperiled Resource, edited by Alan Lufkin, University of California Press, 1991. How pollution and habitat loss have decimated fish populations and efforts to restore them.

Saving Our Ancient Forests, Seth Zuckerman, Living Planet Press, 1991. Easy-to-read primer on protecting old-growth forests.

Saving Water in the House and Garden, Jonathan Erickson, TAB, 1993. Practical ways to conserve water around the home.

Seeds of Change: The Living Treasure. Kenny Ausubel, Harper San Francisco, 1994. Passionate story of the growing movement to restore biodiversity and revolutionize the way we think about food; recipes included.

Silent Spring, Rachel Carson, Houghton Mifflin, 1962. Classic work about the harmful effects of pesticides.

The Social and Environmental Effects of Large Dams, Edward Goldsmith and Nicholas Hildyard, Sierra Club Books, 1984. Thorough discussion of the problems and effects of large dams.

The Social Impact of the Chernobyl Disaster, David R. Marples, St. Martin's Press, 1988. Discussion of victims and environmental and economic impact of Chernobyl.

Soil Conservation in the United States, Frederick R. Steiner, Johns Hopkins University Press, 1990. Recent efforts, initiatives, and recommendations for soil conservation in the United States.

The Solar Electric House, Steven J. Strong, Sustainability Press, 1993. Practical guide for solar energy use in homes.

Statewide Wetlands Strategies, World Wildlife Fund, Island Press, 1992. Thorough discussion of policies to preserve wetlands in states.

Story Earth: Native Voices on the Environment, compiled by IPS, Mercury House, 1993. Visions of the environmental crisis and solutions by traditional peoples around the world.

Superpigs and Wondercorn, Michael W. Fox, Lyons & Burford, 1992. Thorough discussion of biotechnology applications and concerns that its risks and benefits have not been thoroughly weighed.

Surface Water Quality: Have the Laws Been Successful? Ruth Patrick, Princeton University Press, 1992. Thorough review of the state of surface water quality and legislation supposed to protect it.

Sustainable Development: Constraints and Opportunities, Mostafa Kamal Tolba, 1987. How development can occur in a sustainably sound manner.

Sustainable Harvest and Marketing of Rain Forest Products, edited by Mark Plotkin and Lisa Famolare, Island Press, 1992. Examines use, conservation, and marketing of rain forest products.

The Threat at Home: Confronting the Toxic Legacy of the United States Military, Seth Shulman, Beacon Press, 1992. Discussion of U.S. military pollution and waste.

Tired or Toxic? Sherry A. Rogers, Prestige Publishing, 1990. Thorough discussion of the sources of toxics, the human body's ability to handle toxics, and methods to heal diseases caused by toxics.

Tourism and Development in the Third World, John Lea, Routledge, 1988. How tourism affects developing countries.

Toward a Sustainable Society: An Economic, Social, and Environmental Agenda for Our Children's Future, James Garbarino, Noble Press, 1992. Discussion of all aspects of creating and moving toward a sustainable society.

The Toxic Cloud, Michael H. Brown, Harper & Row, 1987. Cross-country travels and discussion of sources and causes of air pollution in the United States.

Toxics A to Z, John Harte, University of California Press, 1991. Thorough listing of toxics, their effects, and their sources.

Toxic Nightmare: Ecocide in the USSR and Eastern Europe, Gary E. McCuen, Gem, 1993. Easy-to-read discussion about toxics and pollution problems in the USSR and Eastern Europe.

Trade and Environment: Conflicts and Opportunities, Office of Technology Assessment, 1992. How trade and economic growth impact the environment.

Trees of Life: Saving Tropical Forests and Their Biological Wealth, Kenton Miller and Laura Tangley, Beacon Press, 1991. Discussion on destruction, preservation, and citizen action regarding rain forests.

Tropical Forests and Their Crops, Nigel J. H. Smith, Cornell University Press, 1992. Thorough discussion of tropical crops and sustainable harvesting of them.

The Warning: Accident at Three Mile Island, Mike Gray and Ira Rosen, Contemporary Books, 1982. Investigative report of the Three Mile Island nuclear accident.

The Wasted Ocean, David K. Bulloch, Lyons & Burford, 1989. Primer on ocean pollution, sources, and prevention.

Water Pollution, Kathlyn Gay, Franklin Watts, 1990. Primer on water pollution and suggested solutions.

Wetlands: An Approach to Improving Decision Making in Wetland Restoration and Creation, edited by Ann J. Hairston, Island Press, 1992. Technical methods to restore and create wetlands.

Where Did That Chemical Go? Ronald E. Ney, Jr., Van Nostrand Reinhold, 1990. Listing of chemicals and where they end up in the environment.

Wind Power for Home and Business, Paul Gipe, Chelsea Green, 1993. The most authoritative source on small wind turbines.

Who's Poisoning America: Corporate Polluters and Their Victims in the Chemical Age, Ralph Nader, Sierra Club Books, 1982. Corporations' responsibility for pollution they create.

A World in Crisis?: Geographical Perspectives, edited by R. J. Johnston, Oxford University Press, 1988. A discussion of global environmental problems.

World of Waste: Dilemmas of Industrial Development, K. A. Gourlay, Zed Books, 1992. Discussion of waste and related policies and solutions worldwide.

You Can Save the Animals: 50 Things to Do Right Now, Michael W. Fox and Pamela Weintraub, St. Martin's Press, 1991. Easy-to-read guide on saving endangered animals.

Magazines

Alternatives, Faculty of Environmental Studies, University of Waterloo, Waterloo, Ontario N2L 3G1; (519) 885-1211. Canadian journal covering major environmental issues worldwide.

Ambio, Royal Swedish Academy of Sciences, Box 50005, S-104 05 Stockholm, Sweden; (46 8) 673 95 00. Scientific magazine regarding worldwide environmental issues.

American Forests, 1516 P Street, NW, Washington, DC 20005; (202) 667-3300. Publication of the American Forestry Association often expressing the view of timber interests.

Amicus, Natural Resources Defense Council, 40 West 20th Street, New York, NY 10011; (212) 727-2700. Environmental affairs and policies of national and international significance.

Aquaculture, Achill River Corporation, 31 College Place, Asheville, NC 28801; (704) 254-7334. Industry magazine about aquaculture worldwide.

Archives of Environmental Health, Society for Occupational and Environmental Health, 1319 Eighteenth Street, NW, Washington, DC 20036; (202) 296-6267. International scientific articles on effects of chemicals and pollutants on human health.

Audubon, National Audubon Society, 700 Broadway, New York, NY 10003; (212) 979-3000. Magazine covering national environmental issues.

Beef, Webb Division, Intertec Publishing Corporation, 9800 Metcalf, Overland Park, KS 66212; (717) 560-2001. Trade journal for cattle farmers and beef producers.

Biocycle, The JG Press, Inc., 419 State Avenue, Emmaus, PA 18049; (215) 967-4135. Journal of composting and recycling.

Biotechnology: The International Monthly for Industrial Biotechnology, Nature Publishing Company, 65 Bleeker, New York, NY 10012. Information on current research.

Buzzworm, 2305 Canyon Boulevard, Suite 206, Boulder, CO 80302; (303) 442-1969. Articles on environmentally sound living.

Chemical Week, P.O. Box 7721, Riverton, NJ 08077; (609) 786-0401. Trade journal of global news and analysis for the chemical industry.

Clean Air, National Society for Clean Air and Environmental Protection, 136 North Street, Brighton BN1 1RG England; (0273) 326313. Quarterly journal on clean air issues in England; half is devoted to technical articles.

Coal, 29 North Whacker Drive, Chicago, IL 60606; (312) 726-2802. Trade journal for coal industry news and developments.

"E" The Environmental Magazine, Earth Action Network, 28 Knight Street, Norwalk, CT 06851; (203) 854-5559. Clearinghouse for in-depth information, news, and commentary on environmental issues.

E&MJ (Engineering and Mining Journal), 29 North Whacker Drive, Chicago, IL 60606; (312) 726-2802. Trade journal on international mining issues and news.

Energy Engineering, 700 Indian Trail, Lilburn, GA 30247. Bi-monthly technical journal of energy engineering problems and solutions.

Environment, Heldref Publications, 1319 Eighteenth Street, NW, Washington, DC 20036; (202) 296-6267. In-depth articles on wide variety of environmental issues.

Environmental Action, 6930 Carroll Avenue, Sixth Floor, Takoma Park, MD 20912; (301) 891-1100. Activist-oriented magazine on environmental issues.

Environmental Affairs Law Review, Boston College Law School, 885 Centre Street, Newton Centre, Boston, MA 02159. Quarterly review of environmental law.

Environmental Law, 10015 SW Terwilliger Boulevard, Portland, OR 97219; (503) 768-6700. In-depth Northwestern School of Law Journal on environmental legal issues.

EPA Journal, New Orders, Superintendent of Documents, P.O. Box 371954, Pittsburgh, PA 15250-7954; (202) 260-2080. Official EPA journal, covering a wide variety of issues the agency works on.

EPRI Journal, Electric Power Research Institute, P.O. Box 10412, Palo Alto, CA 94303; (415) 855-2000. Founded by the nation's utilities to develop and improve electric power production, distribution, and utilization; energy/electricity-related articles.

ES&T (Environmental Science and Technology), American Chemical Society, Department L-0011, Columbus, OH 43268-0011; (614) 447-3776. Chemical industry views on environmental issues; latter half of the magazine is devoted to scientific papers.

Farm Journal, 230 West Washington Square, Philadelphia, PA 19106. Journal for American farmers.

FDA Consumer, Superintendents of Documents, Government Printing Office, Washington, DC 20402. U.S. Food and Drug Administration journal updating current health issues.

Friends of the Earth, 218 D Street, SE, Washington, DC 20003; (202) 544-2600. Newsletter on current issues Friends of the Earth is working on.

Garbage, Dovetale Publishers, 2 Main Street, Gloucester, MA 01930; (508) 283-3200. Practical articles on the environment; attempts to take an evenhanded approach to any issue.

The Gene Exchange: A Public Voice on Genetic Engineering, National Biotechnology Policy Center, National Wildlife Federation, 1400 16th Street, NW, Washington, DC 20036. Newsletter on current status and concerns of biotechnology.

GeneWatch, Council for Responsible Genetics, 19 Garden Street, Cambridge, MA 02138; (617) 864-5164. Newsletter with current information and concerns regarding biotechnology.

Greenpeace, 1436 U Street, NW, Washington, DC 20009; (202) 462-1177. Newsletter of current campaigns and concerns.

Health and Environment Digest, Freshwater Foundation, Spring Hill Center, 725 County Road 6, Wayzata, MN 55391; (612) 449-0092. Newsletter on environmental health issues.

Independent Energy, Marier Communications, Inc., 620 Central Avenue N., Milaca, MN 56353; (612) 983-6892. Information for power and utility industry, particularly for owners and managers.

International Wildlife, National Wildlife Federation, 1400 Sixteenth Street, NW, Washington, DC 20036; (202) 797-6800. International news on wildlife, pollution, and habitat.

Journal of Forestry, 5400 Grosvenor Lane, Bethesda, MD 20814; (301) 897-8720. In-depth articles on forest management and preservation.

Journal of Soil and Land Conservation, Soil and Water Conservation Society, 7515 NE Ankeny Road, Ankeny, IA 50021; (515) 289-2331. In-depth journal about soil conservation issues.

Modern Plastics, 1221 Sixth Avenue, New York, NY 10020. Trade journal on global issues and developments for the plastics industry.

Mother Earth News, Sussex Publishers, 24 East Twenty-third Street, New York, NY 10010; (800) 234-3368. Country magazine with practical articles on organic farming and renewable energy and lifestyles.

National Green Pages, Co-op America, 1850 M Street, NW, Suite 700, Washington, DC 20036. Yellow pages listings for environmental products, services, organizations, and resources.

National Parks, National Parks and Conservation Association, 1776 Massachusetts Avenue, NW, Washington, DC 20036; (202) 944-8530. Articles on defending, improving, and promoting the national park system, while educating readers.

National Wildlife, National Wildlife Federation, 1400 Sixteenth Street, NW, Washington, DC 20036; (202) 797-6800. Articles on wildlife and related environmental problems.

Nature, 1234 National Press Building, Washington, DC 20045; (202) 737-2355.

International journal on natural science; 1/3 general content news, 2/3 technical science news.

Nature Conservancy, 1815 North Lynn Street, Arlington, VA 22209; (703) 841-5300. The Conservancy's efforts to preserve wildlands and wildlife.

Nature Study, 5881 Cold Brook Road, Homer, NY 13077. Journal promoting environmental education.

New Scientist, c/o Virgin Mailing and Distribution, 10 Camptown Road, Irvington, NJ 07111. Articles on current science, with some environmental articles.

Oceanus, Woods Hole Oceanographic Institute, 9 Maury Lane, Woods Hole, MA 02543; (508) 457-2000. Wide ranging and in-depth articles on ocean and coastline environmental problems.

Oil & Gas Journal, PennWell Publishing, 1421 South Sheridan Road, Box 1260, Tulsa, OK 74101. Trade journal for international petroleum industry.

Organic Gardening, Box 7320, Red Oak, IA 51591; (800) 666-2206. Practical articles and advice on organic gardening.

Pollution Engineering, 1350 East Touhy Avenue, Des Plains, IL 60018; (708) 635-8800. Trade journal on pollution engineering methods, equipment, and politics.

Ranger Rick, National Wildlife Federation, 8925 Leesburg Pike, Vienna, VA 22184. Wildlife magazine for children.

Science, American Association for the Advancement of Science, 1333 H Street, NW, Washington, DC 20005; (202) 326-6400. Half devoted to current science reports, half to technical scientific studies.

Scientific American, 415 Madison Avenue, New York, NY 10017; (800) 333-1199. Wide range of science topics, with some environmental coverage.

Sea Frontiers, P.O. Box 498 Mount Morris, IL 61054; (800) 786-SEAS. Articles on the sea and its creatures.

Sierra, Sierra Club, 730 Polk Street, San Francisco, CA 94109; (415) 776-2211. Articles covering environmental issues.

Solar Energy, Pergamon Press, 660 White Plains Road, Tarrytown, NY 10591; (914) 524-9200. International Solar Energy Society journal with technical articles on current solar energy science.

The Battery Man, Independent Battery Manufacturers' Association, 100 Larchwood Drive, Largo, FL 34640; (813) 586-1408. Industrial trade journal about recent developments on starting, lighting, ignition, and generating systems.

The Circle, 1530 East Franklin Avenue, Minneapolis, MN 55404; (612) 871-4749. Indigenous issues covered from around the world.

The Ecologist, MIT Press Journals, 55 Hayward Street, Cambridge, MA 02142; (617) 253-2889. In-depth articles and commentary on global environmental issues.

The Journal of Environmental Education, 1319 Eighteenth Street, NW, Washington, DC 20036; (202) 296-6267. Articles on methods and impacts of environmental education on children.

The Journal of Wildlife Management, The Wildlife Society, 5410 Grosvenor Lane, Bethesda, MD 20814. Quarterly publication of scientific wildlife studies.

The New Farm, Rodale Institute, The New Farm, P.O. Box 7306, Red Oak, IA 51591; (800) 365-FARM. Practical articles for farmers on methods of regenerative/sustainable agriculture and relevant legislation.

USA Today: The World of Science, Society for the Advancement of Education, 99 West Hawthorne Avenue, Valley Stream, NY 11580; (516) 568-9191. Newsletter with a wide variety of short science articles, many relating to new technology that affects the environment.

Waste Age, 1730 Rhode Island Avenue, NW, Suite 1000, Washington, DC 20036; (202) 861-0708. Trade journal of the National Solid Waste Management Association on all facets of waste management.

Wilderness, The Wilderness Society, 900 Seventeenth Street, NW, Washington, DC 20006; (202) 833-2300. Magazine with articles concerning protecting wilderness areas and wildlife.

Wildlife Society Bulletin, The Wildlife Society, 5410 Grosvenor Lane, Bethesda, MD 29814. Quarterly with articles on wildlife research.

Whole Earth Review, 27 Gate Five Road, Sausalito, CA 94965; (415) 332-1716. Quarterly examining how we live on the planet and encouraging individual action.

World Health, World Health Magazine, World Health Organization, CH-1211, Geneva 27, Switzerland. Magazine about health issues in developing countries.

Worldwatch, Worldwatch Institute, 1776 Massachusetts Avenue, NW, Washington, DC 20036; (202) 452-1999. Tracks key indicators of earth's well-being by looking at economic systems and environmental problems.

The CERES Principles are a set of 10 principles held by a coalition of environmental groups, labor unions, religious organizations, and investors controlling more than $150 billion, including the pension funds of New York City and the state of California, and investments of Franklin Research and Development and the Calvert Group. About 70 companies with $20 billion in reserves have thus far pledged to abide by the CERES Principles and submit yearly, standardized reports that are used by investors to judge their actions. The principles, formally called the Valdez Principles, were developed after the oil tanker, *Exxon Valdez*, ran aground in Alaska's Prince William Sound, ruining vast areas of shoreline, destroying fishing areas, and killing large numbers of birds and aquatic life.

Companies pledging to adhere to the CERES Principles are interested in voluntarily shaping their policies, are trying to attract an environmentally conscious public investing group, and wish to join a credible organization. Thus far companies range from Ben and Jerry's, Aveda, and Timberland to Fortune 500 companies like Sunoco and Sun Oil (a new member). CERES has a counterpart in Japan and is expanding in Europe.

Signatories to the CERES Principles agree to pursue the following actions:

1. Protect the biosphere by minimizing pollutants and the effects of activities on environment.
2. Use natural resources sustainably while conserving nonrenewable resources.

3. Reduce and dispose safely of waste.
4. Maximize energy efficiency and conservation.
5. Reduce environmental and employee risks.
6. Market products and services that are safe for consumers and the environment.
7. Pay for any cleanups or other created problems.
8. Disclose any environmental hazards or environmental damage they have created, and protect employees who report them.
9. Appoint a minimum of one director to oversee these principles within the company.
10. Make public the yearly record under the CERES Principles.

For more information, contact CERES—Coalition for Environmentally Responsible Economies, 711 Atlantic Avenue, Boston, MA 02111; (617) 451-0927.

Acknowledgements

Dozens of organizations and even more people contributed information and assistance for this book; I am indebted to all of them. I would especially like to thank the following groups and people for providing information: Dawn M. Martin of the American Oceans Campaign, Verne Achtermann of the American Water Works Association, Paul Gipe of the American Wind Energy Association, the Center for Global Change, Mark Tulay of CERES, Robyn Roberts of Clean Water Network, Pat Olsen of *Coal* magazine, Conservation International, the Earth Island Institute, the Environmental Protection Agency, Kirk Randall of the Federal Energy Regulatory Commission, Greenpeace, Tom O'Brien of the Hawaii State Energy Division, the Institute for Food and Development Policy, the International Institute for Energy Conservation, Dr. Ronald I. Orenstein and Teresa M. Telecky of the Humane Society, Henry Drewes of the Minnesota Department of Natural Resources, the National Coal Association, the National Oceanic and Atmospheric Administration, the National Recycling Coalition, the Nature Conservancy, the Population Council, Sierra Club, Paul Reich of the Soil Conservation Service (USDA), the Union of Concerned Scientists, the U.S. Bureau of Mines, the U.S. Department of Agriculture, Mary Jane Wisenbaker of the U.S. Department of Energy, the U.S. Department of Transportation, Diana Cross of the U.S. Fish and Wildlife Service, John A. Combes of the U.S. Forest Service, the World Bank, the Worldwatch Institute, the World Resources Institute, and Zero Population Growth.

Special thanks to Tom Christensen, Dave Peattie, Kirsten Janene-Nelson, Heidi Fritschel, Janet Mowery, and everyone else at Mercury House for their guidance, enthusiasm, suggestions, and careful attention to detail. Thanks to Mel Saign, Mary Saign, John Sheehan, Jay Handelman, Loras Holmberg, and Marleah Jex for help in gathering information, and to Tim Hartz for use of his fax machine. And thanks to Scott Edelstein for his early suggestions and guidances of the manuscript.

Index

-C-

-D-

Dams, 182
see also chapter on:
    Hydroelectric energy
and deforestation, 58
and earthquakes, 184
and environmental damage, 185
and fish extinction, 111
and indigenous peoples displace-
    ment, 197
and soil degradation, 344
and wetlands, 417
worldwide, 184–85
Dandelions, 103
Danube, 408, 420
DDT, 272, 421, 440
Debt-for-nature swaps, 65, 66–57, 119,
    440
Deciduous forests, 56
Decomposition, 23, 25
Decontamination of nuclear plant, 247
Deep ecology, 440
Defoliants, 302, 441
Deforestation, *56–67*
see also chapter on:
    Extinction
    Livestock problems
and biomass burning, 24
and eutrophication, 57
and indigenous peoples' homelands,
    196, 437
and species loss, 115, 196
and tree diseases, 58
Delaney Clause, 308
Denmark, 425
Department of Agriculture (USDA)
    biotechnology, 39–40
    Federal Animal Damage Control
        Program, 121
    National Seed Storage Laboratory,
        118
    Soil Conservation Service, 208
Department of Defense, 169, 409, 414
    biotechnology, 40
    hazardous waste, 167, 168, 169
Department of Energy, 409

and biotechnology, 40
    Clean Coal Technology Program, 54
    hazardous waste, 167, 169
    Institutional Conservation Program,
        89
    lead in coal, 214
    mercury in coal, 226
    Weatherization Assistance Program,
        89
    wind energy, 425, 426–27
Deposit laws, 358
Des Moines, IA, 26
Desalinization, 137, 441
Desert cacti, 117
Desertification, 217, 340, 344, 441
Desert Storm syndrome, 411
Desulfovibrio sulfuricans, 255
Developing countries
    and asbestos regulation, 20
    biomass crop production, 28
    biotechnology and farming, 37–38
    and coal burning, 55
    disposable products in, 75
    and ecotourism, 72
    electricity use, 86
    energy needs, 26, 125
    and environmental disasters, 98
    and fish protein, 281
    and population pressures, 313
    and solar energy, 350
    and use of banned pesticides, 46
    wildlife sanctuaries, 75
Diesel engine emissions, 268
Dioxin, 8, 43–4, 47, *68–71*, 272
    see also chapter on:
        Chlorine
        Incineration
        waste sites, 69
Discrimination, 38
Disease. *See also chapters on:*
    Air pollution
    Exotic species
Disease, waterborne, 135, 138, 187
Disease-resistant animals, 32
Disinfectants, 302
Disposable products, 75, 362
DNA, 30

Environmental disclosure laws, 214
Environmental Protection Agency.
    See EPA
Environmental racism, 443
EPA (Environmental Protection
    Agency)
  and air pollution, 10
  and asbestos, 20–21
  Bioremediation Field Initiative, 172
  biotechnology, 39–40
  biopesticides, 34
  biotechnology, 39–40
  Chemical Manufacturing Rule, 338
  dioxin as carcinogen, 69
  dioxin standards, 48, 71
  EMF studies, 79
  and exotics, 106
  Green Lights Program, 89
  hazardous waste sites, 169
  and military wastes, 168
  National Acid Precipitation Project, 3
  National Estuary Program, 422
  oil waste disposal, 126
  pesticides, 308–309
  radiation exposure limits, 254
  radon exposure, 322, 324
  sewage, 136, 141
  and wetlands degradation, 420
  wetlands division, 422
Environmental sinks, 160–61, 229,
    230, 444
Erosion, 340–41, 343, 444
Erythropoietin, 32
Eskimos, 196
Estrogen
  effects of EMFs, 78
  mimicking by toxics, 45, 47
Estuaries, 256, 444
Ethanol, 24–26, 28
Ethiopia, 408
Ethylene, 12
Ethylene dichloride (EDC), 46
Europe
  biotechnology, 34

catalytic converters, 14
crop yields, 11–12
energy efficiency programs, 86
forests and pollution, 4, 11
hazardous wastes, 168
organic farming, 374
peat land disappearance, 27
use of tropical timber, 60
European beech scale, 103
European corn borer, 103
European pine shoot beetles, 103
European purple loosestrife, 103
European starlings, 104
European Wind Energy Association,
    425
Eutrophication, 31, 57, 207, 417, 444
Everglades, 104, 416, 420
  hurricane damage, 96
Exotic species, *101–7*, 417
    see also chapter on:
    Ecotourism
    Extinction
  and death and displacement of
    indigenous peoples, 197,
    199–200
  and extinction of native species, 111
  pets and products, 111, 117
Explorers and exotic species, 102
Extinction, *108–22*
    see also chapter on:
    Overfishing
    Population pressures
    Sustainable development
  and biotechnology, 32
  and global warming, 150
  and introduction of exotic species,
    101
  naturally caused, 112
  of Norwegian salmon, 3
  rate of, 112
  result of deforestation, 60
  of West Coast salmon, 59
Extractive reserves, 202
Exxon Valdez, 269

-F-

Fallout, nuclear, 444
  health problems, 253
  in Nevada, 410
Family planning, 313, 315
Famine, 315, 317
Farm Bill, 375, 422, 423
  see also Conservation Reserve
    Program
Farm losses, 37
Farm workers, 307, 309–10
Farming
  see also chapter on:
    Sustainable agriculture
  biological controls in, 32
  crop destruction, 4, 11, 105
  crop diversity, 114
  crop rotation, 358
  crop waste, 25
  crop yields, 316
  deforestation, 58
  habitat loss, 110–111
  integrated pest management, 372–73
  intercropping, 368, 450
  methane production, 230
  multinational farming corporations,
    36, 37, 42
  and pesticide use, 304, 305
  poor agricultural practices, 342
  and population explosions, 316
  regenerative, 369
  and water pollution, 132, 133, 134,
    136, 137
  wetlands loss, 418
Fathometers, 280
Federal Animal Damage Control
  Program, 121
Federal Compliance Act, 414
Federal Insecticide, Fungicide, and
  Rodenticide Act, 308
Federal subsidies
  fossil fuel, 6, 16
  tobacco, 16
Feedback loops, 444
  effect of global warming, 152

Feedlot practices, 219, 444
Feldspar, 235
Fermentation, 23
Fertilizers, 445
  fish use for, 281
  phosphate, 235
  and radon, 321
  and soil degradation, 345
  used in landscaping, 204
  and water pollution, 132
FIFRA (U.S. Federal Inseticide,
    Fungicide, and Rodenticide Act of
    1947), 308
Fiji and ecotourism, 73
Finland, 420
  biomass studies, 27
Finning, 283
Fire ants, 102
Fires
  causes, 94
  environmental impact, 93, 98
  part of temperate ecosystem, 93
Fish
  see also chapter on:
    Overfishing
    Freshwater degradation
    Ocean degradation
  acid rain effects on, 1
  aquaculture, 282
  and coral reefs, 260
  and hydroelectric plants, 183
  mercury in, 224
  and toxics, 8
  ladders, 188
  need for wetlands breeding grounds,
    417
  threatened extinctions, 113
Fishing
  see also chapter on:
    Overfishing
  commercial, 113, 419
  illegal, 280
  tuna, 281, 284
Fission, 246, 445
Flooding
  causes of, 58, 94, 95, 345

-G-

Hawks, 25
Hazardous and Solid Waste
    Amendments, 172
Hazardous waste, *165–174*, 421
    see also chapter on:
        Heavy metals
        Incineration
        Nuclear energy
        Toxic chemicals
        Warfare effects on the environment
    Clark Fork River, 236
    collection sites, 213
    international shipping, 169–70,
        173–74
    ocean dump sites, 258
HCFCs (hydrochlorofluorocarbons),
    277, 291, 292, 294–95
HCH. *See* lindane
Health problems, causes
    *See also "Human Impact" section for
        each chapter*
    air pollution, 13
    biotechnology, 36
    coal, 53
    extinctions, 116
    hazardous wastes, 165–66, 170
    heavy metals, 178
    incinerators, 192
    lead, 212–13
    mining, 237
    natural disasters, 98
    nitrogen oxides, 243
    oil burning, 269
    particulates, 298–99
    pesticides, 307
    polluted oceans, 262
    population explosion, 317
    smog, 337
    synthetic organic chemicals, 275
    toxic chemicals, 392
    warfare effects on the environment,
        412
Heat trapping chemicals, 160
Heaters, stoves, and fireplaces
    and air pollution, 14
Heavy metals, 8, *175–181*, 417
    see also chapter on:

Incineration
Lead (Pb)
Mercury (Hg)
Mining
    and biomass burning, 25
    and deforestation, 57
    incinerators and, 191–92
    leaching, 2
Herbicides, 272, 302, 448
    resistance to, 31
    use in warfare, 406
Heritage Farm (IA), 118
High-grading, mining, 239
High-intensity farming, 371
High-level radioactive waste, 249
Himalaya and ecotourism, 73
Hiroshima, 411
Hoffman-La Roche, factory explosion,
    69
Home energy use, 85
Honduras, 202
Hooker Chemical Company, 172
Hoover Dam, 186
Hudson Volcano eruption, 96
Human gene mapping, and discrimina-
    tion, 38
Human Genome Project, 36
Human protein production in animals,
    32
Humane Society, 38
    and exotics, 106
Humus, 368, 448
Hungary
    air pollution deaths, 12
Hurricane Andrew, 96
Hurricanes
    causes of, 94
    environmental impact of, 93, 98
Hybrid crop
    and pesticide use, 304
    problems, 116, 316
Hybrid seed selection, and natural
    variety extinction, 34, 110, 111
Hybrid vehicles, 399
Hydro-Quebec, 186
Hydrocarbons, 8, 272, 449
Hydrochloric acid, 43, 47

Integrated waste management, 450
International Atomic Energy Agency
    (IAEA), 413
International Decade of Natural
    Disaster Reduction, 99
International Institute for Energy
    Conservation, 88
International Joint Commission (IJC)
    (on the Great Lakes), 45, 48
International Mussel Watch Project, 140
International Whaling Commission, 114
Inuit tribes, 186, 198
Iodine, 235
Ionizing radiation, 450
Iowa, Seed Savers Exchange, 118
IPM. *See* Integrated pest management
Iraq, 411, 413
Ireland
    peat as fuel in, 27
    potato famine in, 114
Iron, 175
Irradiation, food, 253, 254–55
Irrigation, 450
Israel
    breast cancer rates, 44
    pesticides, 44
Italy
    biodiesel fuel, 26
    monuments eroded by acid rain, 4
Ivory trade ban, 115
Ixtoc 1 oil spil, 237

-J-

Japan
    and carbon dioxide research, 162
    biotechnology, 34
    catalytic converters, 15
    fishing, 281, 285
    and hydroelectric power, 186
    mercury poisoning, 224
    nonsustainable economy, 377
    nuclear energy, 250
    ocean energy, 330
    solar energy, 329
    use of tropical timber, 60
    whaling, 115

Japanese beetle, 103
Johnston Atoll, 409
    incinerator, 192
Junk mail, 360

-K-

Kakadu National Park, 74
Kanaf, 66
Kariba Dam, 185
Kauai, 26
Kayapo, 199–200
Kelp forests, 111
Kenya
    and ecotourism, 73
    fuelwood use in, 27
    wetlands in, 420
Kerosene heaters, 243
Keystone species, 111, 450
Killer bees, 102, 105–6
Killer fogs, 337
Koala bears, 109
Krill, 109, 285, 451
Kudzu, 103
Kuwait, 411
    burning of oil wells, 9

-L-

Lake Baikal, 140
Lake Nasser, 183
Lake Victoria, 104
Lakes and mercury contamination, 224
Land reclamation after strip-mining, 54,
    55
Land rights. *See also* Land tenure
    and indigenous people, 195, 197,
    202
Land subsidence, or sinking, 139
Land tenure, and deforestation, 62, 67
Landfills, 354–59, 451
    and asbestos, 18
    and coal ash, 125
    ash from incinerators, 190
    source of methane, 23, 230
Landscaping problems, *204–9*
    see also chapter on:

Mount Saint Helens, 2, 96
Mozambique, 408, 421
Municipal solid waste, 167, 354–59
Mururoa, Polynesia, 410
Myxomatosis, 104

-P-

Rail, 398
Rain forests
  see also chapter on:
    Deforestation
  and biodiversity, 112
  boycotts against destruction, 65
  and tribal peoples, 196, 201
  effects of acid rain on, 4
  medicines from, 200–1, 22
  preservation of through biotech-
    nology, 37, 39
Rare Breeds Survival Trust, 118
Raw materials
  obtained by mining, 235, 239–40
  use, 236, 379
RCRA. *See* Resource Conservation and
    Recovery Act
Recombinant DNA (rDNA) technology,
    30
  see also Biotechnology
Recycling, 355–62, 462
  air pollution reduced, 14
  energy savings, 86
  products using heavy metals, 179,
    180
Red meat, 201, 202
Red Sea, 411
Red tides, 257, 462
Reforestation, 62
Refuse-derived fuel incinerators, 191
  see also Waste-to-energy burn plants
Renewable energy, *326–332*
  see also chapter on:
    Biomass energy
    Geothermal energy
    Hydroelectric energy
    Solar energy
    Wind energy
  tax incentives for, 6
Reprocessing, 246
Reproductive problems
  in animals, 68, 166, 175, 210, 222,
    247, 272, 303
  and EMFs, 80
  in humans, 170, 178, 213, 224, 252,
    275
  and toxics, 45

and wetlands degradation, 417, 421
Reservoirs, 182, 183
Resource Conservation and Recovery
    Act (RCRA), 172, 393
Respiratory disease
  and air pollution, 13
  and fossil fuels, 127
  and fuelwood burning, 27
  and particulates, 299
  and smog, 337
  and sulfur dioxide, 365
Rhinocerus, 408
Rice-Anderson Indemnity Act, 255
River basins, 256
River pollution, 136
Rock mining, 236
Rocky Flats Nuclear Plant, 409
Rocky Mountain Arsenal, 409
Rocky Mountain Institute (RMI) study
    on energy conservation, 85, 88
Rodenticides, 302
Royal Society of London, 318
Run-of-river plants, 182
Runoff, 431
  manure, 216, 217, 218
  pesticide, 205
Russia, 411, 414
  coal and oil reserves, 126
  freshwater pollution, 136, 137
  hazardous wastes, 168
  lack of pollution controls, 12
  livestock, 217, 218
  mercury production, 224
  nuclear policy, 410
  peat as fuel, 27
Rwanda
  ecotourism, 73
  warfare, 408

-S-

Safe Drinking Water Act, 141, 142
Salamanders, and acid rain, 1, 4
Salinization, 136, 137, 183, 234, 342
Salmon
  and dams, 185, 280
  and hydropower dams, 185

-U-

Universal Declaration on the Rights of
Indigenous Peoples, 198
Univ. of Illinois Swine Research Center,
27
Unleaded gas, 214
Upper atmosphere research satellites
(UARS), 294
Uranium, 235, 246, 248, 320, 468
Urban areas
landscaping problems, 205, 209
smog, 334–35
water pollution, 132–33, 134
U.S. Army Corps of Engineers, 99–100
U.S. Bureau of Reclamation, 106
U.S. Fish and Wildlife Service, and
exotics, 10, 106
U.S. Forest Service, 59, 66
aid to Venezuela, 65
biotechnology, 40
policy, 64, 66
U.S. Geologic Survey, 115, 306
U.S. Department of Justice, 115
U.S. Office of Techology Assessment,
259

-XYZ-